P9-DNZ-886

GALILEO

GALILEO

WATCHER OF THE SKIES

DAVID WOOTTON

YALE UNIVERSITY PRESS
NEW HAVEN AND LONDON

Published with assistance from the foundation established in memory of
Oliver Baty Cunningham of the Class of 1917, Yale College

For information about this and other Yale University Press publications, please contact:
U.S. Office: sales.press@yale.edu   www.yalebooks.com
Europe Office: sales@yaleup.co.uk   www.yaleup.co.uk

Set in Arno Pro by IDSUK (DataConnection) Ltd
Printed in Great Britain by MPG Books, Bodmin, Cornwall

A catalogue record for this book is available from the British Library.

Wootton, David, 1952–
   Galileo: Watcher of the Skies/David Wootton.
      p. cm.
   ISBN 978–0300–12536–8 (cl:alk. paper)
   1. Galilei, Galileo, 1564–1642 2. Astronomers—Italy—Biography.
I. Title.
   QB36.G2W66 2010
   520.92—dc22
   [B]
                                                                      2010027620

10 9 8 7 6 5 4 3 2 1

For Alison

. . . For, all the night,
I heard the thin gnat-voices cry,
Star to faint star, across the sky.

Rupert Brooke, 'The Jolly Company' (1908)

Contents

Part three: The eagle and the arrow

Part four: Prisoner to the Inquisition

Illustrations

Acknowledgements

Numerous Galileo scholars, historians of science and historians of the late Renaissance have responded to my enquiries, amongst them Ugo Baldini, Silvio Bedini, Domenico Bertoloni Meli, Mario Biagioli, Christopher Black, Horst Bredekamp, Massimo Bucciantini, Michele Camerota, Linda Carroll, Miles Chappell, David Colclough, Pietro Corsi, Nicholas Davidson, Peter Dear, Simon Ditchfield, Germana Ernst, Dinko Fabris, Federica Favino, Maurice Finocchiaro, Steve Fuller, John Henry, Michael Hunter, Mary Laven, Peter Machamer, Ian McLean, Edward Muir, Ronald Naylor, Paolo Palmieri, Isabelle Pantin, Pietro Redondi, Eileen Reeves, Jürgen Renn, John Beldon Scott, Richard Seargentson, Michael Shank, Michael Sharratt, William Shea, A. Mark Smith, Roberto Vergara Caffarelli and Nicholas Wilding. I thank them for their patience. I also thank Rick Watson of W. P. Watson for generously lending me his copy of Giovanni Battista Stelluti's *Scandaglio*.

I am also grateful for a number of opportunities to try out my ideas before an audience: to the British Association for the History of Philosophy meeting in York; to David Cram and Robert Evans of the History Faculty of the University of Oxford; to Jim Bennett and Stephen Johnston of the Oxford Museum of the History of Science; to David Lines of the Centre for Renaissance Studies at the University of Warwick; to Knud Haakonssen of the Centre for the History of Ideas at the University of Sussex; and to the Renaissance Society of America meeting in Venice in 2010.

Mordechai Feingold, Paula Findlen, Peter Machamer, Alison Mark, Paolo Palmieri, Pietro Redondi, Michael Shank, Michael Sharratt and Nicholas Wilding were kind enough to read various drafts. My text has been greatly improved as a result of their suggestions. My copy-editor Laura Davey did a remarkable job catching at least some of the remaining errors.

Alison Mark, the University of York and the Leverhulme Foundation have generously supported this research – I thank all three for the pleasures of a year of research leave. Thanks to Matthew Patrick who has discussed the problems of biography with me interminably.

Finally, I would like to acknowledge a debt to Stillman Drake. Many years ago (in 1981, I think), when I was a young scholar starting out in academic life, Stillman Drake spent an afternoon showing me his wonderful collection of Galileo books. I am delighted to have an opportunity to record this kindness now.

Introduction: Conjectural history

Now the past, by virtue of being *past*, remains forever inaccessible to us, it is no more, we cannot reach it and it is only from its traces and *remains*, from its debris which are *still present* – works of art, monuments, documents which have escaped the ravages of time and of mankind – that we attempt to reconstruct it. But objective history, the history men make and suffer, is not concerned – or hardly – with the history of historians: it allows the survival of things of no value to the historian and mercilessly destroys the most important of documents, the most beautiful of works, and the most impressive of monuments. What it leaves, or has left behind, are mere fragments of what we should need. Accordingly, historical reconstructions are inevitably fragmentary, uncertain, even doubly uncertain – poor little conjectural science, so has Paul Valéry described history.

Alexandre Koyré (1961)[1]

Our knowledge of Galileo's life we owe largely to three people. One of these is Vincenzo Viviani (1622–1703), his last student, who was sixteen when he met Galileo and not yet twenty when his master died.[2] Viviani wrote Galileo's first biography (which was not published until 1717), preserved his papers, and planned an edition of his complete works. When Viviani died, his property passed first to his nephew and then, in 1737, to the nephews of his nephew. These last did not share their great-uncle's preoccupation with Galileo – despite the fact that 1737 was the year in which Galileo finally became respectable enough to be given an honourable tomb in a Florentine church. They emptied the cupboards in which Viviani had stored Galileo's manuscripts, and turned them to what seemed to them a far better use: storing the household linen.

One day in the spring of 1750, Giovanni Battista Nelli (1725–93), a Florentine man of letters, made a detour to buy some cold meat from a butcher he did not normally frequent. The excellent mortadella was to be his contribution to a picnic. When, out in the countryside with his friends, he unwrapped the meat, he noticed that the waste paper it was wrapped in appeared to be something written by Galileo. Returning with all possible haste to the butcher (but taking the precaution of concealing his discovery from his friends), he

1

eventually traced the butcher's waste paper to its source: a large bin overflowing with documents in Viviani's old house. His grand-nephews were selling off the paper in small parcels as wrapping paper, but were happy to sell the whole bin to Nelli.

Nelli used his treasure trove to write a life of Galileo, and the papers he had acquired eventually ended up in the Florentine archives.[3] There they were to be used by the greatest of all Galileo scholars, Antonio Favaro, who produced what is known as the National Edition of Galileo's works (1890–1909), in twenty volumes printed in twenty-one large tomes (volume iii spans two separate tomes). Favaro was meticulous and indefatigable, and almost everything about Galileo that we know now, Favaro already knew.[4] But Favaro was not an impartial scholar – nobody is. He laboured to defend Galileo's reputation as a scientist, as a man, and as a pious Catholic; inconvenient details were buried, brushed aside, or put in the best possible light. But they were never suppressed.

When Galileo died in 1642 he had been living in Florence for more than thirty years, accumulating letters and papers. But when he moved from Padua to Florence in 1610, at the age of forty-six, he must have packed his clothes, books, papers, telescopes and lens-grinding machinery to be loaded onto the backs of donkeys or mules – and he presumably threw a great deal away. In Favaro's edition of the *Opere* the letters to, from and about Galileo occupy nine volumes. One of these, volume x, contains all the letters for the first forty-six years of his life; another, volume xv, contains the letters for the single year 1633 – the year of Galileo's trial. This imbalance is particularly frustrating in that Galileo made no major scientific discoveries after the age of fifty – and yet the vast bulk of the evidence that survives comes from the last decades of his life. In addition, when Galileo went blind in 1637 he lost control over his own papers: things he would once have thrown away were now faithfully kept.[5]

Thanks to Viviani, Nelli and Favaro we know a great deal about Galileo, and it is easy to be overwhelmed by the bulk of this knowledge. It is also easy to forget that our most important source consists of a bin full of papers. We have no way of knowing how many documents had already been extracted from the bin and used to wrap meat. Unfortunately that is not our only problem. All the documents in the bin had first been through the hands of Viviani. From Galileo's death in 1642 until his own death in 1703 Viviani was engaged in a campaign to restore his master's reputation, blighted by the Inquisition's condemnation of him in 1633. He wanted to see Galileo (who the Church had insisted should be buried without a funerary monument) given a proper burial under a tomb worthy of his status as a great scientist; he wanted to see his works, including his scientific correspondence, published. Because he could achieve neither of these objectives he turned the facade of his own house into a monument to Galileo, recording at length and in stone all his scientific achievements, but making no mention either of Copernicanism or of the Inquisition trial and condemnation.

An essential element in Viviani's campaign to recuperate Galileo's reputation was his insistence that Galileo was a good Catholic who had acknowledged his error in advocating Copernicanism and obediently submitted to the Church's decree. Galileo was portrayed by Viviani as being dismayed by the publication of his works abroad after 1633, in defiance of the Inquisition: the evidence to the contrary, which was and is incontrovertible, was quietly suppressed.[6] It has been suggested that Viviani genuinely believed that Galileo's Copernicanism represented a tragic intellectual error, yet the evidence rather suggests that, as Galileo's faithful disciple, he skilfully combined outward conformity to the requirements of the Church with private dissent.[7]

How far did Viviani's single-minded preoccupation with Galileo's reputation amongst Italian Catholics lead him to falsify the historical record? That he falsified it to some degree we can be sure. He planned to publish an exchange of letters between Galileo and Élie Diodati (a Genevan who was nominally a Protestant but had close relations with a number of unbelievers) on the subject of measuring longitudes, telling Diodati that he would alter or omit some sentences that might provoke hostility to Galileo and make it difficult to obtain a licence to publish.[8] What did he plan to leave out? We do not know. Viviani separated out thirteen letters on longitude, so that they survived to be published in 1718, at which point they were destroyed. We cannot tell how extensively they would have been revised if Viviani had been able to publish them – although there is another case in which we can catch Viviani in the act of rewriting a letter to protect Galileo's reputation.[9] Viviani had in his possession, as far as we can tell, around a hundred letters between Galileo and Diodati. None of these survive; forty are completely lost, and the rest are preserved only in partial copies made by Viviani, or in the edition of 1718, or, in two cases, in copies made by Pierre Dupuy – we do not know how far those copies censor and falsify the originals (even the edition of 1718 appears to be marred by some deliberate omissions).[10] It is theoretically possible that every one of the letters that did not survive until 1718 went to wrap mortadella, but it is more likely that they were destroyed by Viviani himself because they contained proof that Galileo had advocated Copernicanism after 1633.[11]

The charge is a serious one: if Viviani was prepared to falsify the record then we must assume that he destroyed any evidence which cast doubt on Galileo's Catholic piety. On 24 July 1673 he wrote to the scientist and diplomat Count Lorenzo Magalotti in Flanders saying that he had heard that an edition of Paolo Sarpi's letters was shortly to be published in Amsterdam, and that some letters between Galileo and Sarpi might appear in it. Sarpi – who had been a close friend of Galileo's – was notorious throughout Europe as the author of the *History of the Council of Trent* (1619), widely regarded as the most effective piece of anti-Catholic propaganda to have been published since the days of Luther and Calvin. The publication of such a correspondence, even if it was concerned only with scientific matters, would, Viviani felt, be fatal to Galileo's reputation.

In the 1656 edition of Galileo's works, published in Bologna (which omitted the *Dialogue* condemned in 1633 and much else), the censors had required that a passing reference to Sarpi be excised on the grounds that Sarpi had died excommunicate. There was thus no doubt that Viviani would be prevented from publishing any letters between Sarpi and Galileo. But this does not prepare us for his instructions to Magalotti: he must try to get hold of the originals of these letters, if they exist, and of any copies; if printing has begun he must buy up the printed sheets – Viviani will provide the funds. No evidence of such a correspondence must be left in untrustworthy hands. Viviani says absolutely explicitly that publication must be prevented at all costs; he does not say that the letters should be destroyed – rather that he is sure Magalotti will be enraged and will know what to do. 'I do not know what I am saying', he continues, 'Have pity on me as a faithful disciple.'[12] It is clear that Viviani was prepared to go to any lengths to protect Galileo's reputation. Not surprisingly, only two early and innocuous letters from Sarpi to Galileo survive, and only two (which never fell into Viviani's hands) from Galileo to Sarpi, although it was said in 1616 that they were known to be corresponding.[13]

History has to be written from the documents that survive – we cannot recover those sold as wrapping paper or those destroyed by Galileo's 'faithful disciple'. But we will do a better job if we bear in mind that the gaps in our knowledge may be the result not just of happenstance but of deliberate destruction. Even if we had every document that was available to Viviani, we would have no straightforward way of knowing if there were things that Galileo thought but never said, and things that he said but never put in writing. For we can be sure that whenever Galileo put pen to paper he was aware – as every Italian intellectual of his day was aware – that an inquisitor, searching for evidence of heresy, might one day read what he had written. What hope, then, have we of getting past the layers of censorship that came between Galileo's private thoughts and the texts that survived to appear in the Favaro edition? The answer is simple: we must pay particular attention to those sources that bring us closest to Galileo's conversation with his disciples, and to those texts (the margins of books, for example) that Galileo least feared would be read by anyone else. We may not be able to speak with the dead, but we can sometimes listen in on their conversations, and even catch them thinking aloud.[14] In this book I deliberately give such snatches of overheard conversation much greater weight than they have had in previous studies of Galileo.

The first proper book Galileo published was *The Starry Messenger* (1610), in which he announced the discovery of the moons of Jupiter, a discovery which provided new evidence in favour of Copernicus's bold hypothesis that the earth moves and is not fixed at the centre of the universe. A few months after it appeared the poet John Donne published *The First Anniversary*, in which these lines are to be found:

And new philosophy calls all in doubt,
The element of fire is quite put out;
The sun is lost, and th'earth, and no man's wit
Can well direct him where to look for it.
And freely men confess that this world's spent,
When in the planets and the firmament
They seek so many new; they see that this
Is crumbled out again t'his atomies.
'Tis all in pieces, all coherence gone,
All just supply, and all relation;
Prince, subject, father, son, are things forgot,
For every man alone thinks he hath got
To be a phoenix, and that there can be
None of that kind, of which he is, but he.
This is the world's condition now . . .

We know Donne had read *The Starry Messenger* – he referred to Galileo in *Ignatius His Conclave*, written around the same time, where he made clear his own support for Copernicanism. Donne writes in *The First Anniversary* as someone who understands at once the logic of Galileo's position. What are the alternatives to the old philosophy of Aristotle and Ptolemy? The denial of any difference between the superlunary and sublunary worlds (this is what Donne means when he writes, 'The element of fire is quite put out' – the sphere or element of fire marked the boundary between the superlunary and sublunary worlds); Copernicanism ('The sun is lost, and th'earth'); atomism; egalitarianism; individualism (Galileo did indeed think of himself as a phoenix); the unending quest for the new.[15] In 1610 Galileo had only cautiously committed himself to the first two of these, but Donne apparently saw at once the scale of his ambition and the direction in which he was heading. Nobody else grasped, as Donne did, the true extent of the revolution implied by Galileo's first publication.

How did Donne come by this extraordinary insight? There is good evidence that he had been in Venice in 1605 and perhaps 1606, staying with the English ambassador Henry Wotton. It seems that he met Paolo Sarpi, who had – or was certainly soon to have – close links to Wotton.[16] Galileo and Sarpi were close friends, and there were a number of English-speaking students in Padua who knew Galileo.[17] One particular student was giving Wotton especial cause for concern through much of 1605: Thomas Seget, a Scot, had been thrown into a Venetian dungeon for having sex with a nun, and had then faced further charges trumped up by her relatives. Seget's timing was exceptionally unfortunate: during the year in which he was held, the Venetian authorities introduced new legislation making it a capital offence to be in a nunnery without good reason.[18] They were in no mood to accept Wotton's account of Seget's behaviour as a youthful indiscretion. Wotton nevertheless campaigned tirelessly for Seget to be released into his custody, and in October his efforts were at last successful. If Donne was in Venice at the time one can hardly doubt that he would have taken an interest in Seget's case, and he may well have been present in the embassy when Seget (as we may conjecture) said goodbye to his good friends – one of whom was Galileo – before Wotton arranged, as he had promised, for Seget to depart from Venice and its territories for ever. It is perhaps to these meetings that we should attribute Wotton's own interest in Galileo: in 1610 he managed to acquire a copy of *The Starry Messenger* and sent it to James I, even though the book sold out within a few days of publication.

So Donne probably met Galileo; and if not he may have learned from Sarpi or from Wotton or from Seget about the new philosophy, atomist and irreligious, that Sarpi and Galileo had in common. Such conversations would have been confidential – Protestants such as Wotton and Donne had no desire to get their Catholic allies into trouble. When Donne read *The Starry Messenger* he already knew what Galileo thought – hence the confidence with which he went beyond anything that was to be found in the text. Indeed it is likely that Donne knew Galileo better than we ever will because he had heard him talk frankly about his private beliefs. Our task – our impossible task – is not just to read what Galileo wrote, but also to rediscover what he thought. To do so we need to catch the echoes of long-lost conversations. Galileo's conversation with Donne is echoed, perhaps, in *The First Anniversary*; fortunately, there are other conversations that we know more about. If we pay attention to the conversations between Galileo and his friends, we soon find a very different Galileo from the pious Catholic described by Viviani and Favaro.

PART ONE
THE MIND'S EYE

Methinks I see my father . . . in my mind's eye.
Shakespeare, *Hamlet*

A professor teaching in Padua thought he had made some new discoveries and was dismayed that no one had painted up graffiti honouring him. So, being bold, he decided to go out one night with a ladder and paint up in all the public places, 'Long live Distinguished Professor So-and-So!' Unfortunately the police came upon him while he was up his ladder, and thinking he was a thief, they were about to take him off to prison. If they had not noticed the paint bucket tied to his waist and the paint brush in his hand, which pretty clearly indicated the nature of his madness, things might have turned out rather badly for him. But nowadays the discoverers of new things are virtually deified.[1]

Lodovico delle Colombe, *Polite Refutations* (1608)

1

His father's son

Galileo's father Vincenzo (*c.*1525–91) was a Florentine musician. In 1562 he appears in Pisa, where he marries and sets up a music school. We do not know whether he first moved there for love or for money, though we may suspect that neither the marriage nor the music school flourished. The marriage took place on 5 July 1562, and Galileo, the eldest child, was born on 15 February 1564.[2] Surviving documents include accounts supplied by a good friend of Galileo's father, Muzio Tedaldi, who handled his affairs during 1573 (when Galileo was aged nine – Muzio's expenses include the purchase of a football), while Vincenzo was based in Florence, and another document in which Vincenzo made his sister-in-law his attorney in a dispute over payment owing for some silk fabric which had originally belonged to her.[3] This, and the fact that Vincenzo had received part of his wife's dowry in cloth, suggests that he had married into a family that dealt in silk, cotton and wool, and that either he had taken up a role in the business to supplement his income as a musician, or his wife was trading under her husband's name.[4] One contemporary source claims that Galileo was originally intended for the wool business – for his mother's world rather than his father's. All sources agree that the family was, and always would be, impoverished; they were living from hand to mouth. Galileo, the eldest son, would later stress that he had inherited nothing from his father.

Something of the frustration and bitterness which must have overshadowed his early life is reflected in an annotation that Galileo made many years later, in 1619, in a book by Orazio Grassi, a prominent Jesuit critic of Galileo who had published under the pseudonym of Sarsi:

> Sarsi is like someone who wants to buy a piece of cloth made of pure silk and takes it out of the shop into the open air, and there goes over it bit by bit looking for the slightest stain or smallest flaw; and if they find the tiniest defect they run down the whole piece of cloth, find fault with it, and beat down the price, taking no account of the enormous care, patience, and time and trouble that went into making all the rest of it so perfectly; and what is even more barbarous and inhumane, having made a fuss about the slightest stain or pulled thread in that piece of cloth, when they wear it, they have it chopped up,

mangled, cut into a thousand times; and they wear it to a masked ball, or to go gambling, or to the theatre, knowing that before the evening is over it will be completely smeared with mud and torn into pieces.[5]

When Galileo wanted to conjure up a vivid image of someone not being given their due, he went back in his mind to the Pisa of his childhood, to his parents' unsuccessful attempts to make money from the cloth trade.

The young lad quickly showed himself to be interested in making little mechanical models – of ships and watermills. He learnt to play the lute to a professional standard, as one would expect of his father's son. And he showed considerable aptitude for drawing and painting; in later life he spent much time in the company of painters, and said he would have been an artist if he had not become a philosopher.[6] Galileo's willingness to work with his hands (in an age when manual labour was regarded as demeaning) was crucial to his later success: for twenty years his telescopes were the best simply because he ground his own lenses, while other astronomers purchased lenses from spectacle makers.[7]

There is an episode from later in Galileo's life which neatly illustrates the ambiguity in social standing that resulted from his willingness to work with his hands. In 1630 the Florentine ambassador in Madrid promised the king of Spain a Galilean telescope. A Florentine diplomat, travelling to Madrid, was instructed to learn how to use it and bring it with him, but he evidently regarded the matter as a low priority, and was too pressed for time to comply with his instructions. When he got to Spain he was very surprised to find himself pestered by the king for the telescope, and was obliged to write off to Florence to arrange for it to be sent after him. Soon after its eventual arrival the objective lens fell out and broke, and he was instructed to obtain another one as quickly as possible. The letters he wrote to Florence provoked an impatient, even haughty, response. He did not seem to have adequately conveyed to the Spanish government, he was told, that the maker of the telescope was not some sort of artisan, to be ordered about, but a great philosopher and a great man, to be approached with due deference.[8] In Florence, which had a mercantile nobility, the idea of a great man who used his hands was just about comprehensible to some people (though not, it seems, to Esaù del Borgo, the unhappy diplomat who had completely failed to grasp the value of Galileo's telescope for the king of Spain); not in Spain, however, where the king presumably thought Galileo had the same social status as the engineer and painter Cosimo Lotti, who, before moving from Florence to Madrid, had sometimes been employed making fancy cakes for the grand duke's dinner parties.

The young Galileo acquired the standard humanist education of the day – Latin, Greek, Aristotelian logic – from schoolteachers, first in Pisa, and then in Florence, where he moved around the time of his tenth birthday with his mother and his sister Virginia (born a year before, in 1574) to join his father, who was slowly making his way at the Medici court. Young Galileo also spent some time as a novice in a monastery at Vallombrosa, forty miles east of Florence, presumably in order to obtain a free education, and was about to take his vows (so perhaps he was fourteen) when his father, who surely wanted his eldest son to pursue a career rather than abandon the world, came and took him away, claiming he needed urgent medical treatment for an eye infection. When we get to know Galileo's mother Giulia better we will find that she was the sort of woman who would have wanted a monk for a son (and indeed Galileo was never to marry), so here we may guess at a struggle between Galileo's parents over what should become of their child. In later years Galileo must have told his friends how close he had come to a life of poverty, chastity and obedience: the subject surely came up when he returned to teach at Vallombrosa in 1588.[9] One friend, turned enemy, chose 'sfrattato', unfrocked monk, as his preferred insult, scribbling the word repeatedly in the margin of Galileo's testimony in a court case. Perhaps, at times, Galileo regretted his father's decision to rescue him from a monastic life: he was puzzled that his sister preferred marriage to a convent, and he chose to consign his two acknowledged daughters to a nunnery.[10] He may have thought that he had their best interests at heart when he did so. By the time he died, though, he had changed his attitude to the religious life. In his last will he disinherited any descendant who entered a monastery or nunnery.[11]

When Galileo was sixteen his father resolved to find the money to put him through university in Pisa. His plan was for him to become a doctor: indeed he had named him Galileo after a famous ancestor who had been a doctor; this was the original Galileo from whom the Galilei family claimed descent. Galileo was to live in Pisa in the home of his father's old friend Muzio Tedaldi – although there was a rather striking condition: Tedaldi must first marry the woman with whom he was living, or else break off with her. Since Tedaldi's mistress (if she was his mistress – he vigorously denied that theirs was a sexual relationship) was a relative of Galileo's mother, to whom respectability mattered greatly, we can guess that this was Giulia's condition rather than Vincenzo's. Later, when Galileo had a mistress of his own, he did not live with her, and there is no sign that he ever considered marrying her.

Galileo's father must have been an important influence upon the young Galileo. Vincenzo was the author of a significant work of musical theory, the *Dialogue on Ancient and Modern Music* (1581):[12] Galileo was following in his father's footsteps when he wrote dialogues in Italian. In later life Galileo had copies of this book lying around – we find him lending one to a friend.[13] In his study of musical harmony, Vincenzo conducted experiments, and Galileo certainly knew of these and may have helped set them up. Vincenzo showed that it was not just the length of a string that determined the note it played, but its material, diameter and tension; this involved a simple experimental apparatus where weights were hung from strings of different materials. But of course every lute player already knew that you can raise the note of a string either by shortening it – by pressing down on a fret – or by tightening it; Vincenzo would also have known that a new string requires a new tuning. So his experiments, as we now call them, were really ways of illustrating what he already knew. He was impatient with people he called Pythagoreans, who thought that music could be explained by simple mathematical ratios, and insisted that all knowledge must pass the test of experience. We can be confident that he told the young Galileo to distrust abstract theorising. Galileo was always a musician's son. When he wants to describe the reform he is trying to introduce into philosophy he compares it to tuning an organ. But at the same time he was always (at least after 1597) quick to insist that he was a Pythagorean: not only did he believe in the power of mathematical abstraction, but Pythagoras had first proposed the view that the earth went round the sun. Every time Galileo declared himself to be a Pythagorean, Galileo disowned his father.

The elder Vincenzo may well have symbolised failure to his son; he certainly presented him with a model of thwarted ambition. Vincenzo saw himself as a great man who, for all his modest success at court, had never been given his due. Galileo's obdurate refusal to submit to censorship, to sink back into anonymity, may be interpreted as a refusal to relive his father's failure. But Vincenzo's most significant example to his son may lie elsewhere. The culture of late sixteenth-century Italy was one obsessed with the imitation of the ancients. In philosophy, in medicine, in law, in sculpture, the ancient models were still the ones to follow. In music the situation was more complicated. Vincenzo accepted the current view that ancient music was monodic, while modern music was polyphonic. Ancient music, he thought, spoke directly to the soul and was far superior to anything produced by the moderns. But while he experimented with monody in a number of compositions (none survive, but they are thought to have been precursors to recitative in opera), for the most part he wrote 'modern' music.

It has been claimed that Vincenzo's acute awareness of the unattainability of the classical ideal makes his own music the first that is self-consciously modern, and that his idealisation of what was lost meant that he was trapped in an intellectual world that was confused and contradictory.[14] The young Galileo was to opt for a simpler world. Renaissance mathematicians claimed to be able to do what the ancients had done, only better. They believed, quite straightforwardly, in progress. To the young Galileo, impatient with his father's angst, this must have seemed a deeply attractive idea. Here at last was a world in which you could be proud of what you had achieved.

It might be thought that the idea of progress is anachronistic as applied to the world of Galileo's youth, and certainly, outside the fields of painting and mathematics, the idea was tenuous indeed in the late sixteenth century.[15] But it became stronger as the decades went by. The 1630s saw the publication in Venice of a major two-volume treatise designed to show the superiority of the moderns to the ancients: Secondo Lancellotti's *Nowadays, or Today's Minds as*

Good as Ever.[16] In 1635 Fulgenzio Micanzio, reporting the invention of a new pump to power fire hoses, expressed the view that people were getting cleverer all the time.[17] In 1637 a friend of Galileo's claimed that the discovery of the circulation of the blood would revolutionise medicine, as the invention of the telescope had revolutionised astronomy, the compass had revolutionised navigation, and gunpowder had revolutionised warfare.[18]

And yet, despite placing himself firmly on the side of progress, Galileo's ultimate fate would be to repeat what his father had done. Vincenzo Galileo rejected the Pythagorean notion of a harmony that ran through the cosmos and implied that there could be no music of the spheres: sounds are the imperfect products of particular physical objects. His son was to end up destroying the natural correspondences that were supposed to exist between the microcosm (the little world of man) and the macrocosm (the divinely ordained universe), and was to remap the cosmos so that heaven and hell could no longer be located in physical space. He was to leave many of his readers longing for a past that was no longer attainable, just as Vincenzo's readers must have done.

2

Florence

Galileo was born in Pisa, spent much of his childhood there, went to university there, and got his first proper job there. But Viviani thought it important to stress that his father was a Florentine gentleman (albeit one who had fallen on hard times) who just happened to spend a few years in Pisa and marry there. Throughout his life Galileo referred to himself as a Florentine gentleman, 'nobil fiorentino'.[1] Italians frequently identify people by the towns they come from – Leonardo da Vinci, Pietro Perugino – and the Pisans (and some others too) naturally call Galileo 'il pisano'. This would not have pleased him. Pisa was a sleepy backwater. It had once been a free and independent city, a bustling port on the river Arno; but now the population had declined to a mere ten thousand, the streets were half-empty, the university had only a local reputation. Florence was the capital of Tuscany, with a population of eighty thousand. There power was concentrated and fashions were set. Galileo inherited his father's conviction that Florence was and always would be his true home. He went back to Pisa as a student, to the world of his mother's family, with no sense that he was going home. In the university register he was entered as 'Florentinus', from Florence, not 'Pisanus', from Pisa.[2] In later years it was commonly assumed that he had been born and raised in Florence;[3] but there was gossip to the contrary. An enemy, denouncing him to the Inquisition, said that he represented himself as a Florentine, but that he was in fact a Pisan.[4]

Galileo remained an inhabitant of a small part of northern Italy throughout his life. Take a map and a pair of compasses and draw a circle with Florence at its centre and with a radius of 275 kilometres or 170 miles: Galileo never stepped outside that circle. Genoa (with a population of 60,000) to the north and west, Venice (150,000) and Loreto to the east, and Rome (110,000) to the south mark the limits of his travels. In a telling phrase, he once described someone who had been to Genoa, Rome and Milan as having seen the world.[5] His brother went to Poland in search of work as a musician; Galileo stayed close to home. It is true that he talked of travelling further – to Naples, even to Spain.[6] When he did so he always looked south, not north. The books he wrote made the long journey across the Alps, but he did not. In Padua he taught students from all over Europe, but he never went to visit them. A generation before, the philosopher Giordano Bruno had lectured in Oxford and lived in Paris and London, returning to Italy to be condemned as a heretic and burnt alive. Galileo had no intention of following his example.

But during Galileo's long life the world changed. When he was four the Dutch revolt began. Six years after he died, the Treaty of Münster recognised Dutch independence from Spain. Throughout Galileo's life, the balance of wealth and power in Europe was slowly shifting from south to north, from Catholic to Protestant, although there were periods (the years following the Battle of the White Mountain of 1620, for example) when it looked as though this trend would be reversed. At the end of his life Galileo was dependent on a Genevan go-between and a Dutch publisher, and was engaged in negotiations with the Dutch government. The first proper life of Galileo was published in England in 1664. Galileo did not foresee these changes, nor work to speed them up (as, for example, Paolo Sarpi did). If anything, his move from Venice to Florence in 1610 suggests that he misread the course of history, for Florence looked south for its allies while Venice looked north. Had these changes not taken place Galileo would have been remembered, if at all, as an obscure heretical philosopher whose work had been condemned and removed from circulation. If he is remembered as the founder of modern science it is thanks – despite the efforts of Viviani – to Catholic, not, Protestant Europe.

Galileo seems – on the evidence, at least, of his surviving papers – to have had little interest in European politics. We may also doubt that distant places inspired his imagination. In 1617 his best friend, Gianfrancesco Sagredo, then in Venice, sent him a caged bird that had been brought all the way from Agra in India. Sagredo (who had been Venice's representative in Aleppo) found the thought of the bird's vast journey moving. He decided to give the bird to Galileo because he feared he would be deeply upset to have such a precious creature die in his possession, and it seemed almost inevitable that he should outlive it. The bird was sent – and Sagredo heard nothing. No acknowledgement, no thanks. Eventually he extracted news of it from Galileo: the cat had caught it.[7] There is nothing to suggest that Galileo was distressed (as Sagredo certainly would have been) by this unfortunate turn of events.

The bird, on Sagredo's own description, was indeed rather nondescript. But Sagredo loved all creatures – he described his house as a sort of ark. We know the names of his two dogs (obtained for him by Galileo from Florence, where the finest hunting dogs were bred), we know that the bitch was beautiful and the dog was ugly, and we know that the dog chewed the furniture and scrabbled at the rugs.[8] And Sagredo, in his imagination, had shared the bird's journey from India to Italy. Galileo collected fruit trees. As far as we know he had no dogs. And India, it would seem, lay beyond the limits of his imagination. This is surprising: after all, he travelled in his mind's eye to the planets and the stars. It looks as if he was drawn to places where there was no risk of meeting another human being. This is surely because he thought that human beings are, by and large, rather nasty. They can rarely be bothered to help one another, he noted, but are quick to seize any opportunity to do each other down.[9] Moreover there was a bias in the world of which he had become acutely aware: most people were ignorant, and nothing provoked the hostility of the ignorant as much as people who knew more than they did.[10]

Although at the end of his life Galileo was desperate to have his books published in Latin, the language of international intellectual life, and although this was the language in which he lectured throughout his career as a university teacher, he wrote only one short book in Latin: *The Starry Messenger*. He preferred, like his father, to write in Tuscan, the language of a Florentine, of Dante, of Tasso, of Ariosto. As a young man, lecturing on Dante to a Florentine audience, he apologised for the fact that during his lecture he would have to introduce some words derived from Greek or Latin into 'the pure Tuscan language'.[11] Until his books were banned, for so long as he could have Italian readers, he wanted to be read, first and foremost, not in Oxford or Cambridge, Paris or Prague, but in Florence and Bologna, Rome and Venice, Padua and Pisa. Pisa was far too small for the ambitions of young Galileo, but northern Italy was plenty big enough. And far bigger, we may add, than it seems at first to us: in the circle around Florence within which Galileo travelled there were in 1600 some sixteen cities with a population of more than ten thousand, as against seven in the whole of France and four in England. (York was about the same size as Pisa, even though London was much bigger than Venice.) In this respect Galileo lived out his life within his father's world. He justified his decision to publish his second book in Tuscan by saying that what he really cared about was what his fellow Florentines thought of him.

But there is more to Galileo's decision to publish in Tuscan than a desire to address himself to an Italian audience. Latin was the language of international intellectual life, but that was because it was the language of the universities. And in the universities the Aristotelian science that Galileo was committed to attacking held sway. By publishing in Italian Galileo was appealing against the universities to an educated laity, just as Luther, by publishing in German, was appealing against the theologians to an educated laity. Even before Galileo published under his own name in Italian, in 1612, he had been closely associated with a satirical tract written in the Paduan dialogue (1605) and another pseudonymous work written in Tuscan (1606) – he was already trying to escape the narrow world of university learning.[12]

Of course Galileo's books travelled great distances – even to Moscow. But to read all but the first of them you had to hear in your head a language that we anachronistically call Italian. In Galileo's day, however, there were many different Italian languages, with Galileo himself writing in the language of Florence. When one of his books was published in Rome he was anxious lest the

Roman typesetters should turn his Tuscan Italian into Roman Italian, and when he received letters in proper Tuscan from someone in Augsburg who had been reading his books, he said it made him feel as if he had brought Augsburg within the walls of Florence; indeed it made him feel as if Florence had conquered Augsburg.[13] Thus, in his mind's eye, his books brought the world to him, not him to the world. Even so, his Italian readers mattered far more to him than any others. Towards the end of his life he was told that the Latin translation of his *Dialogue* was converting astronomers throughout Europe to Copernicanism; his reply was that since no one in Italy could get permission to read it he feared that eventually all memory of it would be lost.[14] Dutch readers simply did not count.[15]

3

Galileo's lamp

Now for Galileo's intellectual life. It is worth acknowledging what lies ahead of us. There is far too little information about Galileo's intellectual life in the period from 1581, when he became a university student, to 1610, when he became famous. Much of my argument over the following pages will of necessity be tentative and conjectural. If an account existed on which we could rely I would not hesitate to provide a bald summary and move on, but none of the existing accounts will do.

What the next few chapters attempt to trace is a double process. First there is the invention of a new physics, a physics which is finally codified by Newton. Galileo develops an idea of inertia, discovers the law of the acceleration of falling bodies, the parabolic path of the projectile, and the isochronicity of the pendulum (the principle that every pendulum keeps time, and any two pendulums of the same length keep the same time), and formulates for the first time the idea that motion is a relative concept. These five innovations, taken together, constitute the most important development in science between Aristotle and Newton – or, in the words of a contemporary of Galileo's, the most important advance in philosophy for two thousand years.[1] Second there is a crucial advance in the experimental method. Galileo did not invent the experimental method, and his commitment to it was limited. Nevertheless he put it to more effective use than anyone had ever done previously, and it was his students who became the most effective proselytisers for an experimental science.

The problem with this double process – the new physics and the experimental method – is that it is very easy for us to underestimate the obstacles that stood in the way. It is perfectly possible to carry out experiments without having any grasp of the power of the experimental method. It is equally possible to discover one or another element of the new physics (the parabolic path of the projectile, for example) without having any grasp of the full range of interlocking arguments that were apparent to Galileo by 1610. If we are to do justice to Galileo's achievement we have to see that he advanced step by step without knowing where he was going; only at the end could he look back and see that he had constructed a new science. We can see the implications of each step; so, in the end, could he. But only when he reached the end of his journey.

Galileo was sent to university in 1581 to study medicine. Medicine was not only the path to a well-paid career; it was also the most empirical and practical of all the sciences, the one least tainted by abstract theorising, and so a suitable choice for the anti-Pythagorean Vincenzo's eldest son. All doctors were expected to learn a certain amount of mathematics towards the end of their training, mainly in order to cast the horoscopes of the patients; these would tell them which humour naturally predominated in the patient's complexion, and when it was best to administer a therapy. Astrology was a recognised part of respectable medicine, and astronomy and astrology were not separate disciplines, but two aspects of one enterprise. Galileo, however, was impatient to learn some mathematics, and turned to a Florentine friend of his father's, Ostilio Ricci, who taught mathematics to the young men at the Medici court, and who agreed to teach him without his father's knowledge. No sooner was Galileo exposed to mathematics than he fell in love with it. His father quickly became suspicious that Galileo was no longer applying himself to his medical studies and journeyed to Pisa to find out what his son was up to. But for a while Galileo kept up the pretence of studying medicine, hiding his slim maths books within the fat textbooks of medicine to conceal what he was reading from his father.[2]

It is worth pausing over this image of Galileo hiding his real thoughts and feelings from his father, for it points to the future. When Galileo was forbidden to write in support of Copernicanism he reacted in the same way: he postponed outright rebellion for as long as he could, but tried to create his own private world in which he did as he pleased. Later, after his condemnation, he appeared to obey the Church, even while arranging for his manuscripts to be smuggled abroad. The social scientist Albert O. Hirschman has identified three ways in which people respond to affiliations (to political parties, institutions, suppliers of goods and services, spouses, even Churches): loyalty, voice and exit.[3] I love my wife, and will stick with her. I have problems with my wife, and want things to change. I am leaving my wife, and want nothing more to do with her. Galileo practised outward loyalty, inward exit. This is resistance rather than rebellion, subversion rather than revolution, conspiracy rather than compliance.

According to Viviani, Galileo made an important discovery even before turning to mathematics. One day when he was attending a service in the cathedral he noticed a lamp dangling from a long chain. It had recently been lit and, as a result, was swinging slowly back and forth. Counting the length of each swing against his pulse or against the beat of the liturgical music, he realised that even though the arc of each swing was shorter than the one before, every swing took the same amount of time. Viviani says that he went on to confirm this by experiment. Galileo had discovered an accurate way of counting time. (Before the invention of the pendulum clock, clocks were driven by weights, and were far from accurate.) Initially, as a medical student, he used his pendulum to measure the pulse rate of patients.[4]

Viviani's story is sometimes interpreted as implying that Galileo, as a teenager, had discovered what is now called the law of the pendulum. If this were true, then Galileo would already have founded modern physics. Aristotelian philosophers had paid no attention to the pendulum, but if they had they would have been quite unable to make sense of it. In the first place a pendulum seems to have a tendency to go on almost for ever, while in Aristotelian physics movement is supposed to stop as objects approach their natural resting place. Secondly, it is clear that a pendulum slows down and speeds up as it swings. Medieval philosophers had come up with various sophisticated accounts of why a falling body might pick up speed as it fell and why a body thrown upwards might lose speed as it rose, but none of them could account for the perfect symmetry of a pendulum's alternating acceleration and deceleration. Merely to see what a pendulum does, let alone to understand it, you need to stand on the threshold of modern physics.

Viviani's story has long been dismissed as a myth. In Galileo's own papers there is no evidence of an interest in pendulums until twenty years later. The lamp in Pisa cathedral that, in the nineteenth century, was said to be the one he had observed swinging – Galileo's lamp, it was called – turned out to have been installed soon after Galileo's student days. In 1614, Santorio Santorio, a friend of Galileo's, claimed to have invented the *pulsilogium*, a pendulum device for measuring the pulse, and Galileo, who was not someone who would allow even a good friend to take credit away from him, never contested this claim. Nevertheless, Stillman Drake, the greatest of English-language Galileo scholars, who wanted to claim that Galileo had learnt the experimental method from his father, pointed out that he may well have been introduced to the pendulum by his father's experiments with weights on strings, and was willing to concede that he may have started experimenting with pendulums as a young man.[5]

We will never know if the young Galileo sought to measure the duration of a pendulum's swing, but he could easily have concluded that each pendulum swings at a specific rate without discovering the law of the pendulum. According to the law of the pendulum, what determines the time taken by a pendulum's swing (at least in a vacuum) is the length of the line, not, for example, the weight or the specific gravity of the bob, or the material from which the line is made. You cannot reach this conclusion without constructing an elaborate new physics, which was certainly beyond Galileo's capacity as a medical student – the very idea of a new physics had yet to occur to him. At a simpler level, the law of the pendulum entails that any two pendulums with lines of the same length will swing at the same speed – they will be synchronous. But set up a number of pendulums with lines of the same length and you will discover that they swing at different speeds (at least they will if the lines you are using are hemp or linen threads, which stretch after being measured); only if Galileo already knew the law of the pendulum to be true would he have concluded that the lengths were not exactly the same and adjusted them until the pendulums swung in time.[6] It is thus possible to use an empirical test to confirm the law of the pendulum, but you would never discover the law through an empirical test. Rather you would discover that you can tune pendulums, as you can tune the strings on a musical instrument. This is an example of two principles that have become of fundamental importance in the history of science: first, there is always more than one viable interpretation of the outcome of an experiment, so there is more than one way of explaining why a particular pendulum always keeps time; second, replicating an experiment is never straightforward, so it is very unclear what counts as setting two pendulums of the same length swinging. Viviani's story closes the gap between a discovery and its interpretation, while modern history of science wants to emphasise the size of this gap.[7]

Thus the young Galileo may well have reached the conclusion that any particular pendulum has a specific, characteristic rate of swing, just as any particular organ pipe sounds a particular note. He may well have seen an analogy between a swinging pendulum and the vibration that produces a musical note. But he would have interpreted his discovery in the light of his father's work on vibrating strings, and this would have led him to conclude that there could be no simple mathematical formula governing the time taken by a pendulum to complete a single swing, just as his father had shown that there was no simple mathematical rule governing the notes made by lute strings. He may well have seen what a pendulum does, but he would still have been a long way from understanding it.

There is an important difference between a pendulum and a lute string. A musician can hear if a plucked string sounds the note of C; but you cannot see if a pendulum marks the seconds. To measure time you need already to have what you are looking for: a reliable reference point; the reliable ones that nature supplies – the sun and the night sky – measure days, not seconds, and it was not until much later in his life that Galileo devised a pendulum that would count seconds. The *pulsilogium* was presumably set to mark the same time as its inventor's resting pulse, since it was invented long before there was a reliable unit for measuring short periods of time.

4

Eureka!

After three-and-a-half years studying medicine in Pisa, Galileo returned to Florence in 1585 to study mathematics openly. He was now working only for his own satisfaction. He quickly progressed from Euclid to Archimedes, whose key works had been available in print only since 1544. Archimedes was a revelation. Throughout the rest of his life, Galileo's admiration for Archimedes, the Greek mathematician, engineer and scientist of the third century BC, knew no bounds. 'It is all too obvious that no one else has ever been nearly as clever as he was.'[1] He is the 'superhuman Archimedes, whose name I never mention without a feeling of awe', 'the most divine Archimedes', someone Galileo reads with 'unending astonishment' ('infinito stupore'), the works of Archimedes becoming for him the paradigm of what science should be.[2] In so far as Galileo remained to the end of his life a disciple of Archimedes, he never became, and never could become, a 'modern' scientist. For modern science, knowledge begins with facts about the world, often facts established by experiments. And although Archimedes worked with balances, levers and pulleys, although he studied bodies floating in water, he aspired to produce a deductive science of weights and forces, one which would start from indisputable premises and reach incontestable conclusions. He aspired to a physical science which would be modelled on geometry. Galileo's goal throughout his life was a similar deductive knowledge. In such a science a real object, such as a pulley, or a real experiment with floating bodies, might be used as an example to help one think, just as someone studying geometry will draw circles and triangles, but this knowledge is no more grounded in facts than the claim that the angles of a triangle add up to 180° is based on measuring angles.[3] In later life, Galileo's devotion to Archimedes was an obstacle to his reflecting on, or even reporting, his own experimental practices, and the tension between Galileo the disciple of Archimedes and Galileo the founder of a new experimental science must lie at the heart of any account of Galileo at work.

Galileo's relationship to Archimedes is immediately apparent in a work Viviani tells us he wrote in 1586, and that survives in several manuscript copies: *The Little Balance*.[4] Archimedes is, and was, famous for crying, 'Eureka!' as he leapt from his bath and ran naked through the streets of Syracuse. He had solved a problem that had seemed insoluble: Hiero, king of Syracuse, had given a goldsmith some gold from which a crown had been made. The crown weighed the same as the gold out of which it was supposed to have been made, but the king came to suspect that the goldsmith had adulterated the gold with silver, and had stolen some of the gold. He was, however, reluctant to melt the crown down in order to find out whether his suspicion was justified.

One day Archimedes noticed that as he got into the bath the level of the water rose. By putting the crown in a tank one could measure the amount of water it displaced. Silver weighs less than gold, so a crown of silver would be larger than a crown made of the same weight of gold. By measuring the volume of the crown one could tell, approximately, the proportion of silver and gold in it.

This has always been seen as an example of Archimedes' brilliance as a scientist. Not in Galileo's view: the proposed solution was too crude; the measurements would be merely rough and ready. It was not worthy of Archimedes. What he must have done is take a very precise balance and use it to weigh the crown both in air and in water – for it is much easier to measure weights exactly than it is to measure volumes exactly. By doing the same with both gold and silver he could work out exactly how much heavier than water gold, silver and the crown were. What Galileo was introducing was the concept of specific gravity, and he goes on to provide a table of the specific gravities of different materials. Once you knew the specific gravity of gold, silver and the crown, you could tell exactly, not just approximately, what the crown was made of.

The concept of specific gravity is important, and Galileo was putting it to sophisticated use. But Archimedes had the concept. The difference between Galileo and Archimedes is that Galileo wants to measure more exactly, and having measured one substance he wants to go on and measure others. In order to do this he has to devise a new method of measurement. To measure the difference between what an object (a lump of gold, for example, weighing the same as Hiero's crown) weighs in air and what it weighs in water, Galileo balances that object against an equivalent weight in air, and then lowers the object into water. He then marks on the arm of the balance with a fine wire the point to which he has to move the weight in order to restore equilibrium. He does the same with a lump of silver. He now has two points marked on the arm of his balance. Between them he winds the finest brass thread, each thread pressed against the next as tightly as possible. He then balances the object whose content he wishes to analyse (Hiero's crown), first in air, then in water. He only has to count the number of threads on each side of the point from which the weight is suspended on the second weighing to know the exact proportion of gold to silver in the crown. The threads, if they have been wound properly around the arm, are too fine and too close together to be counted by eye. You count them by running a very sharp knife gently and slowly ('adagio, adagio') across them. You can then both hear and feel the number of threads that pass under the knife's blade. Galileo has thus achieved a new precision in measuring the difference between two weights.

The Little Balance shows that from the very beginning Galileo was preoccupied with precise measurement.[5] But what is also important here is his claim that the standard account of Archimedes in the bath is not worthy of so great a man. He is claiming that he has a better understanding of Archimedes than anyone else – and there can be only one reason for this. It is because he alone is Archimedes' equal. In this, the first scientific text we have from his hand, Galileo declares that his ambition has no bounds. And he declares that his confidence is justified because he has already outstripped all his predecessors, all but the great Archimedes. Implicitly, of course, he is claiming to have outstripped even his hero – for there is no evidence that Archimedes had invented the little balance.

5

Seeing is believing

Shortly after writing *The Little Balance*, in the last months of 1587, Galileo sent to the leading mathematicians of northern Italy his solutions to a problem that derived from Archimedes' work.[1] He had been working on this problem for a couple of years. It had apparently been presented to him by the Marchese Guidobaldo del Monte, who was to become Galileo's first patron, and whose work was to have a significant influence on Galileo up to the time of Guidobaldo's death in 1607.[2] The issue was how to calculate the centre of gravity of various solids, and Galileo remained sufficiently fond of his solutions to publish them at the end of his life, although he abandoned work on the topic when he discovered that another mathematician, Luca Valerio, had formulated solutions which, if different from his own, were comparable in quality.[3]

The first problem Galileo addressed was that of determining the centre of gravity of a yardarm which has spread along it at equal distances a series of five weights, weighing from one to five units. His solution to this problem puzzled both Guidobaldo and Christopher Clavius, the leading professor of mathematics at the Jesuit College in Rome:[4] surely he had defined the problem in such a way that his solution was already included in the definition (what logicians call *petitio principii*)?[5] Galileo's response to these queries was not to restate his proof in different terms. It was not to argue that the proof was valid and did not include the logical flaw they attributed to it. It was to send both of his correspondents a new drawing of the problem in which the five weights were bunched up together instead of being spread out along a rod. If they looked at this drawing, he said, they would understand his solution.[6]

Guidobaldo found Galileo's response satisfactory ('patet sensu' – I can see you are right);[7] but Clavius repeated his view that this was no proof.[8] Strictly speaking, Galileo's reply was no reply. His response to a question about logic was to offer a different way of visualising the problem – his claim was that you could *see* that his answer was correct.

Galileo's argument may not be logical, but we are likely to find it convincing because we live in a culture in which seeing is believing. Almost all the information that most matters to us, the information that determines whether we live or die and whether we prosper or fail, comes to us through our eyes – from books, from computer screens, through the car windscreen, from the dials on the dashboard, from X-rays and CT scans. Ours is a visual culture, and Galileo is one of the people who constructed that culture. One of the great puzzles of history is why it took three hundred years to invent the telescope: spectacles were invented in 1284, the telescope – essentially a combination of two spectacle lenses – in 1608.[9] One answer is that the telescope and the microscope are profoundly problematic because they require you to rely on one sense, and one sense alone – the sense of sight.[10] Sight had always been regarded as the sense which was most easily deceived: what is perspective painting but a deception of the eyes? The apostle Thomas did not believe that Jesus had risen from the dead when he saw him – belief came only after he touched him. Machiavelli presumably had Thomas in mind when he complained in chapter 18 of *The Prince*, the chapter on dissimulation, that most men judge with their eyes rather than their hands. Scepticism was particularly appropriate when it came to the use of lenses. What most lenses do is magnify and distort: they are deceptive rather than veridical. The fundamental realisation that our own capacity to see depends on a lens within the eye came only with Kepler's *Optics* of 1604, just before the invention of the telescope. And so sight, particularly when dependent on a lens, was to be trusted only when it could be confirmed by touch or sound or smell. To take the idea of a telescope seriously you have to entrust yourself to your eyes; turning it to the heavens you had to trust it to provide information on a world you would never touch, hear or smell. Galileo was already willing to entrust himself to his eyes: hence his conviction that the telescope could tell him all he needed to know. In his work the old adage that sight is the most noble of all the senses takes on a new meaning.[11]

The scale of the cultural resistance to purely visual information is easy to illustrate, because astronomers had always relied on such information. According to the astronomers the planets moved along complicated paths through the skies – epicycles on circles, effectively spirals. According to the philosophers these were merely hypotheses to 'save the appearances', workarounds to produce results that matched the data. In reality, they insisted, the planets were attached to crystalline orbs that performed perfect circles in the heavens. The result was mutual incomprehension, beautifully illustrated in 1616 when Cardinal Bellarmine argued that Copernicanism was a hypothesis just like epicycles – no one thought they were real, so why should anyone imagine that Copernicanism was true? Galileo's reply was simple: every astronomer thought that epicycles were real, and every Copernican thought Copernicanism was true.[12] But the astronomers could not see their epicycles: they deduced their existence from compiling tables in which they recorded the locations of planets in the sky. What they could see and what they believed in were two very different things, even if the astronomers' knowledge was grounded in visual information.

The astronomers, by 1616, not only had the new evidence of the telescope; they were working in a culture which was increasingly receptive to visual information. It is a nice coincidence that the first recorded instance of the use of the phrase 'seeing is believing' in English dates to 1609, the year of Galileo's first telescopic observations. There is a telling indication from Galileo's own world that his reliance on visual information was in step with the times. In 1611 he was invited to join the Accademia dei Lincei, the Academy of the Lynx-Eyed, a learned society dedicated to the advancement of 'free' (i.e. anti-Aristotelian) philosophy founded in 1603 by the Marchese Federico Cesi (who became Prince Cesi in 1613).[13] The symbol of the Linceans was a drawing of a lynx's eye, representing acuteness of vision; from that moment on Galileo not only refers to himself as 'Linceus', a member of the society; he adopts the lynx's eye as one of his three personal symbols (the others being the parabola of the projectile and the Medicean stars).[14] We should not be surprised to find an opponent complaining of the Linceans that they placed too much reliance on the acuteness of their eyesight.[15]

Galileo was well aware that the revolution he was seeking to bring about required a new attitude to the senses, and to vision in particular. We, who take that attitude for granted, have great difficulty in imagining the obstacles Galileo and his friends had to overcome. In the *Dialogue*, he (or rather Sagredo, a sympathetic character in the dialogue) tells the following story:

> One day I was at the home of a very famous doctor in Venice [probably Galileo's friend Santorio Santorio, the inventor of the *pulsilogium*], where many persons came on account of their studies, and others occasionally came out of curiosity to see some anatomical dissection performed by a man who was truly no less learned than he was a careful and expert anatomist. It happened on this day that he was investigating the source and origin of the nerves, about which there exists a notorious controversy between the Galenist and Peripatetic doctors. The anatomist showed that the great trunk of nerves, leaving the brain and passing through the nape, extended on down the spine and then branched out through the whole body, and that only a single strand as fine as a thread arrived at the heart. Turning to a gentleman whom he knew to be a Peripatetic philosopher, and on whose account he had been exhibiting and demonstrating everything with unusual care, he asked this man whether he was at last satisfied and convinced that the nerves originated in the brain and not in the heart. The philosopher, after considering for awhile, answered: 'You have made me see this matter so plainly and palpably [*sensatamente*] that if Aristotle's text were not contrary to it, stating clearly that the nerves originate in the heart, I should be forced to admit it to be true.'[16]

For an Aristotelian philosopher seeing was not the same as believing. For Galileo seeing *was* believing. Interestingly the stress on seeing is Galileo's: when Santorio tells his version of this story – which is presumably Galileo's source – he writes of 'sensing' rather than 'seeing', for the anatomist feels as well as sees what he is doing.[17] But of course that seeing should be believing is too simple a rule: Galileo praises Copernicus for having held fast to his theory even though the apparent size of Venus and Mars did not alter as his theory implied it should. Now, looking at Venus and Mars through a telescope, one could *see* that Copernicus had been right all along.[18]

In Galileo's 1587 argument about centres of balance lies the first indication of his way of working: he thinks with a pencil in his hand, and he uses his pencil to sketch out diagrams until he draws a diagram in which he can *see* a solution. Then, with the diagram to hand, he tries to turn the solution he has drawn into a logical argument.[19] He used the same method when teaching. His second biographer, Niccolò Gherardini, who, like Viviani, had known him personally, tells us that Galileo made his students *see* what he was talking about.[20] He did not just speak, he demonstrated, just as in medicine doctors displayed the dissected body to their students when teaching them anatomy. His friend the artist Lodovico Cardi known, after his birthplace, as Cigoli, who had watched Galileo at work, said that a mathematician without a diagram was only half a mathematician, as disabled as a blind man.[21] And in *The Assayer* Galileo tells us that sight is the supreme sense, the one we rely on for knowledge.[22]

There are various ways of interpreting Galileo's method of thinking with diagrams. It was a personal peculiarity: this is a man who could have been an artist, whose visual sense was highly developed. It was an instance of a major shift in early modern culture, a shift ultimately caused by the printing press, which had made illustrations reproducible for the first time. But if he had been asked, Galileo might well have said that he was merely following in the footsteps of Archimedes.

In 212 BC, Syracuse, which Archimedes had helped defend with burning glasses and other advanced, even fabulous, technologies, was captured by the Romans. The Roman general, Marcellus, ordered that Archimedes was not to be harmed. A Roman soldier found Archimedes on the beach and ordered him to come to meet the general. Archimedes replied that he would be happy to come later, but that he was a bit busy right now. The soldier promptly executed him for disobeying orders – a classic example of that well-known oxymoron, military intelligence. What was Archimedes busy doing? Drawing diagrams in the sand.[23]

If we take Galileo's two earliest (and roughly contemporary) pieces of work, *The Little Balance* and the theorems on centres of gravity, we find in them two features that make him different from other mathematicians of his day that give him a peculiar advantage when it comes to constructing a new science: a preoccupation with exact measurement, and a belief that a sketch can enable you to see a mathematical theorem before having worked out how to put it into words – even when perhaps you cannot put it into words.

Galileo's theorems on centres of gravity were intended as a sort of portfolio to show what he was capable of doing. He circulated them as evidence that he should be considered for university positions in mathematics. But it would be entirely wrong to take this as a measure of his ambition. In his correspondence with Guidobaldo he apologises for not having spelled out the details of his simpler proofs; he has omitted these, he says, because he is following the example of the great mathematicians, who often skip the details.[24] Just as in *The Little Balance*, where he places himself in competition with Archimedes, Galileo is making clear that nothing less than greatness will satisfy him.

In the autumn of 1589, on Guidobaldo's recommendation, Galileo was to obtain his first appointment – at the University of Pisa, the only university in the territories of Florence and Galileo's own *alma mater*. But first we need to consider the strange case of Giovambatista Ricasoli.

6

A friend in need

We know all too little about the early part of Galileo's life, before he became famous with the publication of *The Starry Messenger*. Even after 1610, what we see of Galileo is mainly what he wanted us to see: those thoughts and feelings he chose to conceal have long seemed forever lost. There is, however, one source which enables us to, as it were, spy on Galileo and watch how he responds to an extremely difficult and morally demanding situation.[1]

In early 1589 Galileo was trying to establish himself as a mathematician, and doing some tutoring on the side – it seems he had taught mathematics in Florence and in Siena, and that in the spring of 1589 he was living in Pisa, having taught in the monastery of Vallombrosa during the previous autumn.[2] He returned from Pisa to Florence (a journey of some sixty miles) to celebrate Easter, and fell in with an old friend, Giovambatista Ricasoli, who was a wealthy young man. They studied philosophy, mathematics and poetry together. As friends did in the Renaissance, they slept in the same bed. One night Giovambatista woke him – a year later Galileo could still remember Giovambatista reaching for him, and getting his arm around his neck – and assured him that he, Giovambatista, had been condemned to death.[3] What would happen to him, did Galileo think? Would he be executed by having his head chopped off, or would he be burnt alive? His crime was to have given a funeral address for the grand duke when the grand duke was still alive. He had in fact given such an address to the Accademia degli Alterati, an organisation of young intellectuals, half-serious, half-playful (we might translate its name as 'the drinkers' academy'), but the grand duke really was dead, having died on 17 October 1587.[4] Night after night, day after day, for thirty days and nights, Galileo struggled to persuade his friend that no one had condemned him to death, but without success – Giovambatista was mad, and was going about in public wearing mourning for his own funeral. Eventually, hoping to escape justice, he ran away to Pistoia, and was brought back a prisoner by his relatives. It was clear, however, that he was going to set off again, and so it was agreed by the family that Galileo and a relative, Giovanni Ricasoli, would travel with him to ensure his safety. So Giovambatista, together with Giovanni, Galileo and a servant, set off across northern Italy, wandering from place to place, travelling sometimes by night, sometimes off the beaten track, as Giovambatista fled his imaginary enemies.

A couple of days before Whit Sunday (seven weeks after Easter) they arrived in Genoa, where eventually they persuaded Ricasoli to seek medical advice.[5] Twelve days later Galileo returned to Florence to take care of business (presumably his application for a job at the University of Pisa). Some eight weeks had now passed since Giovambatista had woken Galileo in the middle of the night to discuss his coming execution. They next met near the end of September, when Galileo set out with some relatives to try to persuade Giovambatista, who in the meantime had travelled to Milan and Rome, to return to Florence. He found Giovambatista still wearing the clothes in which he had set out on his travels in the spring, and unable to sleep at night.[6]

By this time, Giovambatista had donated all his worldly goods to Giovanni, and his other relatives took Giovanni to court, claiming that Giovambatista was not of sound mind. When Giovambatista died in January 1590 the case continued: his relatives were now claiming his property for themselves. There followed another trial, where they sought to set aside an earlier will, also made, they claimed, when Giovambatista was not of sound mind. Galileo, inevitably, was caught up as a witness in these trials, and subjected to hostile cross-examination by those seeking to prove that Giovambatista was indeed sane. The impression he gives as a witness is entirely favourable: he answers with great care and precision, and if he often claims to be unable to answer the question ('Is it expensive to live a courtier's life? Replied he had never lived a courtier's life, and so had nothing to say'), it never seems to be because he is concealing evidence.[7]

Throughout the weeks that Galileo was in Giovambatista's company, his friend was spending freely, as one might if one thought one had only days to live. None of this money ended up in Galileo's hands. When Giovambatista lost money to Galileo playing cards, Galileo (described in the court documents as 'poor') tried to return the money to him. He paid towards his own board and lodging in Genoa. It is true that Giovanni accused Galileo of stealing money from Giovambatista in Genoa, but no one seems to have believed him, and it is easy to imagine why. On one occasion Giovambatista had caused someone to mistake Galileo for a bandit and take aim at him with an arquebus – but fortunately the powder had been damp. Galileo had been greatly distressed by this incident, but even so he had not abandoned his friend.

It is evident that Galileo was genuinely fond of his companion. In trial documents he always speaks of him with respect, and insists that there is no disgrace in lunacy, for disgrace comes only from those things over which we have control. He was in a difficult position throughout, acting both as Giovambatista's friend and as an agent of his family, who were trying to prevent him from coming to harm. We may guess that he was strongly opposed to those who wanted to solve the problem by tying Giovambatista up like a veal calf and holding him prisoner.[8] But he seems to have kept the respect of all concerned. It is true that this little adventure was not without its benefits – it gave Galileo an opportunity to see what he called 'the world'. But only a patient and benevolent person – or a scheming one like Giovanni – could have stood Giovambatista's company month after month. It is worth stressing this, because Galileo was not a particularly good father, and in later life he rarely showed any interest in the welfare of others. His work was soon to become more important to him than his loved ones. If we are to regard him as a man of exceptional good character, it must be on the strength of these months in 1589 that he spent day and night in Giovambatista's company, trying to ensure that his friend came to no harm.

Juvenilia

This detour through a Florentine court case brings us to the most vexed issue in modern Galileo scholarship. It is an issue to which William Wallace OP devoted seven entire books, while unfortunately his most vigorous opponents, Alistair Crombie and Adriano Carugo, disagreed with everything but the central, mistaken premise of his argument.[1] So let us first make that central premise clear. At the very end of his life, in 1640, Galileo described himself to an opponent, Fortunio Liceti, as a good Aristotelian. Liceti was politely incredulous – and he felt confident that the rest of the world would share his surprise. Galileo replied that he was a good Aristotelian in that he admired Aristotle's logic and shared Aristotle's respect for sensory experience – a redefinition of what it meant to be a good Aristotelian that left Galileo free to disagree with the whole of Aristotle's science.[2] Wallace's enterprise (broadly shared by Crombie and Carugo) consists in rediscovering Galileo the good Aristotelian – an undertaking that ought to be met (let me be frank) with the incredulity first expressed so politely by Liceti. For Galileo's enterprise – the enterprise he shared with his fellow Linceans – was well described by his friend Federico Cesi: 'What we are engaged in is the destruction of the principal doctrines of the philosophy which is currently dominant, the doctrine of "il maestro di color che sanno" ' – quoting Dante's description of Aristotle as the master of those who know.[3]

Wallace's enterprise is not merely quixotic, however. Amongst Galileo's surviving papers there are three Latin philosophical texts. One consists of two treatises on Aristotle's *Posterior Analytics*, a logic text; the other two are scientific, one being a commentary on Aristotle's *On the Heavens*, the other on Aristotle's *On Coming-To-Be*.[4] Each comes to around one hundred pages. Each seems to derive from a Jesuit source, and Galileo's first known contact with the Jesuits is when he visited Christopher Clavius in Rome in 1587 – a visit we know of from their subsequent correspondence.

The commentary on *On Coming-To-Be* appears to derive from a course taught at the Jesuit College in Rome by Paulus Vallius, perhaps in 1586. It is written on paper with Florentine watermarks. Now paper is bulky and therefore expensive to transport, so it can be safely assumed that the paper on sale in each Italian city would have been made locally, and that when Galileo travelled from one city to another (as a poor man, he would normally have travelled on foot – indeed there is no record of Galileo ever travelling on horseback) he would not have bothered to carry more than a few pages of blank paper with him. So the watermarks are a good guide to Galileo's whereabouts when he wrote each manuscript. Moreover the Latin spelling of this particular manuscript is corrupted by Italian influences, which suggests a beginner's work. It has been argued that these facts imply an early date of composition, perhaps 1588.

Wallace showed that the two treatises on the *Posterior Analytics* also derived from a course taught by Paulus Vallius at the Jesuit College, a course which ended in August 1588. Galileo must have obtained a set of notes made during Vallius's lectures, and from that copy he made his own summary – presumably early in 1589, for the watermarks on the paper show that it was of Pisan manufacture, and Galileo was in Pisa prior to Easter 1589, after teaching mathematics in Vallombrosa between the beginning of September and the end of November 1588. Finally, the commentary on *On the Heavens* may derive from a course taught at the Jesuit College by Antonius Menu in 1580. It has Pisan watermarks, and Wallace argues that it belongs to Galileo's first year of teaching in Pisa, 1589–90. This is where Carugo and Crombie disagree with Wallace (mistakenly in my view): they are not convinced that Galileo was working from manuscript sources, and their search for printed sources leads them to date these texts to much later, after 1597.

The important thing is to recognise that these three texts are not unique. They are copies and summaries of other people's lectures. Such summaries are not rare: lecturers effectively dictated their lectures, so that every student came away with a summary.[5] They represent the best of contemporary philosophy, but they are Galileo's record of other people's ideas, not his own.[6] Compare them to the copy he made, at some point between 1592 and 1601 (and before a somewhat different version of the text appeared in print in 1602), of a text on optics by Ettore Ausonio, the *Theory of the Spherical Concave Mirror*, or to the manuscript copies of Galileo's *Two New Sciences* that were circulating in Florence before the book appeared in print.[7] The question this raises is straightforward. Why would Galileo spend a great deal of time copying and summarising either Jesuit lectures or Ausonio? One answer is that he wanted to familiarise himself with the latest developments in philosophy: the commentary on *On the Heavens* in particular may have served as a prelude to his own work on falling bodies. But there is no necessary connection between texts like these and Galileo's own research: he was not working on optics when he copied Ausonio.

There would seem to be an even simpler answer. We know that Galileo was earning a living as a private tutor, and every tutor has to give his students something to read. Reading Jesuit lectures would have helped to prepare Galileo for teaching philosophy, but copying them may have served a quite different purpose. It would have provided him with a textbook from which to teach, a textbook he could have lent to his students and reclaimed when they ceased to work with him. Amongst Galileo's papers we have a text that is precisely of this sort: a copy of the *Mathematical Games* of Alberti in the handwriting of Galileo's own teacher, Ostilio Ricci; as in the case of Galileo's copies of Jesuit lectures, Alberti's text, reduced to a textbook, has become anonymous.[8] Copying the latest Jesuit lectures would have enabled Galileo to claim he was providing a better course than other tutors could provide, just as later his copy of Ausonio may have been used as a textbook in the courses on optics that he taught to private students in 1601.

Alas, Wallace's assiduous scholarship and his indefatigable productivity were largely wasted. The juvenile works tell us almost nothing about the development of Galileo's thinking; they tell us only what he could get paid to teach. In Galileo scholarship they represent a blind alley. Wallace (himself an Aristotelian) wanted to use the three texts to argue that Galileo's logic was always Aristotelian, that there is space within an Aristotelian philosophy for experimental knowledge, and that Galileo is to be understood as an Aristotelian natural philosopher, even though on actual questions of physics he parted company with Aristotle. These arguments, unlike his robust arguments about the sources and dates of the three Latin commentaries, are based on the flimsiest of foundations, and involve sliding past the elementary fact that these texts are not a record of Galileo's own views, any more than his copy of Ausonio can be taken as a record of his views on optics in the 1590s.

8

The Leaning Tower

In the autumn of 1589 Galileo became a professor of mathematics at the University of Pisa. He had taken the first step on the road to a professional career and to financial security. But there should be no illusions about his situation: his annual salary was sixty florins, significantly less than the income of a good stone-mason.[1] It was not possible to live as a gentleman on such an income. There were professors of philosophy in prestigious institutions who were earning fifteen times as much; Galileo's own friend Jacopo Mazzoni was earning seven hundred florins at Pisa.[2] And this was not just because Galileo was starting out; it was also because mathematics was not a prestigious subject. As far as other professors were concerned it had two main functions: to introduce philosophy students to basic geometry, and thus to the idea of a demonstration; and to provide medical students with the skills required to make astrological charts. Mathematics was regarded as intellectually uninteresting, a purely technical training, and the salary of mathematicians reflected this.

Galileo had been a student at Pisa as recently as 1585, and he had spent time there at the beginning of 1589, probably doing some freelance tutoring and perhaps lobbying for an appointment. Most of the professors, his new colleagues, had been teaching when he was a student. He would have had old friends, as well as relatives, to make him feel at home. And it is likely that he found himself picking up on intellectual discussions that had interested him as a student. Pisan intellectual life was dominated by two philosophers: Girolamo Borro, a follower of Averroes (1126–98), the great Arab commentator on Aristotle, famous for denying the immortality of the soul; and Francesco Buonamici, who followed the Greek commentators (the one thing on which the two were supposed to be agreed was the foolishness of revealed religion).[3] In 1575 Borro had published a book on the movement of heavy and light bodies: Galileo owned a copy.[4] In 1589 Buonamici had on his desk an enormous manuscript on the same subject; although the book (over a thousand large pages) did not appear in print until 1591, Galileo had quite probably attended Buonamici's lectures as a student, so the book's main arguments would have been familiar to him even before he could read it.

Borro and Buonamici disagreed radically on movement, as on most other things, and their colleagues and students must have been aware of these differences, particularly as students were required to engage in 'disputations', debates in which they would take one or other side of a standard academic controversy. Disagreement thus lay at the very core of intellectual life, and students were taught to approach central issues from sharply contrasting points of view. At the same time the permissible range of disagreement was set by the tradition of commentary upon Aristotle.

In order to understand the dispute between Borro and Buonamici we first need a sense of those assumptions that all Aristotelians held in common – assumptions that Galileo was soon to question, at least in part. First, Aristotelians drew a sharp distinction between natural and forced movement. If a ball is thrown, that is a forced movement; if dropped, it moves to its natural place of rest, and this is a natural movement. If it is thrown upwards, the ball decelerates as it rises, and this is because forced movements tend to peter out; if it is dropped, it accelerates, and this is because natural movements tend to speed up. If a ball is thrown, there is an initial forced movement, and then the ball drops downwards. Aristotelians believed that only one type of movement could take place at a time, so they were committed to the view that the ball travelled in a straight line, and then dropped vertically. They also held the view that any object subjected to a forced movement must be being moved by another object (or, as in the case of the hand, by the will of a sentient creature). It appeared to follow from this argument that once the ball leaves the hand it ought in principle to come to a stop. Aristotle sought to solve this problem by arguing that the ball was pushed on its way by the air, which had been disturbed by the movement of the hand. Later theorists argued that the ball had acquired some sort of internal energy or impressed force, just as it might acquire heat from the hand and stay hot even after it had been let go. Galileo would have found an account of impressed force in the textbook of the Jesuit philosopher Benedict Pereira, which he used while working on his first study of falling bodies, *On Motion*.[5] Crucially, all Aristotelians held the view that the heavier a ball was, the faster it would fall.[6]

All this seems rather odd to us, but it is at least an account of something familiar, the movement of a projectile. Aristotelians believed, however, that the universe consisted not only of weight, but also of lightness. If balls naturally move downwards, flames naturally move upwards. Borro and Buonamici were thus debating the natural movement of bodies downwards *and upwards*. Both movements, they held, continued until the body was obstructed or reached its natural place of rest: in the case of a heavy body this would be the centre of the earth; in the case of a light body it would be an invisible frontier between the earth and the moon – beyond this frontier the only natural movement was circular movement, and there was no coming into being or going out of existence. In the heavens what looked like change was simply, if one waited long enough, repetition.

So the heavier a body is the faster it falls. But what is meant by heavy here, and how much faster? Aristotle was clear that speed was proportional to weight: a

body that weighed twice as much would fall twice as fast. The much more difficult question was how to understand weight. The standard view was that weight
was to be understood in the normal sense: a two-pound lead weight would fall
twice as fast as a one-pound lead weight. An alternative view was defended by
Buonamici: heaviness should be understood as density or specific gravity so that
lead weights of different sizes would all fall at the same speed, and they would all
fall faster than wooden balls. Borro held a third view. He thought that weight was
situationally specific, and related to the composition of the material. Aristotelians
held that all earthly objects are made from four elements: earth and water,
which are heavy; fire which is light; and air, which can be either heavy or light,
depending on its location. And they agreed that a wooden ball contained more air
than did a lead weight. Borro's view was that when you weighed a wooden ball the
air in it had no weight, but the moment it was dropped in air, the air became
heavy (in water, by contrast it would become light). Thus if you took a lead
weight and a wooden ball which weighed the same when suspended in a balance,
the wooden ball was in fact heavier when falling, and would fall faster.

Borro had put this theory to the test. He had gathered a group of philosophers
at his house and they had tossed comparable lead weights and chunks of wood
out of an upstairs window. The wood had consistently reached the ground
ahead of the lead.[7] Buonamici was not persuaded: you would first need to weigh
the objects and make sure that their weight was identical. There was also an
obvious problem: the shorter the drop the harder it was to see what was going
on. The solution was to drop objects from a greater height, and it is known that
by 1612 Giorgio Coresio had tried dropping objects of different weights but the
same material from the top of the Leaning Tower of Pisa. The result, he claimed,
vindicated Aristotle, for speed of fall proved proportional to weight.[8]

According to Viviani, Galileo, when teaching at Pisa (thus between 1589 and
1592), had already carried out this test of dropping objects of different weights
but the same material from the top of the Leaning Tower, and had shown that
they reached the bottom simultaneously.[9] This is like Viviani's story of the
pendulum: he wants to present Galileo as an experimental scientist, and to claim
that he was so precocious that he founded modern physics almost effortlessly.
But in this case there is a manuscript on natural motion, apparently written in
the period 1589–92, in which Galileo confidently claims that he knows what will
happen in such a test: at first, he tells us, the lighter object moves faster, but then
it is overtaken by the heavier object, which reaches the bottom ahead of it. He
has, he tells us, repeated this experiment many times.[10]

Thus it would seem perfectly straightforward to say that Galileo was carrying
out experiments with falling bodies while he was at Pisa, and so, clearly, he was
already an experimental scientist. Historians, however, have made the simple
complicated. In 1935, Lane Cooper published a whole book arguing that Viviani's
story was a myth. In 1937 the great Alexandre Koyré, an enormously influential

historian of science, also dismissed the story: this was not a real experiment, but an imaginary or 'thought' experiment. Such scepticism appears well grounded, for if you drop two objects of the same material but very different weights from a high tower you do not get the results Viviani and Galileo describe: the heavier object will hit the ground well before the lighter object, and this is because the resistance of the air has more effect on the lighter object than on the heavier.[11] Galileo simply could not have carried out the experiment and obtained the results described by Viviani. Unfortunately, Viviani, Cooper, Koyré and the modern physicists who write about high tower experiments share a fundamental misunderstanding, one that has implications for our whole understanding of what science is: they think that experiments are straightforward, and that repeating them is unproblematic.

In 1638 Galileo published *Two New Sciences*. There he described a high tower experiment (he never refers explicitly to the Leaning Tower of Pisa) in which a heavy object and a light object of the same material are dropped, one object being a hundred times the weight of the other. The heavy object reaches the ground only two inches before the lighter object.[12] Many scholars dismiss this as another mere thought experiment. But we know that in March of 1641 Vincenzo Renieri set out to replicate Galileo's experiment by dropping a cannonball and a musket ball from the Leaning Tower of Pisa.[13] The cannonball reached the bottom, he told Galileo, a good palm's width ahead of the musket ball: let us say six inches, instead of Galileo's two inches. According to computer modelling, the difference ought to have been 174 centimetres, or 5 feet $8\frac{1}{2}$ inches. No one could mistake the height of a man for a palm's width.[14] Renieri also tested the alternative experiment, dropping objects made of different materials, and here too he failed to get the result that a modern scientist would expect. When he dropped two objects of the same size, one made of wood and one made of lead, the lead object reached the bottom three *braccia* or two metres ahead of the wooden object; depending on the size of the objects, modern calculations suggest that the gap should have been much bigger: in the region of eight metres.

Renieri was not alone in obtaining results that are at odds with modern expectations (that is to say, the expectations of modern physicists – historians of science no longer assume they already know the 'right' outcome of an experiment). In 1632, Roderigo de Arriaga, a Jesuit, claimed that repeated experiments in which objects were dropped from a very high place proved that all bodies fall at exactly the same speed; while in experiments conducted in 1634, in which objects were dropped from the tower of the Jesuit chapel in Ferrara, heavier objects were found to reach the ground very slightly ahead of light objects – or at least so Giambattista Riccioli claimed, relying not on sight but on the sounds generated when one object struck brass, the other wood, while Niccolò Cabeo insisted that they arrived simultaneously.[15]

These repeated replications of Galileo's high tower experiments vindicate Viviani's claim that Galileo was an experimental scientist. The problem is to understand how Galileo and those who followed him obtained what are, in the view of modern scientists, the wrong results. Our sources do not provide much help: the convention whereby experiments are recorded in detail had yet to establish itself, so there is nothing unusual about the bald summary they provide. But fortunately there can be only one answer to our problem: both Galileo and Renieri failed to drop the two unequally sized objects simultaneously. And indeed, if you imagine leaning out over a parapet, holding in one hand a musket ball and in the other a cannonball, it is evident that your grip on the cannonball would have to be much firmer than your grip on the musket ball, and that it would be virtually impossible to release the two simultaneously. There is, it seems, a systematic bias towards releasing the lighter weight first (although Coresio must have found a way of avoiding this error; it is perhaps significant that he uses the first person plural when describing his experiment, which may indicate that he had two people dropping objects, not one).[16] If one person drops both objects one sees exactly what Galileo reports in *On Motion*: the lighter weight appears to fall faster, and then to be overtaken by the heavier weight. If you were to choose a tower of the right height the heavy weight would overtake the light weight only just before they both hit the ground.

Galileo's high tower experiments were much more sophisticated than Borro's tests, which consisted of throwing assorted lumps of wood and metal out of an upstairs window. But they were nevertheless fundamentally flawed. Galileo's eventual solution to the problem of fall involved ignoring one of the results (he substituted a theoretical projection that was at odds with his own experience) and concentrating on the other. In *On Motion*, on the other hand, he was able to accept both results as reliable because he was confident that he could explain them away.

Galileo describes three experiments in his early manuscripts on moving bodies, *On Motion*. There is the pair of high tower experiments, which were a natural development from Borro's test; he claims to have actually performed one of these and to be certain of being able to predict the result of the other. The third is an entirely original experiment designed to measure the speed of a ball rolling down a slope. Galileo wanted to know how much faster a given ball would roll down a slope if you increased the angle of the slope. But he also reasoned that if heavier objects fall faster than lighter objects, then it ought to be possible to identify a slope such that a heavy ball would roll down it at the same speed as a light ball would fall vertically.[17] There can be no doubt that Galileo devised this experiment because he was looking for a way of illustrating his theories, and that he wanted to carry out measurements of falling and rolling bodies under a variety of conditions. But it should be stressed that he was seeking to illustrate rather than to test his theories, for when his measurements did not accord with his expectations he did not reconsider the theories, but merely concluded that it was difficult to illustrate them with practical examples.

What were these theories? Reading Borro, Galileo will have found an account of the views of Avempace (d. 1138), an Arab philosopher whose arguments Averroes had found challenging.[18] And he would have recognised in Avempace an approach that could be made to fit neatly with Archimedes' account of bodies floating in water. Why, according to Archimedes, does a body float to the surface? Because it has a lower specific density than water. It is actually pushed upward or extruded to the surface. From an Archimedean point of view there is no need to think of objects as having lightness – they have only different degrees of heaviness. And the weight of an object is affected by the medium which it displaces: one can thus calculate an absolute weight, the weight it would have in a vacuum, which is its weight in a medium plus the weight of the medium displaced by the object.

According to Aristotle, rates of fall are determined by the ratio between the weight of an object and the density of the medium through which it is falling. Aristotle believed that a vacuum was impossible, but if there were to be such a thing, the rate at which an object fell through a vacuum would be infinite, corresponding to the weight of the object divided by zero. The object would no sooner enter the vacuum than it would emerge the other side – it would be in two places at once. Archimedes had proposed that weight could be established by subtraction: the object's absolute weight, minus the weight of the medium that it displaces. Why not, Avempace had argued, determine rates of fall in the same way? In that case the rate of fall in a vacuum would be determined by an object's specific gravity. (Buonamici favoured this view.) Any lead object would fall in a vacuum at a constant speed ten times faster than any wooden object, and since air weighs very little, the difference would be very nearly the same in air. A simple thought experiment showed the need to think in terms of specific gravities rather than total weights. According to Aristotle, a two-pound weight would drop twice as fast as a one-pound weight. But supposing you attached two one-pound weights together with a metal rod: at what speed would this now fall? From one point of view it is now a two-pound weight; but at the same time it is obviously still two separate one-pound weights. The only logical conclusion is that it will fall at exactly the same rate as before. Consequently all objects of the same material, no matter how much they weigh, will fall at the same speed.

This is Galileo's basic theory in *On Motion*. Note that he rejects the Aristotelian assumption that there are upward-tending objects as well as downward-tending ones; but that he accepts the Aristotelian assumption that, amongst downward-tending objects, heavy objects will fall faster than light ones. Speed of fall should be a constant determined by specific gravity: Galileo takes it as axiomatic that where the same force is continuously at work, the result, in the absence of other factors, will be a constant speed. But if you actually drop an object it starts from being stationary and accelerates towards this notional constant speed. Why is this? Because, Galileo says, it has an impressed force that only gradually wears away – indeed one that wears away faster in the case of some materials than in the case of others (just as iron loses heat more rapidly than wood). And this explains the unsatisfactory results obtained in high tower and rolling ball experiments: these experiments do not measure absolute rates of fall; they measure only rates of acceleration. There is no way of measuring absolute rates of fall: we can only deduce them from first principles.

It is essential to understand the limited role assigned to experiment in *On Motion*. It is the role that would have been assigned to it by any orthodox Aristotelian. Aristotelians insisted that philosophical accounts of the world must correspond to experience. They recognised that experience might oblige one to rethink a philosophical argument, but the fundamental terms of that argument must come not from experience but from reason. (Famously, what is known as the Paduan school argued that experience, suitably reworked, might help in the construction of the premises from which to reason.) It was reason, not experience, that generated science, because science is the study of necessary or causal relationships. Galileo says precisely this in *On Motion* when he maintains that it is necessary 'to employ reasoning at all times rather than examples (for what we seek are the causes of effects, and these causes are not given to us by experience)'.[19]

For the most part Aristotelians thought of experience as something that was passively acquired over time, though they were not averse to the idea of going out and putting an argument to the test of experience, as Borro did when he threw things from his window. But no Aristotelian, and no Archimedean, thought that experience in itself and on its own had ever resolved any philosophical disagreement. Consequently Galileo could quite happily declare in *On Motion* that his high tower and rolling ball experiments did not correspond to his theory.[20]

9

Inertia

We have now looked at Galileo's core argument in *On Motion*. It is an argument that models itself on Archimedes (and openly declares itself to be in opposition to the ridiculous arguments of the Aristotelians) but one that any competent Aristotelian could have followed. From an Aristotelian point of view it contains two major departures from anything resembling orthodoxy: the denial of lightness, or the insistence that everything has weight; and the possibility of movement in a vacuum. But there is a further argument in *On Motion* which is striking and which contemporary Aristotelians would have found profoundly puzzling.

According to Aristotle there are two types of movement: natural movement, which is directed towards an end, and halts when an object arrives at its natural resting place; and forced movement, which continues only for so long as there is a mover acting on the moving object. As others had done before him, Galileo modifies the account of forced movement to include the idea of an impressed force. But he also invents a quite new type of movement, which he calls intermediate movement. Imagine a perfectly round ball standing on a perfectly smooth sheet of ice. The slightest touch will start it moving, and it will continue to move indefinitely. If this seems too much like an impractical abstraction, think of a river: it flows constantly, and yet the gradient is often minute. It seems that flowing water has almost no resistance to movement; otherwise one would be able to identify a slope that was not steep enough for a river to run down it. Aristotle held that the natural condition of all sublunary things is to be stationary, and that all movement naturally ends in the cessation of movement; Galileo was now suggesting that movement (if it is neither upwards nor downwards but sideways) might have no natural end: that it might be interminable.[1]

This argument comes close to a theory that had been invented in the fourteenth century to replace the older idea of impressed force: the theory of impetus. Impressed force naturally dissipates; impetus continues until it encounters some form of obstruction or resistance. According to the impressed force idea, if a ball were to be thrown into empty space, where it was subject neither to gravity nor to air resistance, it would eventually glide to a halt; according to the impetus theory it would continue. Galileo's theory (although he does not know it) is an impetus theory, and an impetus theory which has escaped from any notion of a natural resting place: the slightest touch could send a notional ball sliding across a theoretical sheet of ice for ever and ever.[2] Moreover, once you have an impetus theory, it follows that if a force continues to act on a moving object, what should happen is not that it will continue to move at a constant speed (as Galileo assumed in his account of falling bodies in *On Motion*), but that it will accelerate.

So far I have written about *On Motion* as if it were a finished text. But it is not. It is a series of drafts and notes, which were evidently abandoned unfinished. Why did Galileo stop work on this, his first major scientific (or, in sixteenth-century terms, philosophical) – as opposed to mathematical – text? Galileo does not say, and so we have to speculate. There are two current theories. The first is that the experiments described in *On Motion* represent the beginning of Galileo's commitment to a programme of experimentation. According to this theory *On Motion* was abandoned because Galileo could not generate experimental results that confirmed his theories.[3] This is hard to believe because there is no evidence that in 1592 Galileo saw experimental evidence as crucial: when experimental results were at odds with theory it was easy to say that this was due to accidental factors. The second is that the ideas that Galileo was developing in *On Motion* were inconsistent and contradictory. Having introduced the idea of intermediate motion, Galileo needed to follow through the logic of this way of thinking and replace his idea of impressed motion by an impetus theory. Had he done so he might have reached the conclusion that a few impetus theorists had already reached in the Middle Ages: that falling bodies have no absolute rate of fall, but continue accelerating indefinitely (in the absence of air resistance) as gravitational force adds impetus onto impetus in a never-ending spiral. This would have involved abandoning the cherished idea that heavier bodies necessarily fall faster than lighter bodies: rather all bodies would be subject to the same law of acceleration.[4] But in 1589–92 Galileo was incapable of taking this bold step. He stalled, stopped and gave up.[5] The problem with this theory is that it implies that Galileo lacked intellectual grip, and this is unconvincing. Had he seen a problem, he would have tried to resolve it.

There is a third possibility. In 1585 Giovanni Battista Benedetti published a new theory of motion. Like Galileo's theory this was an Archimedean critique of Aristotle's views on natural motion. Like Galileo's, Benedetti's theory of forced motion was an impressed force theory, although he (confusingly) used the term impetus.[6] For a long time the standard view has been that Galileo's argument is so close to Benedetti's that he must have read Benedetti.[7] But had he? In the course of *On Motion* Galileo mentions only two modern authors: Borro, his Pisan colleague, and the Jesuit Pereira, whose book was a standard textbook. His whole attitude is that of someone who has a profoundly original argument to propound – an attitude hardly compatible with a knowledge of Benedetti's work.[8] In certain respects Galileo and Benedetti are very close – in the thought experiment regarding two joined falling bodies, for example – and yet Galileo never adopts Benedetti's wording, or his vocabulary.[9] In certain respects Galileo and Benedetti differ, and yet Galileo never engages with Benedetti on these points, or explains why his views are preferable. *On Motion* is written as if Galileo had never heard of Benedetti; nor need he have, for he has many of the same sources (Archimedes, Averroes' account of Avempace, Pereira) to draw on.

It is worth remembering that there was no university library in sixteenth-century Pisa. Books were expensive, and Galileo was poor: we can be sure that he did not own many.[10] Benedetti was not a well-known author, and he had published in Venice. Buonamici, for example, who could certainly afford to buy books, makes no mention of him in his vast work on motion. The first mention of Benedetti by a Pisan author comes in a work by Galileo's close friend Jacopo Mazzoni – they had been friends since at least December 1590 – that was published in 1597. It hardly constitutes evidence that Benedetti was being read in Pisa in 1592.

This points towards a possible explanation for Galileo's decision to abandon *On Motion*. In the summer of 1592 Galileo spent time in the company of two men who had certainly been reading Benedetti: Guidobaldo del Monte and Paolo Sarpi.[11] If he explained the argument of *On Motion* to them they would immediately have told him that he needed to read Benedetti and that his core arguments were not original. For Galileo this would have been a repeat of the experience he had had when working on centres of gravity. There he had given up on discovering that Luca Valerio was working along the same lines; now he discovered that Benedetti had already published a similar argument. He would have to look for another way to make his reputation.

Galileo's later work contains frequent admiring references to Valerio, who became a friend, and remained one despite his opposition to Copernicanism. He never once mentions Benedetti. It seems unlikely that he ever read him. And this brings us to what I think is the correct explanation for Galileo's abandonment of *On Motion*. Around the time he learned of Benedetti's existence the intellectual enterprise which had produced *On Motion* lost all significance for him. Galileo abandoned *On Motion* not because he was unable to make his experiments come out right; not because he realised his argument was incoherent; not (or not only) because he had finally heard of Benedetti; but for a very simple reason: he was now committed to giving an account of falling bodies on a moving earth. He had become a convert to Copernicanism, and this meant that he could no longer rely on Archimedes for an understanding of movement.

10

Nudism

Galileo does not seem to have been a success in his first job. Not only was his salary low, but it was reduced by fines for missing lectures and for failing to wear the required academic dress, the toga, a sort of cassock or academic gown (which can be clearly seen in every portrait of Galileo). His initial appointment was for three years, and at the end of it he faced unemployment. According to Viviani he had made a powerful enemy, and knew he was unlikely to have his contract extended. On 2 July 1591 Galileo's father died, but Galileo inherited no money, and it may be that all that his father left him were debts: the family certainly relied on credit to cover the cost of the funeral.[1] In February 1592 Guidobaldo del Monte invited Galileo to come to Venice that summer: he would try to fix him up with something there.[2] That he succeeded is a tribute to del Monte's influence and reputation rather than to Galileo's: Galileo had published nothing, and had been rejected by the University of Pisa, which was a much less prestigious institution than the university in Venetian territory, the University of Padua; the University of Padua was also about twice the size of the University of Pisa, with twelve hundred students to Pisa's six hundred.[3] Galileo, who should have been on his way down and out, was going up in the world. By 3 September, now in Venice, he was trying to finalise his new salary. He had perhaps already left Pisa in search of a job in Venice when he wrote his satirical poem 'Against the Wearing of the Gown,' for the attitude it expresses is very much that of someone who is planning to leave Pisa far behind.

There was an established tradition of satirical poetry to which 'Against the Wearing of the Gown' belongs. The genre required the poet to represent a world in which conventional values were inverted, and it generally entailed, as in this case, a certain cheerful anticlericalism.[4] The fact that the idea of inversion was a requirement of the genre scarcely explains the enthusiasm with which Galileo embraced it. He starts with the quest for the greatest good, which no one knows how to find because no one knows how to look for it. No one knows how to use his imagination, to be inventive, to look into the future. How do you find the good? Find the bad, proceed by contraries, and then you will have no difficulty in identifying the good, for it will be coupled with the bad like two chickens tied leg to leg in the market.

So what is the greatest evil? The wearing of clothes. Consequently the greatest good is going naked. In the golden age everyone went naked, and then young men and women could easily see if their prospective spouses would suit them. In those days there was no need to fear the pox, as you could easily tell if someone was infected. When clothing was introduced, social hierarchy became possible. Without clothes, everyone was equal – as Donne put it, as if he were familiar with Galileo's poetry, 'Prince, subject, father, son, are things forgot'.[5] Men are like bottles of wine: the fancy bottles always seem to contain dreadful wine, while the plainest bottles contain the best. What does it really matter what clothes you wear, whether you are from Turkey or Bergamo, whether you are called sir or mate? Clothes are the invention of the devil, like gunpowder, witchcraft and demonic possession.

Galileo is nevertheless prepared to wear clothes, with the exception of the wretched toga. No one who was in his right mind would be willing to appear in it. What are its disadvantages? The first is that you will not be allowed into the brothel while wearing it. The second is the requirement to keep up appearances, which means wearing a smart toga, whereas if you were dressed normally you could wear old clothes. The third is that everyone can identify you if you are walking alone, and will conclude that you have no status amongst the other professors. If you are not part of a group or a gaggle, everyone thinks you are ignorant and incompetent. You end up creeping down the street, sliding along the wall, trying to avoid everyone, as if you were frightened of being arrested, like a cat that has been kicked out of the house. It is all very well having long skirts if you never go anywhere, but not if at any moment you may have to run to escape the constable. So Galileo swears that he will resign his chair rather than wear the toga again.

Galileo's poem presents us with a portrait of his life in Pisa – alone, unaccompanied, slipping down the street like a cat that has been thrown outside. He must have read with wry pleasure the reports sent to him by his student Benedetto Castelli, a Benedictine monk, when he became a professor at Pisa (taking on Galileo's old job) in 1614: Castelli describes wandering along the banks of the Arno, surrounded by a mob of students, his success visible to all – the opposite of Galileo's experience.[6] The poem also tells us that he values invention, imagination, the ability to look into the future, and that he has, as Viviani said, a spirit of contradiction. It tells us that he is impatient with the established hierarchies, and longs to be treated as an equal rather than an inferior. And it tells us that he is happy to mock religion – not just when he slyly suggests that if you want to know how evil men can be you should get to know some friars and priests (for the reason, of course, that they are the embodiment of goodness), but rather in the fundamental conceit that runs through the poem. According to the Bible, clothes are not an invention of the devil, but the first expression of shame. It is sin which makes it impossible for us to return to the

Garden of Eden, where Adam and Eve went naked. In praising nudity, in condemning clothing, Galileo is not just saying that everyone should be equal. He is also claiming that there is no such thing as sin, and no such thing as salvation. His aversion to wearing the toga is only a pretext for a poem which attacks Christianity itself.

PART TWO
THE WATCHER OF THE SKIES

Then felt I like some watcher of the skies
When a new planet swims into his ken
John Keats, 'On First Looking Into
Chapman's Homer' (1816)

I think the same thing is going on as when Michelangelo began to design buildings which were quite different from those that had been built by anyone else before him. Putting their heads together, people said that Michelangelo had ruined architecture by deviating so greatly from Vitruvius; but I, hearing some say this, said they had got it the wrong way round, because Michelangelo had not ruined architecture; what he had ruined was the [other] architects.[1]

Lodovico Cardi da Cigoli (1612)

11

Copernicanism

It much disadvantageth the Panegyrick of Synesius [in praise of baldness], and is no small disparagement unto baldnesse, if it bee true what is related by Ælian concerning Æschilus, whose balde-pate was mistaken for a rock, and so was brained by a Tortoise which an Eagle let fall upon it. . . . some men critically disposed, would perhaps from hence confute the opinion of Copernicus, never conceiving how the motion of the earth below should not wave him from a knock perpendicularly directed from a body in the ayre above.

Thomas Browne, *Vulgar Errors* (1646)

The years 1592 to 1610, during which Galileo taught at the University of Padua, are undoubtedly the most important in his life as a scientist. During this time (the best years of his life, he would say later) he made all his major discoveries in physics, while the two great last works, the *Dialogue* and the *Two New Sciences*, are in large part polished presentations of work undertaken during this same early period.[2] Yet the record that survives from this period is thin, even meagre: one great book, or rather pamphlet, *The Starry Messenger*, published in 1610; a privately printed instruction manual and a printed protest against a plagiarist; perhaps a pseudonymous pamphlet or two, minor contributions to a long-forgotten intellectual controversy; some manuscript teaching materials, on astronomy, on mechanics and on fortification; almost 350 letters from, to or about Galileo; and a file of rough notes and jottings recording experiments, calculations, conclusions. Perhaps one-tenth of Galileo's surviving texts derive from these eighteen years. For the first ten of these years there is considerably less – not a single publication, no research notes, only ten letters written by Galileo and some teaching materials. Renn and Valleriani's 'Galileo and the Challenge of the Arsenal', an important recent account of these years, maintains that during this decade Galileo almost abandoned philosophy and science, and turned to technology and applied knowledge. The most significant event in the whole of Galileo's intellectual life disappears from view if this approach is taken: his conversion to Copernicanism.

Copernicus had published *On the Revolutions of the Heavenly Spheres* in 1543 – he died the very day on which the first copy was placed into his hands. In it he argued that the assumption held by all educated (and indeed uneducated) people for at least the previous fifteen hundred years, that the earth was stationary and located at the centre of the universe, was wrong. One could make sense of the evidence, Copernicus argued, only by recognising that the earth was a planet, and that it and all the other planets revolved around the sun. By making this shift Copernicus was able to describe the movements of the heavenly bodies entirely in terms of circles – which was what mathematicians had always wanted, believing that circular movement had a perfection no other type of movement could attain in that it could be repeated for ever without variation.

Copernicus did not get rid of the epicycles – circles on circles – that the Ptolemaic system had used to explain why the planets generally move in one direction across the sky and occasionally seem to loop backwards. He reduced the number of large epicycles, in that the movement of the earth around the sun amounted to an epicycle affecting the apparent movement of all the other planets, but he actually increased the number of small epicycles. What he did eliminate, however, were 'equants' – formulas for speeding up and slowing down movement in a circle by defining it as movement which swept out equal angles in equal times as measured from a line between a point other than the centre and the circumference. The whole point of circular movement in the heavens was that it was supposed to be unchanging and continuous, whereas the equant transformed circular movement into a movement that accelerated and decelerated. Equants had thus always been recognised by astronomers as an admission of defeat, and they had never been happy with them. What Copernicus offered was the restoration of true, unchanging, continuous circular movement. That was attractive; but the rest of his system was not. Copernicus's universe had a centre which was near, but not within, the sun; around this centre there was a series of excentric spheres carrying the planets. Because the Copernican universe had to look exactly the same no matter where the earth might be in its orbit, its scale had to be vastly greater than that of the conventional, earth-centred Ptolemaic universe.

Astronomers, when making calculations, were perfectly happy to think in terms of a moving earth within an expanded universe; philosophers, however, absolutely refused to do so, as did nearly all astronomers when considering how the universe was actually constructed. What is now called physics was then a branch of philosophy, and philosophy, as a deductive science which described how things must be, was accepted as superior to astronomy, a descriptive science which described how things seem to be. As long as the philosophers remained opposed to Copernicanism, it could never hope to establish itself as the correct account of the universe. And every calculation that could be made within the Copernican system could also be made, at the price of introducing equants, within the Ptolemaic system. Naked-eye astronomy was never going to produce a decisive argument in favour of Copernicanism.

Copernicus's book aroused a great deal of interest, particularly because it simplified calculations: very helpful tables were soon produced based on Copernicus's calculations. There was a second edition in 1566 (by the end of his life Galileo owned copies of both the first and second editions), and a third only in 1617. But there were almost no whole-hearted Copernicans amongst astronomers, and none at all amongst professional philosophers. It is a remarkable fact that in the first fifty years after the death of Copernicus, only four people published in support of the Copernican system: the English mathematician Thomas Digges, the Italian philosopher Giordano Bruno, the Spanish theologian Diego de Zuñiga, and the Italian mathematician Giovanni Battista Benedetti, whose account of motion was so similar to that provided by Galileo in *On Motion*. Benedetti mentioned Copernicus only briefly in his major work, published in 1585, but he did describe heliocentrism as a 'most beautiful' theory and praise Copernicus's version of it as divine, indicating that it was to be preferred to that of Ptolemy.[3] A modified Copernicanism (sometimes called semi-Copernicanism) had also been advocated by Celio Calcagnino in a work written as early as 1525 but not published until 1544; he proposed a daily rotation of the earth as preferable to the Ptolemaic idea that the heavens turned (at vast speed) around the earth, but he left the earth at the centre of the universe. Galileo's (entirely conventional) lectures on astronomy suggest that he was familiar with this semi-Copernican position.[4] There were, of course, other Copernicans who were too cautious to publish their views. In Germany, Michael Maestlin taught Copernican astronomy, Christoph Rothman defended it in correspondence and David Origanus relied on it for his calculations, though nothing by any of them that was explicitly Copernican appeared in print before 1596.

A crucial reason for the lack of support for Copernicanism lies in the existence of a new alternative to the geocentric system of Ptolemy and the heliocentric system of Copernicus. This was the geo-heliocentric system of Tycho Brahe, first described in print in 1587. According to Tycho the earth was, as Ptolemy and the Aristotelians claimed, stationary at the centre of the universe, and the sun, moon and fixed stars revolved around the earth. But the planets, Tycho argued, following Copernicus, revolved around the sun. These three systems were so fundamentally different that you might at first think that it would be easy to find evidence to settle the question of which was right. But if one allows for adjustments of scale, they are each capable of making predictions for the positions of the moon, sun and planets in the sky which are identical to those made by the others: they are geometrically equivalent. It was soon apparent that naked-eye astronomy was incapable of choosing between them. The fundamental choice was one of plausibility. Which seemed more likely, that the earth rotated once a day, or that all the fixed stars rotated round the earth once a day? Which was more plausible, that the earth was just another planet, or that the vast bulk of the sun circled the earth every day, carrying all the planets with it? When thinking about questions such as these the objections to a moving earth – which were physical rather than astronomical objections – were absolutely central. Matters were to change in 1609 – new evidence from telescopic observations, above all the observations of Galileo, became available; and in the same year Johannes Kepler published his *New Astronomy*, in which he argued that no theory which insisted on circular movement could account for the orbit of Mars, which was, he claimed, an ellipse. Galileo, who was at the forefront of the telescopic revolution, simply avoided discussing Kepler's revolutionary proposals (even though his patron Federico Cesi found them convincing). He may have doubted whether Kepler's measurements were sufficiently reliable; he surely found his arguments implausible; and he probably felt that they offered the opponents of Copernicanism new grounds for rejecting the Copernican system as philosophically ill-conceived. Galileo remained to the end an old-fashioned Copernican.[5]

It seems very likely that when Galileo became a Copernican he had never met a follower of Copernicus: the first Copernican he met was probably Edmund Bruce, an Englishman (he describes himself as 'Anglus' not 'Scotus', and indeed he was on the payroll of the English secret service), a disciple of Copernicus, of Bruno and eventually of Kepler, who was living in Padua when Galileo arrived there.[6] It also seems very likely that Galileo had never read a work by a Copernican other than the work of Copernicus himself and the *First Report* of Copernicus's disciple Georg Rheticus, first published in 1540 (before the publication of *On the Revolutions*) and reprinted in 1541 and 1596. In the 1566 edition of Copernicus, the one likely to have been the first to be purchased by Galileo, the *First Report* appears as a supplement to *On the Revolutions* – although as the work of someone who was listed in the various editions of the Index of prohibited books as a heretic whose works were therefore banned, it was sometimes excised.

One of the greatest Copernicans alive in 1592, and certainly the greatest of those who had already published in support of Copernicanism, was Giordano Bruno. Bruno, who was a philosopher rather than an astronomer, was not just a Copernican; he also believed that the universe was infinite, and that there were other inhabited planets circling other suns. (In arguing for the infinity of the universe he was following a broad hint given in *On the Revolutions*, although Copernicus had avoided committing himself to such a subversive doctrine.) Not surprisingly, he was never able to obtain secure employment as a philosopher in a university. Bruno was in Venice when Galileo arrived there in early September to lobby, successfully, for an appointment at the University of Padua – an appointment that Bruno had hoped to win for himself. But by September Bruno was already in the prison of the Venetian Inquisition; he had been held there since May, and was to be transferred to Rome in February of 1593. (Venice exercised firm control over the Inquisition within Venetian territory, and would not have handed Bruno over had he been a Venetian or a teacher or even a student in Padua. But he was only a visitor, and the state had no interest in protecting him.) Once in Rome, Bruno was repeatedly tortured and eventually executed, burnt alive in the Campo de' Fiori, in 1600.

It is inconceivable that Galileo did not hear about Bruno during these years. He worked in a library which contained copies of Bruno's works. Again and again, his views were compared to those of Bruno. Yet in all of Galileo's surviving works – his books, his letters, his notes and drafts – there is not a single reference to him. Curiously, the exact charges against Bruno were and are unknown; the record of his trial was destroyed in the nineteenth century, and no contemporary account describes the charges in detail, so Galileo would not have known them either. It was thus safest to avoid any mention of him, and to steer well clear of two arguments that were closely associated with his name: the infinity of the universe, and the plurality of inhabited worlds. The trace of Bruno's hidden influence can be seen, however, in Galileo's two chief works. In Plato's dialogues, Socrates is always present to speak for himself. In Cicero's dialogues the views of great philosophers are discussed, and they are named. Bruno had invented a new dialogue form, in which the disciples and opponents of an unnamed philosopher (referred to as 'il nolano', for Bruno was from Nola) discuss his views in his absence. Galileo copies this formal innovation in his own dialogues, in which Sagredo, Salviati and Simplicio discuss the views of someone called 'the Academician' – that is, Galileo himself.[7]

Galileo's first documented contact with a committed Copernican came in August 1597.[8] He wrote a letter to Kepler, who had published the first major development within Copernican astronomical theory, *The Cosmographic Mystery*, in 1595. This argued that the proportions of the solar system were determined by the shapes of the five regular solids (the pyramid, the cube, the octahedron, the dodecahedron and the icosahedron – figures with four, six, eight, twelve and twenty sides, respectively), first described by Plato. Kepler had given two copies of his book to a friend who was travelling to Italy, who ended up giving them to Galileo. Kepler must have been astonished to receive a letter out of the blue in which Galileo, who was quite unknown to him, declared that he had been a Copernican for a number of years.[9] This letter is the most important single document for understanding Galileo's intellectual biography and, in conjunction with two other documents, it obliges us to adopt a very different view of Galileo's intellectual trajectory to that which has been generally accepted.

As a mathematician, Galileo was obliged to teach the different astronomical systems, and it was therefore natural that he should know of Copernicanism. Galileo's mentor, the great Jesuit mathematician Clavius, had expressed admiration for Copernicus, even while insisting that he must be wrong. In *On Motion* there is a reference to a technical detail described in Copernicus's great work, so it seems that by 1592 Galileo had read Copernicus for himself. *On Motion* itself is certainly not a Copernican text: it assumes that the earth is at the centre of the universe, that there are absolute directions downwards and upwards, and that heavy objects, when they fall downwards, fall towards the centre of the universe. This gives us a tentative chronology: Galileo was not a Copernican when he left Pisa in the summer of 1592, but he must have become one very soon thereafter. In Padua (where Galileo started teaching in December 1592) he became an associate of Giovan Vincenzo Pinelli.[10] Pinelli had an astonishing library, which he took great pains to keep up to date. It contained two copies of Copernicus, works by Bruno and a Copernican work by Thomas Digges; and publications by astronomers such as Tycho and Kepler were added as quickly as Pinelli could obtain them. When Pinelli died in 1601 his library was sold and shipped to Naples. The ships into which it was loaded were seized by pirates, who were furious to discover that the cargo consisted of old books, many of which they threw overboard. The ships were then wrecked, and fell into the hands of fishermen, who used what they regarded as waste paper to caulk their boats and light their fires.[11] But from 1593 to 1601 Galileo had the use of what for his purposes was the best library in the world. Since he could not afford to buy many books in these years, Pinelli's library was a treasure trove for him.

When Galileo wrote to Kepler in 1597 he made a most remarkable claim: he not only said that he had been a Copernican for a long time, but claimed that 'with this hypothesis [I] have been able to explain many natural phenomena, which under the current hypothesis remain unexplainable'.[12] He had, he said, written at length in defence of Copernicanism; but had no intention of publishing while Copernicanism remained so generally scorned.[13] Galileo, it is worth noting, was employed, like the rest of the professoriate, through a series of renewable short-term contracts, so he could not afford to adopt in public views which were generally 'scorned'. But by 1597 he appears to have already been working on an early version of what would eventually become his masterpiece, the *Dialogue Concerning the Two Chief World Systems*.[14]

This is the problem: the defence of Copernicanism that Galileo wrote at some time between 1592 and 1597 has not survived. We can only pick up the record of his research in physics and then astronomy, as preserved in his letters, papers and publications, in 1602. The most important intellectual activity of his first decade in Padua, the activity that was in its later development to establish his reputation as a scientist and bring him to a trial and condemnation by the Inquisition, is lost. Insoluble problems are best avoided, and so the standard accounts of Galileo's life pass over this problem – and the letter to Kepler – with indecent haste.[15] Serious scholars argue, against Galileo's own unequivocal testimony, that he became committed to Copernicanism only after 1610. Those who disagree with them have been handicapped by the fact that they have been unable to fill in this missing decade.

It turns out, however, that there is rather good evidence regarding Galileo's thinking when he first arrived in Venice. Indeed, given the general paucity of information on his intellectual life during these years, we are extraordinarily fortunate when it comes to tracing his conversion to Copernicanism. In 1592 his friend and sponsor Guidobaldo del Monte invited Galileo to visit him, and we can securely date to that visit, in the summer of 1592, del Monte's notes on an experiment that he and Galileo probably conducted together, the idea for which may indeed have been Galileo's.[16] The experiment was very simple. Take a hard ball and coat it with a substance that would leave a mark; take a hard, flat surface, and place it at an angle fairly close to the vertical; then throw the ball so that, like a pinball, it flies across the sloping surface, first rising, and then, after it slows, falling: what will be left behind is the track of the ball across the surface, and this track will be very similar to the path of a ball flying free through the air – to the path of a projectile.[17] The experiment showed two things. First, the path of the ball is a symmetrical curve – the projectile never travels in a straight line, and its path while rising is identical to its path while falling. Second, this curve is mathematically legible. Galileo and Guidobaldo felt it was very similar to a parabola (a curve generated by the formula $y = x^2$) or a catenary (a curve generated by hanging a chain from two fixed points). Or rather they believed it was both a parabola and a catenary. Modern mathematicians know that a catenary is not a parabola, but that the two curves can be very similar; Galileo, throughout the rest of his life, was to make repeated efforts to prove that the two were one and the same.[18]

This result was astonishing. First, traditional Aristotelian physics maintained that an object's movement must be governed *either* by nature *or* by force. Thus a cannonball fired at an angle of 45° should start by travelling in a straight line and end by falling straight down towards the earth; in between there might be some sort of curve linking the two movements. Some contemporary theoreticians, Archimedeans rather than Aristotelians (Tartaglia, Benedetti, Sarpi), argued that none of the lines would be quite straight, but if asked to draw the trajectory of a cannonball, they would have drawn a distinctly unsymmetrical path, a sort of fish-hook – Tartaglia drew such paths in his *New Science* (1537), as did Galileo himself in the manuscript of *On Motion*. Nothing would have prepared Galileo and Guidobaldo for a symmetrical path. The path, they would have felt sure, ought not to be symmetrical because the upward path was determined by impressed force and the downward path by weight. Two different forces ought to have produced two different paths, not two paths that were a mirror image of each other.

Second, let us think of the path of a projectile as a graph with an x-axis and a y-axis (the language is anachronistic but will serve). The y-axis then represents height, and on it we see the projectile rising and falling. But the x-axis is a little more confusing, as it can represent either distance or time – the time the projectile is in the air. Either way, the projectile's speed over the ground looks pretty constant.

Any mathematician would know that a parabola is obtained by combining a figure that varies according to an arithmetic sequence ($y = 1$, then 2, then 3) with a figure that varies according to a geometric sequence ($x = 1$, then 4, then 9); x is thus the same thing repeated over and over again (e.g. 1, plus 1, plus 1) – on a graph, it is a straight line. The problem is that this x is the forced movement, while y, which is decelerating upwards and accelerating downwards, is, at least on the downturn, the natural movement. In Aristotelian terms it should be exactly the other way round: forced movements should be constantly changing, while natural movements should be stable and continuing. So the path of the projectile not only implies that the same force is governing deceleration and acceleration, upward movement and downward movement; it also implies some sort of unchanging sideways movement. In an Aristotelian world there are no unchanging movements except in the heavens.

Galileo was to have great difficulty in understanding how to relate time and distance when thinking about projectiles and falling bodies – he was still confused on this question in 1604. He seems quickly to have escaped from another possible confusion. A parabola is the result of two forces working at right angles to each other, a catenary the result of two forces (one is gravity pulling downwards, the other is the tension in the chain which pulls to the side) which are never at right angles to each other, because the chain is never straight. Hence the initial attraction of thinking of the movement of a projectile as a catenary, since a cannonball or musket ball is fired at an angle between the horizontal and the vertical. But in successfully avoiding the mistaken conclusion that the path of a projectile is the outcome of two forces not at right angles to each other, Galileo made the opposite mistake – he concluded that a catenary is the result of two forces at right angles to each other.[19]

In *On Motion* Galileo had argued that in addition to forced and natural movement there was a third movement that was neither one nor the other. Forced and natural movements come to a stop: a rising body decelerates until it starts to fall, while a falling body accelerates until it reaches either an obstacle or the centre of the earth. But imagine a millstone whose centre coincided with the centre of the earth, and which turned without any friction. Set it going with a crank. Once started it would carry on spinning for ever. This, of course, is a step towards the semi-Copernicanism of Calcagnino. It means that if the earth were set rotating it would continue to rotate for ever, but the earth would still have to be at the centre of the universe, not flying through space in an orbit around the sun. Again, imagine a perfectly smooth pebble on a perfectly smooth sheet of ice. Just as the slightest touch would move a perfectly poised balance, so the slightest touch would move this pebble. And if the sheet of ice in fact curved with the surface of the earth, so that the pebble was always moving at right angles to gravity, then in principle the pebble would keep on moving for ever, and keep on moving always at exactly the same speed.[20] In Newtonian physics this is the principle of inertia; in Galileo's physics all real-world instances of inertia have to involve some sort of circular movement.[21]

Studying the path left by his pinball as it slipped across a long, steeply sloping surface, Galileo must have seen two things, one entirely unexpected and one strangely familiar. The first, entirely unexpected sight was the symmetry between the upwards movement and the downwards movement; the second, strangely familiar sight was the constant movement that seemed to be the other element generating the curve – the constant movement of an imaginary pebble on an imaginary ice rink on an imaginary flat earth. What Galileo recognised when he saw that the curve was a parabola – and at the same time what he did not recognise, at least not yet – was that gravitational fall implies some sort of multiplication; and that movement without gravity implies some sort of inertia. It was to be twelve years before he fully grasped the first thing he had seen; and he never fully grasped the second.

The significance of the curve left by the pinball as it rolled along was something Galileo perceived immediately, slowly and never.[22] To understand the rest of his life's work we have to find a way of doing justice to all three elements of his response – the immediate recognition that he had seen something that clarified how bodies moved in space; the slow realisation that there were problems here to unpick; and the insuperable obstacle that prevented him from ever quite imagining movement in a featureless, abstract, geometrical, Newtonian space. Why was Galileo unable to grasp at once the full implications of what he had seen? Perhaps because the law of fall as he was eventually to formulate it involved recognising that a falling object, if not subject to resistance from the medium, would accelerate indefinitely, without any limit. For a long time he must have wanted to preserve the assumption he had made in *On Motion* – that acceleration was a temporary phase, which would be succeeded by a constant rate of fall.

For the moment let us concentrate on the immediate recognition. Galileo thought it was of foremost importance, and laid claim to it. It was a custom in Renaissance Europe for gentlemen to collect in an album the autographs of their friends accompanied by mottoes or drawings. Four such inscriptions by Galileo survive, three of which, one from 1599 (before he had fully grasped the law of fall) and two from 1629, include little sketches that correspond very closely to the drawing of the parabolic shape of the path of a projectile which Galileo published for the first time in 1638.[23] The earliest of these sketches appears in the friendship book of Thomas Seget, who seems to have worked for a while as Pinelli's librarian, and who was soon to be imprisoned in a Venetian dungeon, until Wotton secured his release and introduced him (if my conjecture is correct) to John Donne.

In 1599, then, this was already Galileo's great unannounced discovery. At some point in the future his friends would be able to look back and see that he had already laid claim to his discovery in their albums, while Galileo would be able to adduce their albums as evidence if he ever found himself embroiled in a priority dispute. (The fact that he so clearly regarded the discovery as his own suggests that the experiment recorded by Guidobaldo was Galileo's idea. And his anxiety about a possible priority dispute was not misplaced: a friend of his, Raffaello Gualterotti, described the descending path of a projectile as a parabola in print in 1605.)[24] Galileo's discovery was not just of theoretical significance: it was a crucial contribution to the science of gunnery, proving, for example, that the angle of elevation which would result in the maximum range for a gun was 45°. In 1610 we find Galileo promising to write a treatise on this applied science, and his Venetian friends hoped for its publication soon after he moved to Florence that year.

But what Galileo had seen in 1592 was something more than a contribution to applied knowledge. He had seen that the traditional assumption that one movement must predominate over another was false: an object could be moving in two quite different ways simultaneously. He had also seen something close to visual confirmation of his theory of circular inertia: the curve would not have been symmetrical if the object had not moved at a constant speed at right angles to the force of gravity. The importance of this was that it related directly to a passage in Copernicus's *On the Revolutions*. There Copernicus had identified a number of possible objections to his theory of a moving earth. Why were heavy objects on the earth's surface not thrown off into space? Why, if an object was dropped from a high tower, did it fall at the foot of the tower, not to the west? (In 1596 Tycho Brahe was to publish an updated version of these objections, adding a new one: why did a cannonball not fly further if shot to the west than if shot to the east?)[25] Copernicus's answer to the high tower question was that the object must actually fall along a curved path so that it managed to stay at all times above exactly the same spot. In Copernicus's formulation this was a rather far-fetched claim, because no such curve had ever been seen, and the very possibility of such a trajectory seemed doubtful. What Galileo had now seen was precisely the sort of curved trajectory which would be needed for Copernicus's theory to be correct, and he had also seen evidence in support of precisely the sort of circular inertia that would cause a falling object to remain constantly above the same spot on the earth's surface. He could now explain how – to take the version of the problem that Thomas Browne had encountered, perhaps when studying in Padua – in a Copernican universe, an eagle would be able to take aim and drop a tortoise directly onto Aeschylus's head, even though Aeschylus's head was moving as the earth turned.

When Galileo wrote *On Motion* he was not a Copernican: he did not even address the question of falling bodies on a moving earth because, even though he had formulated an idea of circular inertia, he assumed the earth was stationary. It looks very much as though by the time he left Guidobaldo's he had become a convert – it is even possible that Galileo had converted Guidobaldo, as Guidobaldo left amongst his papers when he died in 1607 a treatise on the movement of the earth.[26] What had converted Galileo to Copernicanism was not some new understanding of astronomy: it was a new approach to the problems in physics that he had been addressing in *On Motion*. Galileo became a Copernican because he now had the conceptual apparatus with which to understand the physics of a moving earth, but his conversion implies an eagerness to embrace Copernicanism, for he had found no new evidence in favour of it – he had merely found a way of refuting one of the standard objections to it. Conversion for Galileo was easy. He was a mathematician, and so could appreciate the beauties of the Copernican system. He was hostile to Aristotelian philosophy, and so had no attachment to the old order. He believed (judging by the poem against the wearing of the toga) neither in heaven nor in hell, and so had no need to worry about their spatial location. He will have been well aware that the arguments that convinced him would not be enough to convince other, more conventional thinkers.

Before Galileo left, he also discussed with Guidobaldo a puzzling problem in materials science: why was a long rope more prone to break than a short rope? They concluded that the length of the rope was irrelevant; what mattered was its strength at its weakest spot. A long rope was simply more likely to have a weak spot.[27] Evidence of Galileo's thinking next appears in a very different source. When Galileo arrived in Venice in September (he stayed for around a fortnight and then returned briefly to Pisa) it seems that he almost immediately made the acquaintance of a friar, Paolo Sarpi. Sarpi was someone with an extraordinary range of interests. In the future he would become a great historian, the author of the *History of the Council of Trent* (published in Italian, Latin and English in Protestant England in 1619). In 1606 he was to be the leading theological adviser to, and polemicist on behalf of, Venice when she was placed under an interdict by Pope Paul V – a role which made Sarpi famous throughout Europe. In 1607 he survived an assassination attempt, which turned him into a symbol for intellectual liberty. After 1606 Sarpi was to become preoccupied with religion and politics, but in the early 1590s much of his intellectual energy was going into scientific and philosophical subjects.[28] He was a materialist, an atomist and an atheist (or at least an agnostic); and although there is no evidence that he was ever completely convinced of the case for Copernicanism, he had long been puzzling over the arguments for and against the new astronomy. He was also actively involved in anatomical research, and may have played a part in the discovery of the valves in the veins.

Sarpi recorded many of his thoughts on philosophical and scientific questions in a notebook, dating them year by year. Under the year 1592 we find an account of Guidobaldo's pinball experiment, spread across four numbered paragraphs; the first of these offers an objection to the principle of symmetry, an objection that Sarpi was to restate in a letter to Galileo in 1604. There is also a paradoxical version of the rope length puzzle: if a rope were uniformly of the same strength throughout, then it would never break, because there would be no weakest spot. The only possible source of these two distinctive sets of arguments is Galileo; indeed, this is the proof that Guidobaldo, who records all this material in his own notebooks, had gone through this material with Galileo, who then went through it with Sarpi.[29]

Between these two arguments that correspond to material in Guidobaldo's notebooks we find other material which is distinctively Galilean and which must also derive from conversations with Galileo – indeed, Sarpi's notebooks enable us to, as it were, listen in on his conversations with Galileo. In one paragraph Sarpi lays out a version of what will become Galileo's argument in the *Two New Sciences,* that all bodies will fall at the same speed in a vacuum – an argument very different from that of *On Motion*, where it was maintained that the speed of bodies falling in a vacuum would be proportional to their specific gravities.[30] In two others Sarpi presents a number of reasons why heavy bodies are not thrown off the surface of the earth by centrifugal force as it rotates: one of them becomes Galileo's (mistaken) argument in the *Dialogue* that the rate of fall towards the earth must always exceed the rate of projection into space.[31]

What Sarpi's notes show is that by the time Galileo arrived in Venice in 1592 he had formulated sophisticated responses to the standard arguments against a moving earth – arguments that went far beyond anything in Copernicus. He could explain why objects dropped from a high tower appear to fall vertically by means of the principles of circular inertia and combined (rather than predomi-nating) movements; and he could explain why objects are not thrown off into space by centrifugal force by comparing estimates of the speed at which bodies fall towards the earth with estimates of the speed at which they would move away from the earth if projected along a tangent by centrifugal force. When he eventu-ally read the debate between Tycho Brahe and Christoph Rothman which Tycho published in 1596, he would have immediately felt that he had already addressed all the key issues raised (the principle of circular inertia could easily be employed to address Tycho's question about the trajectory of cannonballs fired to the west and to the east).[32] In his biography of Galileo, Viviani claimed that he began work on the project that forty years later was to become the *Dialogue* as soon as he arrived in Venice: this claim, apparently extravagant, now seems exactly right.[33]

This being established, it is possible to revisit Galileo's remarkable claim in his letter to Kepler that he had used Copernicanism to explain many natural phenomena which were inexplicable within conventional physics. Kepler imme-diately guessed that Galileo was talking about the tides.[34] This seems to be an extraordinary leap until we realise that a classical source had suggested that the tides might be linked to the movement of the earth. And Kepler was right, for in Sarpi's notes for 1595 we find an account of the theory that Galileo was to put forward in 1616, and publish in 1632, as proof that the earth moves.[35] The basis of the argument is simple. If you are on a boat into which a little water has leaked, the boat and the water within it travel along together quite happily. But if the boat bumps into the bank then the water in the bottom of the boat carries on travelling forward, piling up at the front of the boat. The tides, Galileo argued, must be caused by a similar phenomenon: by some change in the speed or direction of the earth's movement, which causes the water contained in the earth's basins to slosh around. Copernicus attributed three movements to the earth, of which the two most important (indeed in Galileo's view the third was spurious) were the annual movement around the sun and the daily rotation on its axis.[36] The combination of these movements would, Galileo argued, mean that any point on the earth's surface would be travelling faster when it was on the side of the earth away from the sun (annual movement plus diurnal movement) and slower when it was on the side of the earth towards the sun (annual movement minus diurnal movement); in between it would experience acceleration or deceleration. This is not in fact what causes the tides (they are caused, as Newton showed, by the gravitational attraction of the moon and the sun), but it could cause tides in a theoretically possible world: the argument is valid in principle, but not in its application to our particular world.[37] (That it is the wrong explanation for tides in our world was never apparent to Galileo, who refined it to explain why tides vary at different times of the month and the year, and who sought to explain away the fundamental problem that there are two tides every day rather than, as his theory implied, one. Nor did Galileo have an explanation for the fact that the timing of the tides varies from one day to another.)

In his letter of 1597, then, Galileo is referring to his theory of the tides, which he has already explained to Sarpi (or, just possibly, which Sarpi has explained to him). But Kepler's genial interpretation has distracted attention from the fact that Galileo says that he has explained 'many natural phenomena', not just one. What else does he have in mind? One answer, I think, is the principle of the relativity of movement. Copernicus had argued that perception of movement is relative. We have all experienced the moment when two trains are standing side by side and one starts to move; if you are on one train and look through the window at the other you cannot at first tell which is moving – one train moving is visually indistinguishable from the other train moving. Galileo will have experienced the same phenomenon when on a boat casting off from the dock – at first it looks as if the dock is moving, not the boat. But by 1594, and probably in 1592, he had formulated a different principle of the relativity of movement: that movement at a constant speed is indistinguishable, unless there is an external reference point, from standing still. If the boat moves at a steady speed, the water in the bottom of it lies calm and undisturbed. If one person in the boat tosses a coin to another, he does not have to aim off to allow for the movement of the boat, because he, the coin and the other person are maintaining their positions with regard to each other. And if someone drops a coin from the top of the mast (this is a version of Copernicus's problem – hence its importance for Galileo), it will land not behind but at the very foot of the mast. In conventional Aristotelian physics, objects in this world are naturally at rest: they move only if they are forced, or if they are out of place. So an object dropped from the mast of a moving ship will move vertically downwards towards the centre of the earth; it will not continue to move forwards as it falls, keeping pace with the ship.[38]

In the new physics of Galileo, rest and constant movement are indistinguishable, and this alone can explain how objects behave in boats. When Galileo says he has used Copernicanism to explain many natural phenomena he means that he has used his new principle of the absolute relativity of movement, a principle which must be true if Copernicanism is to work, to explain a wide range of phenomena. And since this principle is central to the *Dialogue*, it is now clear that the *Dialogue* was a direct descendant of the defence of Copernicanism which he referred to in his letter to Kepler of 1597.

There is another discovery that Galileo probably had in mind when he used the word 'many' in his letter to Kepler. In his first great work, *The Starry Messenger* (1610), he talks of a discovery made 'not recently but rather many years ago'.[39] This discovery related to the faint light illuminating the dark mass of the moon at new moon. Some had claimed the light came from the moon itself, some that it came from the stars, and some that it was the sun's light shining through the translucent body of the moon. Galileo claimed – correctly – that the light is what we now call earthshine; in other words, that the earth illuminates the moon just as the moon illuminates the earth. The argument he presented was not new. It is to be found in one of Leonardo's notebooks; it had apparently been presented in a work, now lost, published by Maestlin in 1596; and it had certainly been presented by Kepler in a book published in 1604 (the fact that Galileo was unacquainted with this work is an indication of the extent to which his reading had fallen behind with the loss of the Pinelli library) and by Raffaello Gualterotti (perhaps on the basis of conversations with Galileo) in a book published in 1605. Before Galileo had published his account of it, Thomas Harriot had taken this argument for granted in naming the phenomenon, for the first time, 'earthshine'.[40] More to the point, it had been formulated by Sarpi in his notebooks long before he even met Galileo; it therefore seems likely that in *The Starry Messenger* Galileo presented as his own a discovery that he owed to Sarpi.[41] Galileo, Maestlin, Harriot and Kepler were all Copernicans, and the great attraction of the argument for a Copernican was that it suggested that the moon and the earth were fundamentally alike – an argument that is absolutely central to the *The Starry Messenger*.

In *The Starry Messenger* a crucial argument for thinking of the earth and the moon as alike is that the moon is not, as the Aristotelian philosophers claimed, a perfect, crystalline sphere: it has mountains and valleys. The mountains can be identified by the shadows they cast, and by the fact that their peaks are caught by the rising sun, lighting effects that are clearly visible with a telescope.[42] In *The Starry Messenger* the impression is given that this is a completely new argument. But this was a phenomenon that Galileo discovered almost as soon as he turned his telescope towards the moon in 1609, yet he showed no initial urgency to publish this extraordinary finding. The reason is that it was merely confirmation of something he will already have seen with the naked eye, for if you have sharp enough eyesight and know what you are looking for, you can see evidence that the moon has mountains without a telescope. Galileo owned a copy of a book published in Florence in 1606 in which this phenomenon was clearly (if slightly mistakenly) described: 'There are also on the moon mountains of gigantic size, just as on earth; or rather, much greater, since they are [even] sensible to us. For from these, and from nothing else, there arise in the moon scabby little darknesses, because greatly curved mountains (as Perspectivists teach) cannot receive and reflect the light of the sun as does the rest of the moon, flat and smooth.'[43] There is good reason to think that this book, published pseudonymously, was by a pupil of Galileo – indeed, it may well have been by Galileo himself – and it is quite possible that Galileo had reached the conclusion that there were mountains on the moon, and that this proved that the moon and the earth were alike, as early as 1597.

These arguments – the argument from the relativity and combinability of movement, the argument limiting the impact of centrifugal forces on a turning earth, the explanation of earthshine, and the explanation of the tides – were all at Galileo's disposal by 1597, perhaps together with the argument from the existence of mountains on the moon, and they certainly justify his claim of that year that he had explained many natural phenomena on Copernican principles.

There exists one other Copernican text from 1597 – a long letter from Galileo to his friend in Pisa, Jacopo Mazzoni (a copy of which was to be found in Pinelli's library at the time of his death, showing that Galileo's Copernicanism was no secret from his friends in Padua).[44] Mazzoni was a Platonist, and Galileo's letter contains a much more unequivocal declaration of their shared Platonism (in other words, presumably, of their agreement that the structure of the world was fundamentally mathematical) than of his own Copernicanism.[45] As far as that goes, he says merely that he has 'a bias' ('qualche umore') in favour not only of Copernicanism, but also of some arguments that are inseparable from Copernicanism (arguments, we may assume, on the physics of a moving earth). Mazzoni had just published a book on Platonism containing a criticism of Copernicanism: the subject was relevant because the Pythagoreans had held that the earth went round the sun, and Pythagoras was the major alternative to Plato for those concerned to read the universe in mathematical terms. To this criticism Galileo feels obliged to respond, his response providing indirect confirmation that he had not been a Copernican when he was living in Pisa, since the issue is clearly not one he has discussed with Mazzoni in the past.

Mazzoni's argument took for granted the notion that we can 'see' the heavens – that is, the sphere to which the fixed stars are attached. If Copernicus is right, he argued, then when we look upwards at the heavens during the daytime the part of the heavens we are looking at is further away than the part we see at night by the diameter of the earth's orbit around the sun. The result should be that in the daytime we are aware that the dome above our heads represents more than half of a sphere, and at night we are aware that it represents less than half of a sphere. This argument might be reformulated (since all that can be seen during the daytime is a featureless blue sky) by saying that it ought to be possible to see at midnight that the stars on the horizon are appreciably further away than the stars overhead. Galileo rebuts this argument by putting some elementary geometry to work. His argument depends on the fact that even in a Ptolemaic system we do not see the universe from its centre, but from the surface of the earth, which is at a distance from the centre. Taking the measurements of the universe accepted by the Aristotelians, he maintains that displacing the earth from the centre to the location of the sun would be equivalent to seeing 176°56' of the sphere of the heavens at night and 183°04' during the day – a difference equivalent, if the earth was at the centre of the universe, to seeing the heavens from the top of a mountain one mile high. It is easy to see, he argues, that this difference might be imperceptible; and we certainly will not be aware that the stars above our heads are closer than the ones on the horizon if the scale of the universe is increased as the Copernican theory requires. What Galileo is trying to formulate (rather clumsily it must be said) is a simple, fundamental claim: if we look at the heavens with the naked eye, either at a single moment in time or over the course of a day, nothing that we can see can possibly help us choose between the Copernican and Ptolemaic systems.[46]

12

Money

Between 1591, when his father died, and 1610 Galileo was engaged in a constant struggle to make ends meet. It seems likely that his sister Virginia married in 1591 on terms negotiated by his father. Galileo inherited the obligation of paying for her dowry, and in 1593 Benedetto Landucci, his brother-in-law, was threatening to have him arrested if he came to Florence.[1] Galileo borrowed two hundred ducats in an effort to sort things out.[2]

In 1601 the marriage contract for his sister Livia provided for a dowry of 1,800 ducats, to be paid over a period of five years by Galileo and his brother Michelangelo.[3] But no money was forthcoming from Michelangelo – indeed Galileo had just spent roughly 60 ducats on equipping Michelangelo so that he could take up a job in Poland.[4] In 1602 we find Galileo borrowing 250 ducats, which he was unable to repay. He had to turn to his friends for help after his employers agreed to pay him only a year's wages in advance when he had asked for two. A year later he obtained a second year's pay in advance.[5] Nevertheless, in March 1605 he was being sued for money owing on Livia's dowry, in May he was facing legal action for money owing on Virginia's dowry, and in 1608 he was yet again receiving a year's salary in advance.[6]

We can see how difficult Galileo's position was when we realise that Livia's dowry represented close to thirty years' salary at the rate at which he had been paid at Pisa, and eight years' salary at the rate he was then being paid at the University of Padua. At Pisa he had been paid sixty ducats a year; at Padua he started on 180 florins (a Venetian florin being worth approximately 70 per cent of a Florentine ducat), which rose to 320 florins when he was given a new contract in 1599 and to 520 florins when he received a third contract in 1606.[7] His improved telescope led to his being offered a contract for life at a thousand florins a year, but this was not to start until 1612, and it was stipulated that there would be no further increases.[8]

Before 1606 Galileo simply could not meet his obligations out of his university salary. Moreover he soon had a mistress, Marina Gamba (about whom little is known, and whom Galileo left behind when he moved to Florence in 1610), and children of his own to support (Virginia was born in 1600, Livia in 1601, Vincenzo in 1606).[9] He supplemented his income by giving private lessons in the mathematical skills required by young gentlemen who might aim to pursue a military career. He also taught the use of a calculating and measuring device known in contemporary English as a sector (the Italian name is a 'compasso', or compass); this required an instrument maker, hired in 1599, to supply sectors, and a copyist to copy the instruction manual, until it was privately printed in an edition of sixty copies in 1606.[10] (Galileo's instruction manual is now so rare that a copy sold at Christie's in 2008 for more than $500,000.) Young gentlemen who came to study with Galileo often boarded with him, bringing with them either friends, who attended the lessons, or servants, who did not. The peak year for which there is evidence is 1603, when Galileo was struggling to pay Livia's dowry without the help he had expected from Michelangelo. In that year he had about a dozen students, and sold a dozen instruments; with students, their friends and their servants, he had at any time around ten boarders, and a staff of five (plus his mistress) to supply their needs.[11] He had of necessity a large house, with a vineyard which will have supplied wine for the household.[12] Teaching brought in 660 ducats, which will have been mainly profit (although he had a copyist working full time to supply texts to read); boarders paid a thousand ducats, much of which would have been eaten up by expenses; the profits on the sale of instruments will have been small – perhaps five ducats each. We can safely say that Galileo earned three times more in that year from private lessons, boarders and the sale of instruments than he did from lecturing at the university. It seems likely that he was exaggerating when he claimed in 1610 that he could make at least another two thousand ducats a year from private teaching and boarders, though he had indeed managed to more than triple his salary in that way in the past.[13] But by 1603 he was certainly earning at a rate which allowed him to begin to make headway against his obligations. The Interdict crisis of 1606–7, when Venice's policy of asserting the rights of secular states brought it into conflict with the papacy, evidently caused a short-term fall in his income, but by 1610 he was in a position to lend four hundred ducats (a very large sum – a year's salary for most professors) at 6 per cent to his friend the philosopher Cesare Cremonini, money which he had great difficult recovering. Cremonini was the highest paid professor in Padua, but he lived, we are told, like a cardinal: he had not one but two carriages, and six fine horses, so that it is not entirely surprising that he needed to borrow money.[14] With Galileo safely out of the Venetian territories, Cremonini even discovered that he had a principled objection to lending at interest.[15] (Like many of Galileo's Venetian and Paduan friends, he saw Galileo's departure for pro-papal Florence as a betrayal, and his failure to pay may have been a sort of protest.)[16]

These were years, however, in which Galileo was carrying out an active programme of experiments, and the long hours of teaching must have seemed an unbearable distraction. He came to hate teaching so much that he began to think of it as equivalent to prostitution. It is not surprising that he began to long for a better life, and to make enquiries as to whether he could find employment elsewhere. What he was looking for was not more money but more time. When he finally moved to Florence in 1610, the salary (allowing for the exchange rate) was about 40 per cent more than he was to have received if he had remained in Padua – but a good deal less than he could have made if he had stayed in Padua and continued providing lessons. The big difference was that in Florence he was not going to be required to do any teaching at all – not even the sixty half-hours a year that were required under his current university contract. This, what he and his correspondents called 'ozio', or leisure, not money or status, was (officially at least) the irresistible attraction which caused him to leave Padua.

There was another attraction. In Padua Galileo had been dependent on the patronage of a small circle of powerful young noblemen. They had looked after him, ensured his salary was regularly increased and arranged for it to be paid in advance, and they had lent him money when he was in need. But in return they expected to be able to make demands on his time and his energies – they expected him, for example, to accompany them on holiday.[17] The demands of such friends and patrons came to seem so burdensome that satisfying them involved what Galileo saw as 'ceaseless labours'.[18] It seems clear that he had come to resent this unequal relationship. In 1606 he tried to obtain a job in Florence for his friend the great Paduan professor of medicine Acquapendente. Acquapendente, he felt sure, would be happy to move – he would be delighted to give up lecturing, and keen to escape from the network of dependency that held him imprisoned.[19] It is hard to avoid the suspicion that Galileo was describing not so much Acquapendente's likely response to the offer of a job in Florence (he insisted he had not mentioned the matter to him) as his own.

After he moved to Florence, Galileo had a good income, and he was never in financial difficulty again. He had, it is true, to arrange for his salary to be paid in advance in 1614 and 1628–9: on the first occasion he was collecting the funds required for his daughters to take their vows as nuns – the equivalent of a dowry was payable to the nunnery – and on the second he was recuperating the money he had laid out to marry off an unrecognised illegitimate daughter.[20] But there is nothing to suggest that he regretted the reduction in overall income that followed his move, or that earning more money was ever a significant objective for him again. Between 1591 and 1608 he had been running fast in order to stand still, like a hamster on a wheel. In 1610 he stepped off the wheel, and wanted never to get back on it. He had no need: on his Florentine salary he was able to save substantial sums of money, to the extent that his nephew thought of his uncle as a rich man.[21]

13

Fields of fire

When Galileo arrived in Padua there was, some maintain, a major reorientation in his intellectual interests.[1] *On Motion* is a purely philosophical work; by contrast his very first surviving letter from Padua is concerned with a practical problem in naval technology: how long should the oars of a galley be? There could be no more basic question in Mediterranean naval warfare in the period, for fighting was entirely conducted, as in the classical era, by boats powered, not by the wind (there were too many calm days), but by the bodies of human beings – usually convicted criminals condemned to the galleys. This new interest in technology was soon reflected in a number of different areas of activity. Galileo wrote courses of lectures on machinery and on fortification; he designed and sold his sector or compass; and he designed a drainage pump of some kind (perhaps some sort of Archimedes' screw).[2] His primary interest now was in applied science, and this new interest went hand in hand with an interest in developing experimental procedures. The birthplace of modern science was not Galileo's lecture room in Padua but what Venetians called the Arsenal – the vast shipyard where vessels were built and repaired for the Venetian navy.[3]

This story sounds persuasive, but I am not convinced that it is right: it cannot survive a reading of Galileo's first Paduan letter and Giacomo Contarini's reply.[4] Galileo is answering a question asked by Contarini, who has responsibility for naval architecture. The question is whether it is better to have the oar pivot on the gunwale or on an outrigger; later it becomes clear that Contarini is confident he knows the answer to the question, so his purpose is probably to test Galileo. Galileo replies that the most efficient oar will be the one in which the rowlock is as close as possible to the oarsman. If the handle section of the oar is long the oar will be efficient at moving seawater; if the blade section of the oar is long the oar will be efficient at moving the boat.

Contarini was not impressed by this analysis, and one can see why: if Galileo's theory were right, no one would ever build a Venetian gondola. Contarini wrote back with barely concealed impatience. Galileo, he protested, had not considered that oars have to be raised out of the water and lowered into it, as well as being pulled through it. He had not considered that the oarsman needs to use the whole weight of his body, and therefore needs to be able to work standing up. The oar thus has to be proportioned to the oarsman's arms, for moving it up and down, and to his whole body, for pulling on it. Galileo's proposal would render the outboard weight of the oar too heavy for the oarsmen to manage (on a galley there would be several rowers on each oar); indeed, even with existing oars (where the handle is one-third the length of the whole) great lead weights have to be added to the oar on the inboard side to bring it almost into balance, and this would be impossible if the inboard section were too short. However, the angle at which the oar enters the water is also important: the steeper the angle the more energy will be wasted cutting downwards and upwards through the water rather than driving the boat forward.

Contarini's reply amounts to telling Galileo that he cannot design a galley just by applying the theory of the lever, which is precisely what Galileo had claimed to be doing. He needs to look at some boats and some oars, to watch rowers at work, to ask skilled craftsmen why they proceed as they do. In order to claim that Galileo had become an applied scientist one would need to be able to show that he promptly learnt the lesson that Contarini was trying to teach him. Contarini, for his part, seems to have decided that Galileo had failed the basic test he had been set: there is no evidence that he consulted him again.[5]

At some point, it is true, Galileo does seem to have learnt this lesson: the *Two New Sciences* begins with an account of a conversation with shipbuilders in the Arsenal. It is from talking to them that Galileo learns that the weight of a large boat may be such that it is in danger of breaking up when it is in dry dock, even though it is perfectly sound when afloat. He was already interested in the strength of materials before he arrived in the Venetian state – he was puzzling over the breaking points of ropes. But we may wonder if he ever spent much time studying naval architecture: he writes as if the beams on wooden boats are supported at one end only, when they are nearly always supported at both ends.

It is even true that Galileo placed a higher value on applied knowledge than on merely theoretical knowledge: he thought there was nothing more admirable than the invention of the alphabet or of weaving. His own great contribution to this type of knowledge was, he believed, a reliable method of establishing longitude.[6] But his work on longitude provides decisive evidence that he never grasped the idea that applied science is different from theoretical knowledge. In 1635 and 1636 he was claiming to be able to build a clock that would be accurate to within one minute, even to within one second, in a month. In making this claim he was relying on the principle of the isochronicity of the pendulum and the fact that he thought it would not be difficult to devise an escapement for a pendulum clock. But he had never built a pendulum clock, and he had absolutely no idea how accurate such a clock would be in practice, when subjected, for example, to variations in temperature, let alone when on a ship at sea.

Thus Galileo never fully grasped that theory has its limits, and that one needs to learn from experience. He relied on experience only when theory was defective and incomplete. He had already conducted one experiment, the pinball experiment, which had transformed his understanding of a fundamental philosophical problem: but there is no evidence that this had led to any wider shift in his intellectual interests. His failure to grasp that the pinball experiment provided him with a law of acceleration shows that he had not understood how much he might be able to learn from an experiment. There is no text of Galileo's that can be dated to before 1600 (we know of the pinball experiment only indirectly, through the accounts of Guidobaldo del Monte and Paolo Sarpi) in which experience is used to supplement, correct or substitute for theory.

It might seem odd to claim this when we have manuscript textbooks associated with two early lecture courses by Galileo – one on machines and the other on fortification – that would seem to be contributions to applied technology rather than to theoretical physics. These courses were not part of the university curriculum: like many mathematicians, Galileo supplemented his academic salary by providing private tuition to young noblemen who had come to Padua not to take a degree but to acquire useful skills. Crucially, they needed the technical knowledge that would enable them to command armed forces in warfare, beginning with elementary topics such as the use of the magnetic compass. How could Galileo have lectured on these subjects without having acquired a considerable respect for experience?

Not only did Galileo lecture on applied subjects; he also designed his own very sophisticated version of a basic scientific implement used by soldiers: the military compass or sector.[7] The sector made it possible to speed up calculations involving ratios or proportions by reading off figures from one scale to another. In a world in which few people were capable of carrying out multiplication and division reliably, the sector greatly reduced the risk of error. Someone designing a sector had to decide which calculations he wanted to simplify: by providing lots of scales he could simplify lots of calculations, but he also increased the cost of the instrument, and made it more and more fiddly and complicated to use. Thus, to use a simple example, a sector might provide scales that enabled a merchant to see the value in florins of a certain number of ducats without multiplication or division. Another scale might enable him to translate volumes into weights, so that given the height of a pile of grain he could use the sector to work out its weight, and then to calculate its value. With a plumb line attachment a sector could be used to calculate the angle of elevation of a cannon's barrel. With a sighting device (which Galileo supplied) it could be used to calculate ranges or in surveying.

The sector was the sixteenth-century equivalent of a slide-rule or a calculator, and like slide-rules and calculators it provided in-built formulae for the most common calculations. In Renaissance warfare a sector was an invaluable tool for anyone in charge of artillery: it could be used to calculate, for example, how many barrels of gunpowder would be needed to fire off a thousand cannonballs, or to turn an estimate of the range of an enemy fortification into an angle of fire.

Galileo's sector was the most sophisticated available in northern Italy at the time: other people were soon producing copies, and his instruction manual was plagiarised. (There was a court case in which Galileo demonstrated that the plagiarist was not, as he claimed, the inventor of the instrument by showing that he did not know how to use it.)[8] It was Galileo's sector that brought him to the attention of Christina of Lorraine, the wife of the grand duke of Tuscany, Ferdinando I – she employed Galileo to teach its use to her son Cosimo in the summer of 1605.[9]

There is no doubt that Galileo taught applied or technological sciences. There is no doubt that he thought hard about them, and that his teaching was innovative and brilliant. But it is a fundamental fact that his approach to them was theoretical rather than practical. It was through his clarifying and simplifying the exposition of theoretical principles that Galileo thought that he would make it possible for his students to acquire practical competence.

Galileo's manuscript textbook on fortification is a brilliant little book because it focuses narrowly on the issues raised by fighting in the era of gunpowder.[10] It comes with extremely helpful illustrations so that the argument can be followed with ease. Since artillery can knock down walls, defence depends on the ability to prevent the enemy from occupying positions from which they can attack – both positions at a distance, from which cannon can fire on the walls of a fort or city, and positions at the foot of the wall, where ladders can be set or tunnels dug in order to lay explosives. The first requirement is met by mounting artillery in a high place. The second is more complex: it requires that every point of the defended perimeter should be reachable by both flanking and raking fire. Raking fire enables you to hit the enemy as they approach the wall, but flanking fire enables you to hit them (without destroying your own walls) when they are right at the base of the ramparts. If a castle is a simple cube, then artillery can be mounted on the top of the cube to fire out over the countryside, and sharp-shooters can be set on the walls to pick off approaching troops. But if the enemy actually reach the foot of the wall there is very little that can be done: it is impossible to hit them with your artillery, and you have to expose yourself to their fire in order to shoot a musket at them. What shape of structure will improve the field of fire? The basic solution is the bastion: a fortified projection which makes it possible to fire along the flank of the fort, and which often takes the form of a mini-fort at each corner of the larger fort. Of course the same problem is replicated within the bastion itself: how can one achieve flanking fire along its sides? The answer is that no side should be parallel to the main defence walls. The science of fortification thus becomes fundamentally a problem in geometry: how, in any particular location, to design a structure such that every point on the perimeter can be reached by flanking and raking fire from within the structure. Where the ideal shape of a medieval castle was a cube, the ideal shape of a defensive perimeter would now be a star, were it not for the fact that a star-shaped castle would be expensive to build and would enclose only a small area.

Galileo is perfectly happy to discuss the proper height and depth of walls – the base needs to be a massive earthwork which can absorb the impact of cannonballs, while the top has to provide shelter for marksmen. But he addresses practical problems of construction only when describing the trenches and banks that need to be dug out and thrown up by an attacking force as it tries to advance as close to the walls as possible. Here an officer will have to take direct command, rather than simply hiring engineers and masons, and so Galileo provides illustrations of the tools and the tactics required. But these are secondary issues: for Galileo, fortification, in the gunpowder era, is primarily a problem in applied geometry because guns shoot (at least when seen from above) in a straight line, so that all that is needed to study fields of fire is a ruler and a pencil.

Galileo's manuscript textbook on mechanics is a similar exercise in theoretical simplification.[11] The machines he is interested in are what are now called the simple machines: the balance and the yardarm, the lever, the wedge, the screw, the wheel and axle, the pulley, the inclined plane. His fundamental theoretical claim (which he seems to have been the first to formulate) is that you never get out more work than you put in, so that a device that enables a one-kilo weight to lift a ten-kilo weight through one metre will always require that the one-kilo weight travels ten metres: it only seems that the machine is giving back more than is being put in if the weights are considered and the distances are forgotten.[12] If there are ten bricks, they can all be put in a hod and carried together, or they can be carried one at a time: the amount of work involved is exactly the same. Not nearly as much strength is needed to carry them one by one, but you have to walk ten times as far. It should immediately be apparent that Galileo thinks that all the simple machines work on the same principle: each represents a method for dividing a weight by multiplying a distance. If one of them is understood, it becomes easy to make sense of the others.

It should also be apparent that Galileo is not talking about real machines. In real machines there is always some loss of energy: think of the friction involved in turning a screw. There is no such thing as a frictionless pulley, no rope that bends without resistance and never stretches, no lever that does not flex. Of Galileo's idealised machines the one that comes closest to being realisable in practice is the yardarm, and Guidobaldo had carried out experiments to see if one could design a balance that would weigh exactly. But Galileo is not interested in improving the design of these machines. His purpose is to clarify the principles on which they work, and this involves applying to them all the conceptual tools that Archimedes had employed to understand the yardarm. In the modern world Galileo's arguments could be modelled with a computer simulation; in Galileo's world you cannot illustrate them by playing with real machines. He supplies geometrical sketches, and they have to be animated in the mind's eye. For Galileo, mechanics is, like geometry, a purely theoretical discipline. In *On Motion*, describing his first experiments with sliding bodies on inclined planes, he had written, 'We must assume that the plane is, so to speak, incorporeal.'[13]

One might go further than that and say that mechanics has become his template for how to think scientifically. Galileo can tell you in theory what work a frictionless pulley will do, and anyone can see that a real pulley is an imperfect version of this theoretical model. But when Galileo imagined a puck sliding across perfectly smooth ice, or concluded that the path of a projectile is a parabola, he was engaged in exactly the same sort of abstraction, the difference being that he was treating the real world as if this too were simply an imperfect realisation of a theoretical model.[14]

14

The experimental method

When Galileo caused balls, the weights of which he had himself previously determined, to roll down an inclined plane . . . a light broke upon all students of nature. They learned that reason has insight only into that which it produces after a plan of its own, and that it must not allow itself to be kept, as it were, in nature's leading-strings, but must itself show the way.

Immanuel Kant, *Critique of Pure Reason* (1781),
trans. Norman Kemp Smith

Was Galileo an experimental scientist? Few subjects in the history of science are more contested. There are a number of possible sources of confusion to be resolved before proceeding. First, what is an experiment? In modern usage, the idea of an experiment involves the reproduction of some naturally occurring phenomenon in a carefully controlled and therefore artificial setting – what Kant called reason proceeding 'after a plan of its own'. We might distinguish between an experiment, which is designed to play a part within an argument, perhaps even to test a hypothesis, and an experience: I have plenty of experience of thunder, but I would not have the first idea how to devise an experiment to establish the cause of thunder. Observation can be placed somewhere between experience and experiment: having observed that I see lightning before I hear thunder, and having hypothesised that they both have the same cause, I would be a good step on the way towards devising an experiment that would make it possible to compare the speed of sound with the speed of light. In order to avoid unnecessary complications I have used the concept of experiment rather loosely so far. I have talked about Galileo's pinball 'experiment', but it might be more accurate to speak of his pinball *observation*: Galileo had simply devised a way of recording the path of a projectile. (He might, equally, we might add, have looked at the path of a jet of water from a fountain, although the recognition that this too formed a parabola came only after he had published his *Two New Sciences*.) He had not attempted a series of controlled observations, studying, for example, what would happen if he doubled or halved the speed of the projectile.

These distinctions between experiments, observations and experiences might need refinement, but they are commonplace concepts and terms. They may be difficult to define, but everyone knows how to use them. This was not the situation in Galileo's world. Intellectual debate was mainly conducted in Latin, and classical Latin, the Latin of Cicero, had no word for either experience or experiment: the words 'experientia' and 'experimentum' appear soon afterwards, but they are always rare, 'experimentum' being the rarer of the two. Galileo only twice uses this word in Latin, and he rarely uses its equivalent in Italian, 'esperimento': for him 'esperienza' usually does the work of both.[1] Indeed, he assumes there are only two types of knowledge: knowledge based on proof (mechanics, for example), and knowledge based on experience.

He has, however, a Latin phrase for testing a hypothesis: 'periculum facere', to put something to the test.[2] Here it might be helpful to distinguish tests according to whether they are open-ended or (to coin a term) closed-ended – that is, merely confirmatory. In an open-ended test it should be possible to get a result, but the result will not be known in advance. So if a timber beam is tested to establish its breaking strain, and the wood the beam is made of is previously untested, then the result will provide new information. It will be impossible to predict the result in advance, although it might well be possible to make a guesstimate. At the other extreme, if I have just learnt Pythagoras's theorem (that in a right-angled triangle the square on the hypotenuse is equal to the sum of the squares on the other two sides), I might well proceed to draw a triangle on a piece of graph paper and measure the size of the squares, knowing full well what the result ought to be, and concluding, if I failed to get the right result, that there was an error in my drawing or my counting, rather than in Pythagoras's theorem. The testing of a hypothesis would lie between an open-ended test and a confirmatory test: the result cannot be safely predicted in advance, but the range of possible results has been carefully limited to 'yes' or 'no'. This can be called a binary test, and is what is meant by 'periculum facere'.

It should now be clear then that there is something inherently anachronistic about asking if Galileo was an experimental scientist; it is much more straightforward to ask if he argued by deduction, or by induction from evidence – although 'evidenza', in this sense, is another word he lacked. Even a deductive science will be supplemented by references to evidence – as in the case of drawings illustrating Pythagoras's theorem. An empirical science will rely on open-ended tests or observations; an experimental or theoretical science will rely on hypotheses and binary tests; a deductive science will supply confirmatory examples.

There is a third difficulty. It is easy to assume that, given some basic understanding of science, or some relevant experience, we can predict the outcome of experiments. After all, we have the advantage of a familiarity with modern scientific principles. Two problems of this sort have already been discussed. Can we reliably predict what will happen if someone drops a heavy weight and a light weight from a high tower? It turns out that simultaneously dropping two weights is far more difficult than might be expected. Similarly, can the supposed law of the pendulum be established through experiments? It turns out that it is much easier to construct two pendulums of the same length by tuning them to swing in time than it is to show that two pendulums of the same length will always swing in time.

A classic example of our tendency to overestimate the value of our own knowledge is provided by a passage in the *Two New Sciences*. Galileo tells us to fill a glass container with wine and to seal it with a stopper containing a very small hole – the hole needs to be so small that there will be no flow of wine out of the container if we hold it upside down. Now place this container upside down in a basin of water. Slowly, very slowly, the wine will drift down out of the container and fall to the bottom of the basin while the water will seep up into the container – the water and the wine will exchange places. A great historian, Alexandre Koyré, read this passage and thought it obviously ridiculous. Everyone knows that you can dilute wine with water, and so he reasoned that in this case you would end up with both a basin and a container full of dilute wine. So, he concluded, this must be a thought experiment. Galileo could never have tried it out in practice. Koyré's conclusion is a remarkable one, as he can hardly have imagined that Galileo did not know that you can dilute wine with water. But no one actually repeated the experiment until 1973, when it turned out that Galileo was right and Koyré was wrong.[3] There is an important general conclusion to be drawn from this: there is no way of testing whether Galileo was an experimental scientist, other than by trying to replicate the experiments he describes. It used to be thought that he had laid claim to a significant number of experiments that he had never actually conducted: I can find only one example of his representing a guesstimate as if it were an actual experiment.[4]

Now to the fourth and final difficulty: was there a tradition of experimental science prior to Galileo? The argument of the next few chapters will be that Galileo effectively invented the modern idea of an experimental science. It is worth pausing for a moment to consider two alternative accounts of the birth of experimental science. The first is that the experimental method was invented by Ibn al-Haytham, now also known as Alhacen (but known in Renaissance Europe as Alhazen). Alhacen lived from 965 to 1039 AD, but his work was translated into Latin only in the early thirteenth century and this translation became generally available in the West when it was published in 1572. The translator unwittingly shaped our modern vocabulary, adopting the obscure word 'experimentum', and perhaps being the first to use the verb 'experimentare' and the noun 'experimentator'. Galileo's occasional use of this latter word is an indication of the influence, both direct and indirect, of Alhacen's Latin translator.[5]

There is no doubt that Alhacen employed an experimental method and it is unfortunate that we lack a good study of the impact of his work, particularly after 1572. What is missing is evidence that Alhacen was seen in the Renaissance as providing a model for how science in general or even optics in particular should be conducted. Nor is there at present any evidence that Alhacen had a crucial influence on the pioneers of the experimental method in the early seventeenth century. Here it is important to see that there was nothing particularly problematic about the idea of a science founded on experience: anatomy and astrology were two such sciences. What was needed were two further steps: the formulation of the logic of experimental enquiry, and a claim that experiment had a part to play in all the sciences of nature. There is no evidence that the Renaissance readers of Alhacen (and Galileo was certainly one) learnt these steps from Alhacen, or were provoked by their reading of him to take these steps for themselves. Alhacen's text seems to have been what we might call a sleeper: looking back we can see the influence it might, indeed should, have had; but Galileo's contemporaries assumed that it was of only minor importance. For them it was an old text that had been read by generations of experts. Since it had had only a very limited impact in optics, let alone on scientific methodology, the idea that it could provide a revolutionary new way of understanding the world would have seemed deeply puzzling, even implausible.

We cannot be sure exactly when Galileo read Alhacen, but it was probably for the course on optics he taught in 1601.[6] It is absolutely clear that Galileo failed to grasp one of his central arguments, namely that vision depends solely on rays of light entering the eye, not (as had traditionally been assumed, and as Galileo went on assuming) on something going out from the eye to the perceived object – a failure that must have been a basic obstacle to Galileo's understanding of the workings of the telescope. He did not grasp the importance of Alhacen's work, or of Kepler's *Optics* (1604), which developed Alhacen's line of argument – indeed Galileo may never have read Kepler's book.[7]

So although Alhacen *ought* to have been a crucial figure in the development of modern science, in reality the scientific revolution was not the result of reading Alhacen. A second and more persuasive contention is that it was the result of reading a book whose importance has been consistently underestimated: William Gilbert's *On the Magnet*, published in 1600. Gilbert's book consists of a long series of experiments – and this is the word he uses – with magnets. Everything he knows about magnets comes from these experiments, and he helpfully assists the reader by employing a star system in the margin to indicate their importance: a four-star experiment indicates something important that we would not otherwise know.

Gilbert's starting point is an interest in the mariner's compass and in two phenomena associated with it: the difference between magnetic north and true north, and the dip of the compass needle, which points somewhat downwards. His hypothesis is that the compass needle is attracted to the magnetic north because the earth itself is a magnet. He therefore works with spherical magnets which he calls 'earthlets', and his aim is to reproduce all the phenomena of the compass on the surface of an earthlet. He provides not only cunning experiments, but also exact and careful measurements. And it is worth stressing that his experiments are controlled and artificial: because the earth is so big that it can be used only for observational and not for experimental purposes, he works with an experimental substitute with which he seeks to replicate the phenomena that interest him. And at the end of his book he reaches a remarkable conclusion: the north-south axis is not in the heavens, but in the earth. North and south are not reference points, but real magnetic features of the terrestrial sphere. The earth has an axis, and this makes it inherently plausible that it rotates; indeed it is easy to construct experiments in which magnets rotate around their axes. Gilbert thus ends his book with a vindication of Copernicus (or to be exact, of the semi-Copernicanism of Calcagnino, though Gilbert had read Copernicus, not Calcagnino): the study of magnetism may provide no evidence in favour of the earth's rotation around the sun, but it does provide a new way of thinking about the daily rotation of the earth on its axis because it shows that this axis is something real, not something imaginary.

Magnetism is what Galileo's contemporaries called an occult force: it is hidden from view, and no one knows quite what it is. So too the supposed influence of the stars on our lives depends on occult forces. During the scientific revolution scientists became more and more impatient with occult forces because they could not successfully measure them or predict their behaviour, magnetism being a crucial early exception, and Newtonian gravity, at the end of the scientific revolution, being a new occult force. By turning the study of magnetism into an experimental science Gilbert had produced knowledge that was completely reliable and entirely new. No one, or so it seemed, had done this before.

Galileo had read Gilbert by 1602.[8] He later tells us that he was given a copy by a philosopher (perhaps his friend Cremonini) who did not want to give it space on his bookshelves.[9] He hurried to share the book with others: we find him discussing Gilbert in a series of letters exchanged with his two close friends Sagredo and Sarpi. Sagredo even sent a letter to Gilbert in England expressing admiration for his work; the letter was received with delight, but Gilbert died shortly afterwards, apparently without having yet replied.[10] It may have been reading Gilbert that persuaded Galileo that he could write a book in defence of Copernicanism which would not for the most part be about astronomy at all: he would later date the beginning of his great project on the two world systems to 1601.

It is surely not a coincidence that it is in the very next surviving letter after the one in which Gilbert is first mentioned that Galileo describes an experiment: and indeed an experiment of fundamental importance – the classic experiment establishing the isochronicity of the pendulum.

In order to understand this experiment it is necessary to take a step backwards. In *On Motion* Galileo had described experiments involving bodies rolling down inclined slopes. He had also, in his textbook on mechanics, compared the force required to pull an object up an inclined slope (a perfectly smooth object up a perfectly smooth slope) with the force required to lift it vertically. It was easy to see that an inclined slope was simply a way of slowing down the acceleration of a falling object: indeed it could be hypothesised that the speed at the bottom of an inclined slope (assuming no friction) would be the same as the speed the object would reach if it had fallen the same height, and that the time it would take to descend the slope would be in the same ratio to the time of fall as the length of the slope to the height of fall.[11] If one could devise an exact enough way of measuring time one could confirm this hypothesis by rolling balls down polished surfaces. (Galileo was not to know that rolling balls behave slightly differently to sliding objects because they are rotating.) At some point Galileo carried out experiments to confirm this result, but he will have been confident of it long before he tested it. He knew the results could not be perfect: the ball would never be perfectly round, the slope would never be perfectly polished (he cut a groove in a plank, to keep the ball on track, and then lined the groove with parchment to minimise friction), and the time-keeping would never be perfectly accurate (he used a fast-flowing water-clock and weighed the water to measure small amounts of time). But all in all, the results were entirely satisfactory.

Galileo now asked himself another question. Suppose an object takes one unit of time to fall an ell or arm's length. Is there a formula for calculating the length and steepness of all the possible slopes that an object would slide down in the same unit of time? Clearly the longer the slope the steeper it must be, until a slope of 89° is going to be very nearly an ell in length; the shorter the slope the nearer it must approach the horizontal, until a slope of 1° is going to have almost no length. (It is in the logic of this argument that there are slopes on which objects slide infinitely slowly.) Galileo must have doodled around with lengths and slopes until he realised that if all the slopes ended at the same point, then all the starting points would form a curve. What sort of curve? The simplest curve is a circle, and Galileo was soon able to demonstrate (on the basis of his earlier hypothesis about heights, lengths and times) that if a circle were placed vertically, then an object sliding down any chord to the lowest point would take the same time as an object sliding down any other chord. There was no need to test this: it simply had to be true if the initial assumption was true, and any test of the initial assumption would confirm this new law. Galileo was constructing a science of falling bodies which was just like his mechanics: it was about idealised, abstract, imaginary bodies, and it was deductively true given an initial hypothesis or definition.

If every chord drawn to the lowest point of the circle represents an identical time of fall, what about every arc? Surely all pendulums of the same length should take the same time to swing through different arcs, whether wide or narrow? Galileo could see that this ought to be right, but try as he might he could not find a geometrical method of proving it – he had no mathematical procedure for handling a constantly changing angle of descent. He took a step in the right direction by showing that if an object fell from the halfway point of the circle to the lowest point along not one but two contiguous chords, then it would take exactly the same time as if it had fallen the length of the diameter or the length of a single chord. It would fall faster on the first leg than it would have done if it had been falling along a single chord, but it would then gain speed more slowly on the second, and the two would exactly cancel each other out. The object would travel further, but it would take exactly the same amount of time. In principle an arc was just an infinite series of chords, but Galileo could no more make a calculation for an infinite series of chords than he could for a constantly changing angle of descent. He was stuck.

This was presumably the point Galileo had reached when he read Gilbert. Gilbert had shown that you could prove theories through experiments. Could Galileo show through an experiment that all pendulums of the same length will swing through differing arcs in the same time? Could he prove the law of the isochronicity of the pendulum in fact, if not in theory? It is precisely this experiment that Galileo describes in the first of his letters after he read Gilbert: he writes to Guidobaldo del Monte describing pendulum experiments which he claims demonstrate this law.[12]

These experiments have always been problematic because we know that Galileo's hypothesis is false: the length of a pendulum has to be made to alter very slightly as it swings if every swing, whether wide or narrow, is to take the same time. (Christiaan Huygens was to establish seventy years later, in theory and in practice, how to construct a genuinely isochronous pendulum.)[13] Galileo says that a pendulum swinging through a large arc and a pendulum swinging through a small arc will swing together – never getting a whole swing out of step – for hundreds of swings. With the length of pendulum Galileo describes, and the differences of arc he has in mind, modern theory suggests that the two pendulums would be a whole swing out of step after thirty or so swings. The conclusion seems simple: either Galileo never conducted the experiment – it was, as Koyré would claim, a thought experiment – or he falsified the results. But Paolo Palmieri has recently shown that if you perform the experiment you will indeed get the results Galileo describes, as the theoretical models had miscalculated the effect of air resistance.[14] Galileo, too, was trying to allow for air resistance, and it was not unreasonable for him to claim, on the basis of the experimental evidence available to him, which he reliably reported, that the law of the isochronicity of the pendulum would hold in a vacuum.

Galileo believed he now had two new laws regarding the speeds of falling bodies. His law of chords implied that bodies started to fall infinitely slowly – they clearly accelerated as they fell. He had seen this acceleration at work in the experiment – or rather observation – he had carried out with Guidobaldo del Monte in 1592. At first, what he thought was important about that observation was that it demonstrated the combination of different types of motion that must exist if Copernicanism were to be true. Now he wanted to know if he could find a mathematical law which would describe the acceleration of a falling body. In a letter to Sarpi in 1604 he tried to formulate such a law in terms of the speed attained after falling a certain distance, but he soon realised the results were impossible. He then reformulated it in terms of speeds and times, and the results seemed coherent.[15]

Suppose that in a unit of time x a body reaches a speed of y; that after 2x its speed is y muliplied by the square of the time, or 4y; that after 3x it is 9y; and so on indefinitely. This is a law of acceleration – and it will work whatever units of time and distance are employed. In the first time x, the object gains one unit of speed; in the second it gains three; in the third five; in the fourth seven; in the fifth nine; and so on indefinitely. How far has it travelled? Draw this curve of times and speeds (it is, of course, a parabola) on a piece of graph paper: the distance is now the area within the curve. If x is seconds and y is expressed in metres per second, then we can now calculate the number of metres travelled after any number of seconds.

But is this right? It is mathematically neat and conceptually coherent. The only paradox it produces is the idea that when bodies start to fall they are moving infinitely slowly – and as we have seen, this is a paradox Galileo had already encountered. Galileo could prove the law of chords by appealing to a self-evident principle: the time taken to slide down a slope as compared to the time taken to fall the same height will be in the same ratio as the length of the slope to the height. He could then confirm that experience corresponded to theory. Now he had a theory which was based on principles which were far from self-evident, even if they were simple and elegant; it was crucial that he should test his hypothesis and see if it was correct.

According to Galileo's theories, the same principles of acceleration would apply on an inclined plane. Take a plane that has a slope of 45° and an object that takes one second to slide down: at the end of its slide one can say it has reached one unit of speed. Lengthen the plane so that it takes two seconds to slide down, and it would reach four units of speed. Just as Gilbert had substituted earthlets for the earth, so Galileo substituted inclined planes for vertical falls: the one was manageable where the other was not, but the principles involved were the same. Now set this plane near the edge of a table, so that the object first starts to move horizontally (and stops accelerating) and then flies into space. No matter what the speed of the object, it has the same height to fall, and provided it is projected exactly horizontally into space, this should always take the same time. So the object which has reached four units of speed will fly four times as far before it hits the ground as the object which has reached one unit of speed. All that needs to be done now is to conduct the experiment and measure the distances: the practical difficulty lies in marking points on the slope which correspond to times, and this will require fine adjustment and repeated trials (he probably used a water clock to measure units of time). Galileo performed this experiment with great care and obtained results which were entirely satisfactory: they correspond almost exactly to the results we would expect to get today. Galileo was now an experimental scientist.

After 1602, after he read Gilbert, Galileo believed that scientific knowledge could advance both through deduction (as in the case of the law of chords) and through experimentation (as in the cases of the isochronicity of the pendulum and the law of acceleration). Later, he would say that his own method was very similar to Gilbert's – which was to indicate Gilbert's priority – and describe him as a very great philosopher.[16] Why then claim that Galileo founded the modern method of experimental science, when he was merely following in Gilbert's footsteps? The answer is that Gilbert had shown that experiments could add new knowledge: before Gilbert there was no proper theory of magnetism. But Galileo had shown something much more remarkable. He had shown that experiments could transform existing knowledge. Ever since Aristotle there had been a sophisticated account of the behaviour of falling bodies. Galileo had shown that that account was wrong in every respect. Bodies did not fall at a constant speed: they accelerated. Heavier bodies did not fall faster than lighter bodies: in a vacuum all bodies accelerated at the same rate.

This refutation of Aristotle was crucial because philosophers had previously felt confident that one could assume that, even if Aristotle's meaning was not always clear, he was never simply mistaken. Interpreting Aristotle might be difficult, but one could always learn from him. Here, for the first time, Aristotle's authority had been tested and found wanting. Moreover the crucial evidence came from experiments which showed that the world does not work in the way in which we think it works. Aristotle thought the job of science was to explain what he called 'the phenomena', but the word 'phenomena' included not only one's own direct experience, but also what was generally understood to be the case.[17] Aristotelian philosophy simply assumed that popular preconceptions were always well founded. Galilean science dealt, not with 'the phenomena', but with what would now be called – and what Galileo already called – the facts. And the facts had to be carefully established before they could be taken for granted.

Galileo could now give a complete explanation of the path followed by a projectile – he could *explain* what he had observed in 1592. And if parts of his new science (such as the law of chords) seemed purely theoretical, other parts clearly had practical implications. He could now prove that a cannon would shoot furthest if set at an elevation of 45°; and he could use his pendulum to measure time more accurately than ever before. What he now wanted to do was to get these results into print. They would establish a new science; they would encourage others to adopt the experimental method; they would prove that Galileo was a worthy successor to Archimedes; and they would show that Copernicanism was compatible with the laws of physics. In 1604 Galileo was forty. He had served a long apprenticeship. He was beginning, at long last, to make headway against his debts. He was about to become famous. He began writing – in Latin, of course – the book that he knew would have an impact far greater than that of Gilbert's *On the Magnet*. The evidence suggests, indeed, that by 1608 he had come very close to completing this book, and that we can find its text reproduced in Latin within his Italian dialogue on the *Two New Sciences*.[18]

What is striking about that book is how unlike Gilbert's it is. Gilbert, following Alhacen, presents his new science as a series of experiments; Galileo presents his as a series of axioms. Gilbert, like Alhacen, describes every experiment with great care; Galileo never even reports his table-top projectile experiment – we know about it only from his unpublished notes. If Galileo had become an experimental scientist, it was not as an experimental scientist that he wanted to present himself to the world, but rather as a follower of the deductive science of Archimedes. The true role of experiment, in Galileo's view, was to establish facts that one could then seek to explain and, where possible, demonstrate. Gilbert (like Galileo's father) had been willing to stop when the job was half done, but Galileo was not so easily satisfied. 'You may be sure', he wrote, 'that Pythagoras, long before he discovered the proof for which he sacrificed a hecatomb, had satisfied himself that the square on the side opposite the right angle in a right triangle was equal to the squares on the other two sides. The certainty of a conclusion assists not a little in the discovery of its proof.'[19] In Galileo's mind his experiments were merely preparatory procedures undertaken in the quest for proofs. Even then, the experiments only clarified what was already, in some sense, 'known'. (Galileo could thus play with the idea that he upheld the Platonic doctrine of reminiscence.) In the *Two New Sciences* he insists that his method is to argue from 'reasons, observations, or experiences that are well known and familiar to everyone'.[20]

So far was he from identifying himself with an experimental method that throughout the two great works of his intellectual maturity he never uses either of the two words for 'experiment' that were available in the Italian of his day ('esperimento' and 'cimento'); he uses only the word 'esperienza'.[21] This is true even when he is describing Gilbert's own work, Galileo presenting Gilbert as being, like himself, an *experiential* scientist. Yet it is also important to note a major difference between the use of the English word 'experience' and that of the Italian word 'esperienza': in English an experience is something you passively undergo, while an experiment is something you perform; in Italian you can 'make' an experience ('fare una esperienza'), which places 'esperienza' somewhere between the English words 'experience' and 'experiment'.[22] To one well-informed contemporary, Galileo was 'the founder of the experimental method in all its exactness'.[23] Moreover, after his death his pupils founded the Accademia del Cimento, dedicated to the pursuit of scientific knowledge through experiment.[24]

Had Galileo published his little book on projectiles and falling bodies, and done nothing else, he would have been a significant figure in the history of science: the founder of modern physics, the greatest physicist before Newton, someone certainly greater than Tartaglia or Stevin. In later years Galileo's friends were to urge him repeatedly to publish, yet he did not do so until 1638. This thirty-year delay cries out for an explanation. After all, Galileo had endless opportunities to publish. He could have published at any time after 1616: banned from defending Copernicanism, this was certainly the sensible thing to do. He even had a clean copy of his draft prepared in 1618 so that he could resume work. For this delay there is only one explanation: in Galileo's mind this project was inseparable from what he saw as a much larger and more important one, the campaign to vindicate Copernicus. Prestige and fame were not enough for Galileo. Proving Aristotle wrong was not enough for Galileo. Revolutionising physics was not enough for Galileo. There was something else that mattered to him, something that he thought was much more important. What exactly was that something? This is the central puzzle of Galileo's life; until we understand why Galileo cared so much about Copernicanism we have not begun to understand him at all.

In Galileo's view the answer to our question was obvious. 'The constitution of the universe I believe may be set in first place among all natural things that can be known,' he wrote, 'for coming before all others in grandeur by reason of its universal content, it must also stand above them all in nobility as their rule and standard.'[25] No other subject was as grand or as noble; it came first in precedence and was raised above all others in status. Galileo was surely sincere when he said this. But he must have known that it was odd to claim, as he went on to do, that Ptolemy and Copernicus were undoubtedly the greatest scientists the world had ever seen: he himself admired Archimedes more than anyone else, and most philosophers thought Ptolemy a minor figure compared to Aristotle. Why then privilege cosmology above all other subjects? It is worth bearing in mind that we are ourselves part of the universal content of theories of the constitution of the universe. What is our place in the cosmos? Is the universe friendly or hostile? The law of fall provides no answer to such questions, but Copernicanism implies an answer, or at least a new way of putting the questions, and there are grounds for believing that these implications mattered greatly to Galileo. To prove Copernicanism would be great and noble; at the same time it would be to prove the fundamental insignificance of the human species. For, as Descartes would later state explicitly, the one thing we can now be sure of is that the universe was not made for man.[26]

15

The telescope

In the autumn of 1604 a new star appeared in the sky. News of it spread rapidly amongst those interested in astronomy. Thirty years earlier, in 1572, Tycho Brahe had turned a new star into a *cause célèbre*. According to the philosophers, the heavens were unchanging, eternal, perfect. There could be no change in the heavens, and any change that did take place must occur in the vicinity of the earth – in the sublunary world. In their view, consequently, the choice was simple: either the new star had always been there, or it was not a star at all, but some peculiar phenomenon in the atmosphere, a pseudo-star. Since Tycho's new star soon disappeared the argument was left hanging, to be immediately reawakened by the appearance of the nova of 1604.

As a Copernican, Galileo will have known at once where he stood. If Copernicus was right, the distinction between a sublunary and superlunary world was misconceived: the earth was in the heavens, inseparable from them. If there was change on earth, then equally there could be change in the heavens. As someone who taught the military sciences Galileo also knew how to measure distances. He knew it was not necessary to approach an object to work out how far away it was, as long as there was some other reference point whose distance was known. By looking at an object from different positions you could see how its relationship to other objects altered: simple geometry could turn a measurement of parallax into a calculation of distance. If the new star, when looked at from different places on the earth's surface, had an unchanging relationship to the stars around it, then it was far away – much further than the moon.

Galileo had soon collected the information to prove that the new star was indeed a star – that it certainly was not closer to the earth than the moon. He gave a series of public lectures on the new star: they attracted very large audiences and provoked a vigorous debate within the academic community. Galileo's friend Cesare Cremonini, the leading philosopher in Padua, was amongst his opponents: philosophers were unused to mathematical arguments, and had difficulty understanding an argument from parallax.

Galileo must have considered publishing a version of his lectures on the nova: either in Italian or Latin, they would have sold well, and they would have been his first proper, scientific publication. He seems to have had a manuscript ready for publication, but he held it back from the press.[1] Instead he sent one of his students into the lists to do battle on his behalf against a painfully ill informed pamphlet produced by an associate of Cremonini's. The student, Girolamo Spinelli, writing under the *nom de plume* Cecco di Ronchitti (Dead-End Dick is about as close as one gets to a translation), produced a little pamphlet of his own in which three Paduan peasants debated the new star. It was written throughout in the local dialect, and belonged to a distinct local genre of sophisticated literary works that used Paduan dialogue for comic effect. Such works were aimed at the educated elite, not at the peasants who appeared in them. The implication of this little dialogue, which was published early in 1605, and naturally in Padua, was obvious: any dunderhead could follow Galileo's arguments better than a sophisticated Aristotelian philosopher. It would have been difficult to find a more direct way of expressing irritation with the academic establishment – an irritation that was surely Galileo's as much as Spinelli's and that had not been diminished by his regular increases in salary.[2]

In the autumn of 1604, Galileo could show that the new star's location appeared to be the same when viewed from different European cities. The star then disappeared below the horizon of the night sky, to reappear in the spring of 1605. It was clearly Galileo's hope that when it reappeared its relative position would have changed. If Copernicus was correct he would now be looking at it from a point distant from his previous point of observation by the diameter of the earth's orbit.[3] Galileo must have waited with anxious anticipation for the star's reappearance: but there was no change in its location. This may have been a disappointment, but it will not have been much of a surprise – Tycho and Digges had been unable to measure any displacement in the nova of 1572 (a copy of Digges's work was in the Pinelli library).[4] Spinelli's comic dialogue was reissued in Verona (fifty miles from Padua) later in 1605, but with some marginal references to Copernicus toned down. This has been interpreted as meaning that Galileo had lost confidence in Copernicanism, but there is no reason to think that Galileo himself was responsible for these changes, which may well have been the work either of the Veronese printer or of Spinelli.

A second pamphlet defending Galileo's interpretation of the new star appeared in Florence in 1606. This too was pseudonymous. Its author, calling himself Alimberto Mauri (it is abundantly clear that there was no real person of that name), seems to have been a Florentine living in Padua, and he wrote with approval of Cecco's little dialogue. The claim has been made that Mauri was Galileo himself,[5] and I tend to think (although there seems no prospect of certainty) that this claim is correct. The pamphlet is dedicated to Monsignor Luigi Capponi, the pope's treasurer general, who was later to prove a friend of Galileo's. Mauri writes in the preface to Capponi, 'I should much like to present myself before you and to offer myself with something of greater value and more suited to your merits . . . but being unable, for many reasons, I do what I can.'[6]

In 1606 Venice and Rome were at loggerheads. Venice was under papal interdict and there was a real prospect of war. Nobody living in the Venetian territories would have dared openly to seek patronage in Rome – hence the use of a pseudonym. Galileo was not only unable to present himself before Capponi because of the Interdict crisis; he also had something 'of greater value' in the pipeline – his book on projectiles. The strongest objection to Galileo's being the author used to be that Mauri appears sympathetic to astrology, but we now know that this is true of Galileo in these years.[7] And the strongest argument in favour continues to be that we know of no one else capable of writing Mauri's little book. If Mauri was Galileo, then the decision to dedicate the book to Capponi represents an important political choice: he was secretly distancing himself from his Venetian friends Sarpi and Sagredo, who were bitterly opposed to the papacy, and seeking support from their enemies. Mauri steers clear of any explicit discussion of Copernicanism (like Galileo in 1597, he expresses a fear of being laughed at if he speaks too freely), but he does refer to Copernicus several times, and always with approval. Moreover, as we have seen, he boldly asserts that the moon has mountains, just like the earth – an argument which, if Galileo was not the author, must certainly have caught his attention.

So, after some heated arguments and some good-natured mockery of the Aristotelian establishment, Galileo's first involvement in astronomy came to an end. His attention turned back to physics. Early in 1608, however, a Dutch spectacle maker discovered that if a convex lens was held behind a concave lens you saw an enlarged image. He mounted the lenses in a tube with a sliding mechanism, thereby making the first telescope (although the name, which is constructed out of the words for sight and distance in ancient Greek, was to come later, invented by Galileo's admirers).[8] Soon a number of different people were claiming to be the first inventor, and word of the discovery spread rapidly. In May of 1609 Galileo met Paolo Sarpi in Venice. Sarpi had received a very vague account of the instrument from a former pupil of Galileo's, Jacques Badoer, in Paris (or so Galileo gives us to understand – Badoer's letter does not survive).[9] On his return to Padua, within the space of a few days at most, Galileo 'reinvented' the telescope: that is to say, he made a telescope without ever having seen one, and without having been given an account of how one was constructed. He would later claim that 'reinventing' was just as difficult as inventing. Archimedes, after all, had weapons of war that no one had been able to reinvent. He would also claim that his knowledge of optics was crucial for his being able to reinvent and improve the telescope.[10] Galileo certainly had some knowledge of optics, but it is not clear that this played a significant role in either the reinvention or the improvement of the telescope. All that was involved was intelligent trial and error: Galileo saw at once that the number of possible lens combinations was limited, and all he had to do was try them out.

Galileo's first telescope, using lenses made for spectacles, magnified only three times; the best Dutch telescopes of the day magnified about six times. The lens quality will have been poor – there will have been a halo consisting of the colours of the rainbow around the edges of every object, and the image will have been blurred because the curvature of the lens was irregular; its value will have been limited. Galileo's true genius was that he grasped at once that the telescope could be improved. We live in a world where manufacturers are constantly offering us improved versions of products, so if you or I had been shown a primitive telescope we would have asked immediately what scope there was for improvement. Galileo's world was not like this. Even new technologies – guns, printing presses, compasses – were improved slowly and over very long periods of time. By the summer of 1609 there were thousands of people – many of them mathematicians, scientists, engineers – who had seen and used the new telescopes. But Galileo was the only person who immediately saw in the telescope a challenge: how could one improve it?

Everything had prepared Galileo for this moment. He began his career with the little balance – a precision weighing device. His sector was a significant improvement on previous sectors. His whole view of science was based on the conviction that progress was possible. But one particular experience had trained Galileo to understand that a new technology represented an opportunity – an opportunity for profit. As soon as he knew that he could produce a working telescope he will have felt sure that he could make money from it.

The source of his confidence would have been a bit of business he had done very recently, in 1608. Galileo was in contact with the Tuscan court because he had been providing mathematical instruction to the young prince. He will have known that the grand duke, like many rulers and aristocrats, liked to collect remarkable objects: this was the age of the cabinet of curiosities, in which great men displayed rare and wonderful objects, ranging from ancient Egyptian artefacts to unicorns' horns. Galileo's wonderful object was the most powerful magnet in the world.[11] It belonged, he said, to an acquaintance, but he could acquire it for the grand duke. It would be capable, he promised, of lifting five times its own weight. For this wonderful object he was paid a hundred doubloons, more than two hundred ducats.

There are two little wrinkles to this story. First, Galileo claimed to be working as an intermediary, concealing the identity of the purchaser from the seller, and that of the seller from the purchaser. Eventually he revealed to the grand duke that the seller was his wealthy friend Sagredo, and presumably he told Sagredo that the purchaser was the grand duke. The money itself went to Galileo. Did he pass it all on to Sagredo, as he said he would? Did they share the proceeds? The two were involved in a complicated network of favours and exchanges, so it is certainly possible that Sagredo allowed Galileo to keep all the money, and equally possible that Galileo was repaying previous favours and expected no reward. What was clear was that an extraordinary magnet was worth something.

The second wrinkle is that Galileo enhanced the power of the magnet by encasing it in a metal armature. The technique for doing this had been described by Gilbert. Given a large magnet, anyone who had read Gilbert could have produced a magnet that would lift five times its weight and pass as the most powerful magnet in the world; the value of Sagredo's magnet depended on Galileo keeping his knowledge of Gilbert to himself. It should not, therefore, surprise us that Galileo quickly set about making the most powerful telescope in the world, and it should not surprise us that the enterprise, like his sale of Sagredo's magnet, involved him in what an impartial bystander might reasonably call deception.

Galileo soon had a telescope that magnified eight times. He took his improved telescope to Venice, where he displayed it to the city's rulers.[12] It may have been the first telescope many of them had seen, though some will have known that others were offering to sell the secret of how to make one. Standing on top of the bell tower in St Mark's Square, they looked through it out to sea and saw ships that were invisible to the naked eye – the technology had obvious military applications. Galileo assured them that additional, equally good, telescopes could be made. Supported by powerful friends and associates, including Sarpi and Sagredo, he pressed his claim for a reward, and it was agreed that he should receive an appointment at the university for life, and a salary of a thousand ducats a year. In return, he agreed to spend the rest of his life in the service of the Venetian state.[13]

Impartial bystanders soon started muttering that Galileo had practised deception.[14] Telescope vendors were spreading across Europe carrying packs full of Dutch-made telescopes. Soon one could buy a telescope quite cheaply in St Mark's Square, right where Galileo had demonstrated his new invention. This was the first ambiguity: had Galileo presented himself as the inventor of the telescope, or only of a better telescope? And that led to the second: if Galileo's telescopes were better than the other telescopes that were now generally available, was there any secret to their construction, or could anyone who put in a little time and trouble produce an improved telescope? The Venetian establishment seems to have had its doubts. They could not go back on their word – not unless they were prepared to charge Galileo with a crime. But by the time the Senate came to vote on his reward it seems clear that they felt they had been misled. Galileo's new appointment – and his new salary – was now to begin only when his existing appointment came to an end. And the Senate decreed that his new salary was never to be increased; they might as well have said that they had been taken in once, and did not intend to be taken in again. We can guess that Galileo was offended by this restriction, seeing it as an invitation to seek patronage elsewhere. The new salary was, after all, no better than that currently being received by his friend Cremonini, who went on to double his salary in the years that followed (although without receiving an appointment for life).[15]

In 1609, both Galileo and the rulers of Venice were trying to make sense of a new technology. Galileo was certainly in a position to guess (he had never held another telescope in his hands, so he could not be absolutely sure that other telescopes worked on exactly the same principles as his own) that others would soon be able to produce telescopes as good as his. But it is also the case that Galileo had and kept a lead in the production of high-powered telescopes. By the autumn of 1609 he had a telescope that magnified twenty times; by the beginning of 1613 one that magnified thirty times; and by 1615 one that magnified no less than a hundred times – and he was still ahead of the competition. He managed to stay ahead for twenty years. Galileo really did have something to offer the Venetian state but his telescope was not a new invention nor was it made with any special techniques. It is impossible not to conclude that there was an element of deception involved in his dealings with the Venetian government, as there had been in his dealings with the grand duke of Tuscany over Sagredo's magnet.

16

Mother

In November 1609, Galileo's mother, Giulia Ammannati, wrote a letter from Florence to one of Galileo's domestic servants, Alessandro Piersanti.[1] In it she expresses concern that she has heard nothing from him for several weeks, and reassures him that his letters to her will not be intercepted. She wants him to collect some cloth from the weaver's – and it is important that no one from Galileo's household catches sight of it. She has strong views on what should be done with Galileo's eldest daughter, Virginia; unfortunately Piersanti already knows what they are, so she does not repeat them for our benefit, but she had taken Virginia back to Florence with her, presumably to rescue her from the corrupting influence of her parents. She tells Piersanti to write back, and she expects at least a page on the celebrations that will have been provoked by her departure, for she knows that a page will not be enough to provide even a bare summary.

Three days later she writes to say she has received his letter.[2] She wants details of everything that is said in Galileo's household, and she insists that Marina Gamba must not be allowed to get her hands on her fabric. There are complaints that she has been tricked and lied to – apparently to get her to return from Padua to Florence. On 9 January she writes again, asking Piersanti to steal two or three or four telescope lenses from Galileo – not the eyepieces, but the objectives – and send them to Galileo's brother-in-law Benedetto Landucci (Landucci has repeatedly asked Galileo for them, but he has never replied), concealing them under pills made by the famous doctor Acquapendente.[3] Virginia, it seems, is living in the Landucci household, and Galileo is so ungrateful for all that his brother-in-law has done for him that it is only right that Landucci should have these lenses. (A year later, Galileo records the settlement of his accounts with Landucci.)

Galileo's mother was not alone in thinking she had a right to make money out of Galileo's discovery. Galileo sent a few telescopes to his brother so that his brother could give them to influential people and spread the word of Galileo's discoveries. A while later his brother wrote back: he had sold the telescopes for high prices, but now good telescopes were widely available, so Galileo would have to send him some even finer ones if he was to do as well out of the next lot.[4] Galileo sent his brother no more telescopes.

In 1609 Galileo is forty-five; his mother is seventy-one. We would love to know about Galileo's relationship with his mother during his formative years; instead we must make do with this passing glimpse of their relationship forty years later, for these letters are three of the four in which we hear her speak for herself. It is apparent that she is angry, devious, perhaps a little paranoid, and so self-righteous that she can see nothing wrong with spying or theft. She presumably knows that Galileo has been given a massive pay increase by the Venetian government in return for the secret of how to make a superior telescope, so she can scarcely be unaware that specimens of his lenses have commercial value. She believes, surely with good reason, that her son and his mistress dislike her intensely.

We have another, earlier glimpse of her relationship with Galileo in a deposition made to the Venetian Inquisition by Galileo's copyist – employed to make copies of the textbooks Galileo supplied to his students – Silvestro Pagnoni, in 1604. Galileo's mother had told him that her son had been in trouble with the Inquisition in Florence, that she had given evidence against him, and that he in response had sworn at her, calling her a whore and an ugly old cow.[5] On this occasion too she seems to have set out to turn one of Galileo's servants against him, and Galileo found himself reported to the Inquisition partly as a result of her promptings.

Pagnoni deposed that Galileo never went to church, although he did not make irreligious statements. He spent a great deal of time in the company of Padua's leading philosopher Cesare Cremonini. (Cremonini, who served as guarantor in 1608 for the year's salary Galileo owed the university,[6] was himself in constant trouble for arguing that reason – and Aristotle – established that there could be no life after death. He was such a charismatic figure that he had made religious belief unfashionable amongst the young Venetian nobles who attended his lectures, Gianfranceso Sagredo's father, a leading figure in Venetian politics, complaining that he had educated a generation of atheists.)[7] Specifically, Pagnoni claimed that Galileo was making astrological forecasts – which were perfectly legal. But according to Pagnoni, Galileo claimed that his astrological forecasts were totally reliable – that events would turn out exactly as he predicted. This, if proved, was a heretical denial of divine providence. The Venetian authorities acted effectively to block further enquiries: as far as they were concerned, it was their job, not the Church's, to regulate the conduct of professors at the University of Padua.

What are we to make of Pagnoni's deposition? A Galileo who never goes to church but does not directly attack Christianity sounds like someone who is basically indifferent to religious questions. His mother, on the other hand, appears deeply hostile to her son, and keen to get him into trouble with the authorities.

Finally we have a letter from Galileo's brother, written to Galileo in October 1619, when their mother is eighty: 'I am more than a little astonished to hear that our mother is just as dreadful as ever. But given her physical decline, she won't be with us much longer, so that there will be an end to all these conflicts.'[8] She died in August 1620.

Seven years later her son Michelangelo returned to Florence for the first time in many years. He spent fifty scudi and a great deal of effort in doing up a house in town. And then he decided that he could not bear to live there and left Florence in haste, telling Galileo to sell the property or at least pay him for it.[9] He had no money – he could not have afforded to buy himself a house; his brother was refusing to help him with his money troubles; and you do not spend significant sums of money on a rented property. So what house was this? The only plausible answer is that it was the parental home, which had stood empty and neglected since his mother's death, and which had presumably been left to him in his mother's will. It tells us something, when the parental home is left to tumble down and nobody is prepared to live in it.

Yet if Galileo and his mother were caught up in an unending conflict, there were also times when she clutched him to her. In 1590, when he was lecturing in Pisa, she was seriously ill, perhaps about to die. Galileo, who was visiting Florence, remained there to be by her side because his presence was so important to her – although she had her husband and her other children nearby.[10] He stayed on even when it became clear that the illness was unlikely to be fatal but was going to be long drawn out. Galileo's pay was docked because he failed to deliver the lectures he was contracted to give.[11] It is hard to resist the impression that she made the most of the power she briefly acquired over her son when he thought she was at the point of death.

What effect did it have on Galileo to have a mother who was bullying and devious, who regarded everything that was his as rightfully hers, and who believed she had the right to police his words and his thoughts? We can only guess. Perhaps it discouraged him from marrying, although it is worth remembering that Galileo spent his formative years as a young adult in the Venetian territories, where it was common for the nobility not to marry but rather to take a mistress (Sagredo was typical in this respect). It might be that it made him determined to think for himself, to resist pressure to conform, and to aspire to recognition for achievements that were his and no one else's. And surely it made him guarded and cautious. If so little is known about Galileo's emotional attachments, about his feelings, about his private beliefs and inner convictions, this is not because of some cultural inhibition which prevented him from disclosing himself to others. To read Sagredo's letters to Galileo is to learn much more about Sagredo than we ever learn about Galileo – about his feelings about growing old, about his regimen to maintain himself in good health, about his servants, about his girlfriends. If we know far less about Galileo there is one person above all to blame: Giulia Ammannati.

The Starry Messenger

Me thinkes my diligent Galileus hath done more in his threefold discoverie than Magellane in openinge the streights to the South Sea or the dutchmen that were eaten by beares in Nova Zembla. I am sure with more ease and safetie to himselfe & more pleasure to mee. I am so affected with his newes as I wish sommer were past that I mighte observe the phenomenes also.

William Lower to Thomas Harriot, on 'the longest day of 1610'[1]

In Venice there are approximately sixteen hours of daylight in June and July, and only eight in December and January. In the autumn of 1609, as the days short-ened, Galileo turned his improved telescope on the heavens. He mounted it on some sort of stand, but he still had to learn to slow his breathing; even his pulse seemed to shake the telescope; and as the evening temperature dropped the glasses kept misting up. Early in January 1610 he discovered he could reduce the halo around objects by fitting a circle of masking material with an oval hole in it over his lens – he had, in photographic terms, stopped the lens down, and, in the language of opticians, corrected for an astigmatism in the lens. Still the stars remained mere points in the heavens – except that there were many more of them than before. He discovered that the Milky Way was not a mysterious white band in the sky, but a vast number of small stars, individually invisible to the naked eye. Everywhere there were new stars to be seen. It is easy for the signifi-cance of these new stars to escape a modern reader: it might be thought it hardly matters how many stars there are. But Galileo's contemporaries believed that the universe embodied a rational purpose. The sun, the moon and the stars existed to give light to the earth. Invisible stars were a profoundly anomalous concept: what purpose could a star serve if no one could see it? Only a few Copernicans (Bruno, Digges, Gilbert) had imagined a universe so large that there were distant stars invisible from earth. As for the planets, through Galileo's telescope they were not points but tiny discs floating in space.

Naturally Galileo turned his telescope to the moon, which was so greatly magnified that he could look at less than half of it at a time, even with his twenty-power telescope. Everyone knew that the moon was not perfectly uniform in appearance, but the philosophers insisted that it must be a perfect sphere, even if parts of its surface were more reflective than others. If the moon were smooth, the line between the illuminated half and the unilluminated half should be perfectly regular – but Galileo could see that the line was irregular, a phenomenon which could be accounted for only if the surface of the moon were irregular. Moreover, near the margin between the illuminated and the unilluminated half he could see two anomalies. On the unilluminated side of the margin he could see little flecks of light: the sun was clearly reaching some areas before it reached others; these must be high points. On the illuminated side he could see dark spots which the illumination took longer to reach; these must be shadows. Galileo's interest in painting, and his experience of looking at paintings where tricks of light were used to convey textures and shapes, may have helped him understand what he was seeing. But he also grasped at once that what he was seeing was comparable to a familiar phenomenon on earth: at dawn and dusk, the sun stays on the mountain tops when the valleys are deep in shadow. Galileo had discovered (or rather, since he knew Mauri's book on the nova of 1604, he had confirmed) that the moon had a landscape of mountains and valleys; in this respect it was just like the earth, and this was welcome corroboration of his view that the earth, seen from the moon, would look like an enormous moon.

This provided an elegant confirmation of the Copernican view that there was no fundamental difference between the heavens and the earth: on 7 January Galileo sat down and wrote a letter in which he carefully outlined his discoveries. That evening he turned his telescope to Jupiter and noticed three little stars arranged in a line with the planet, two to the east and one to the west: he assumed these were yet more new fixed stars.[2] The next day he happened to look again: now all three stars were to the west of Jupiter, although Jupiter itself was moving from east to west. Two 'fixed' stars had overtaken Jupiter! Galileo now began to observe Jupiter every night on which there was no cloud cover. By the eleventh he had decided that he was observing three satellites orbiting Jupiter. On the thirteenth he discovered a fourth. Galileo now knew that he had made a momentous discovery.

By 30 January he was in Venice, arranging for the publication of a book on his telescopic discoveries – the mountains on the moon, the nature of the Milky Way, the moons of Jupiter – while continuing his observations.[3] Galileo was in a hurry – other people had telescopes, and if theirs were not yet as good as his, it might not be long before someone else had a telescope capable of resolving Jupiter's moons. He was right to be anxious: Harriot in England and Lembo in Rome had already observed the irregularity of the moon's surface.[4] Presumably Galileo now found a printer and an engraver.

And so, at long last, after years of keeping his thoughts to himself, Galileo (whose forty-sixth birthday fell on 15 February) was writing a book. On 12 February he returned to Padua; it will have been from there that he arranged at the end of that month for the book to be licensed for the press.[5] To obtain a licence he must have produced a complete draft, yet he went on making last-minute additions: the last (sixty-fifth) observation of the moons of Jupiter that he reports is dated 2 March.

As he worked the title of his book changed, and the various titles tell us that Galileo became increasingly confident that he was producing a book without precedent. The most conservative is the title which appears once on the manuscript, and survives as the running title across the top of each opening in the printed book: 'Observat[iones] sidereae recens habitae', Starry Observations Recently Undertaken.[6] The word 'recens' is the only clue that a revolution is being announced.

It was common when printing a book to print the body of the book first, and then to add the prefatory matter. It is clear that this happened in the case of Galileo's observations. The large title on the first page to be printed was 'Astronomicus nuncius' – the Astronomical Message or the Astronomical Messenger (a 'nuncius' can be either a messenger, or the message he carries).[7] In his correspondence Galileo referred to the book in Italian as his 'Avviso astronomico', Astronomical News, or as his 'Avviso sidereo', Starry News. An 'avviso' was generally a news report, which suggests that by 'nuncius' he meant 'message' rather than 'messenger', and that the sort of message he had in mind was one which reports important news.[8] And the final title – finalised only as the last pages went to press on 12 March – was 'Sidereus nuncius', the Starry Message or Messenger – that is, a message about the stars, or a messenger from the stars.[9] The title was a source of difficulty from the beginning: one friend of Galileo's who had no Latin complained that no one had been able to explain to him what it meant.[10] We can guess what Galileo had in mind from a comment he makes twice in his letters: there are some people who are so reluctant to accept his new discoveries that they would be convinced only if the stars themselves came down to earth and spoke to them.[11]

The book was a collection of observations, a message, or a report. It was much closer, judging by its title, to a traveller's report from a foreign country than it was to a work of philosophy. And indeed it lacked a central characteristic of any contemporary work of philosophy: philosophy books discussed other philosophy books, whereas this book did not mention any philosophy books at all – or indeed any books, except catalogues of stars. Galileo was always impatient with books about books. His claim was that what he was reporting was so fundamentally new that all previous discussions of the heavens were irrelevant. This was not yet a new science; rather, these were the raw materials, the raw observations, out of which a new science would have to be constructed. In his final choice of title, Galileo presented himself as a traveller returning from a voyage through the heavens with strange and wonderful things to report. If the book was the message about the stars, then Galileo was the messenger who had come from the stars. He was, thanks to the telescope, the first spaceman – indeed this is how Viviani describes him on the monument he erected to record his achievements.[12] And Galileo's message – made explicit only in his final revisions to the text – was that the earth is itself a spaceship.

The book needed to be carefully illustrated. No one else had a telescope nearly as good as Galileo's, and his extended discussion of the moon required the reader to 'see' the fine details he was describing – the title page says the book 'displays Galileo's observations to the gaze of all'. When Galileo produced a carefully illustrated record of his observations he was innovating: no astronomer, mathematician, physicist or engineer had needed illustrations of this quality. What model did he have in mind? Years later he was to compare the telescope to a scalpel in the hands of an anatomist: it exposed to sight what had always been there but had been hidden from view. As we have seen in the story of the dispute between the anatomist and the philosopher that Galileo recounts in the *Dialogue*, anatomists had pioneered the idea of a knowledge based on direct, sensory experience. So *The Starry Messenger* was comparable to an illustrated anatomy text, the first and greatest of which was Vesalius's *On the Construction of the Human Body* (1543). In producing a factual record of his observations, Galileo was imitating the doctors, and at this moment his years of medical education were suddenly, unexpectedly, relevant.

I have casually referred to *The Starry Messenger* as a 'factual record'. Galileo did not describe it as such. The word 'fact' (and its equivalents in Italian and French) comes from the Latin verb 'facere', to do. A fact is initially a deed, an act. In a legal context, the factum is the deed which is the subject of litigation. John stabbed Thomas: that is the factum which must be established, either by confession or by testimony. And then the law must be applied to the fact. Was it self-defence or murder? In English law appeal was allowed on a matter of law (could one invoke self-defence if one had not first tried to retreat?) but not on a matter of fact: the jury alone could determine whether John had indeed stabbed Thomas.[13]

In the late Renaissance the word 'fact' slowly takes on its modern meaning as 'something that has really occurred or is actually the case'. Bacon uses it in this sense in Latin, and when he is translated into English it appears in English. (The earliest English example recorded by the *Oxford English Dictionary* is in 1632, slightly earlier than the translations of Bacon.)[14] The assumption in the literature has been that the fact is an English invention, but in truth facts were invented first in Italy (the earliest examples of the new linguistic usage that I have found date to the 1570s), which is one reason why the scientific revolution began in Italy, not England.[15] Galileo, like his contemporaries, has a Latin phrase, 'de facto', which he uses when writing in Italian as well as in Latin, and an Italian phrase derived from it, 'di fatto'. Just as he very occasionally uses the word 'experiment', so he occasionally uses the Italian word for a fact. As far as I have been able to ascertain he uses the word 'fatto' in its modern sense, aside from its use in phrases such as 'di fatto' and 'in fatto', two dozen times – most strikingly and most frequently (six times in all) in the *Letter to the Grand Duchess*

Christina of 1615, where he says that the first thing we need to do is establish the facts: 'prima fosse d'accertarsi del fatto'.[16] So it looks as if Galileo *could* have described *The Starry Messenger* as a record of the facts, but I think he would have hesitated before doing so. Used in this sense the word was a neologism or a colloquialism, not respectable in either Latin or Italian. (The *Letter to the Grand Duchess Christina* was initially intended for manuscript circulation, not for publication.) What Galileo offers us are not facts, but something astronomers had always relied on: *observations*. Some of these observations – the new stars, for example – required little interpretation; others – the mountains on the moon, the moons of Jupiter – could be turned into facts only by a process of deduction.

The illustrations of the moon were remarkably accurate; if they are placed side by side with a modern photograph their precision is immediately apparent. Indeed, since the moon's appearance changes not only from day to day but from month to month, it is possible to use them to establish exactly when Galileo recorded what he saw. Their accuracy was necessary, because their purpose was to provide a virtual experience of what one could see if one looked through a telescope. They had one significant flaw, however, which has puzzled later scholars: a single feature, a crater which Galileo describes as being as large as Bohemia, appears enlarged – so big that critics have complained that if the illustrations were accurate it would be visible to the naked eye.[17]

The simple answer is that Galileo was persuaded that this single feature was far larger and more noticeable than any other because he had *already* seen it with his naked eye. Mauri knew in 1606 that his claim that one could see mountains on the moon would be met with scepticism: 'And for proof of this I shall adduce an easy and pretty observation that can be made continually when she [the moon] is in quadrature with respect to the sun; for then the semicircle is not smooth and clean, but always has a certain boss in the middle. For this what more probable cause will ever be adduced than the curvature of those mountains?'[18] Galileo was not a perfectly detached observer: he saw the 'boss' enlarged because he already knew that it was there, and was already convinced that it was more noticeable than any other feature.

In later books, with one exception, Galileo did not provide beautiful, careful illustrations. The reason is straightforward: he expected his readers to have, or to be able to obtain, a telescope of their own. They could look for themselves. The exception is his book on sunspots, where it was important to be able to see how sunspots changed their shape and position over time. Galileo could not assume that his readers would have the patience or the expertise to see this for themselves, and so provided extraordinarily exact illustrations.[19]

I feel obliged to stress the quality of Galileo's illustrations for *The Starry Messenger* because shortly after it was printed a pirated copy appeared, published in Frankfurt. Since the publisher of this edition did not have access to the original plates, and since his only interest was in making a quick profit, he substituted crude woodcuts, quickly and carelessly produced. Readers of that edition will have known that the illustrations had not been approved by Galileo.

In 1975 Paul Feyerabend published a now famous book called *Against Method*.[20] In it he argues that the success of *The Starry Messenger* was unwarranted. Galileo did not know how the telescope worked, so he could not tell if it represented the heavens accurately. He had not accurately represented what he had seen, and so his reports should not have been taken on trust. He had, in short, bamboozled his contemporaries into accepting his claim to have made new discoveries in science.

Feyerabend's first criticism would have had some force if *The Starry Messenger* had been concerned only with the topography of the moon. It was easy to show that the telescope accurately represented objects on earth, but since what was at stake was whether objects in the heavens were comparable to objects on earth, one could hardly argue that because the telescope was accurate on earth it must also be accurate when pointed towards the heavens. But Galileo's discovery (a word that he himself does not use – he says the new stars were 'unknown to anyone until this day') of the moons of Jupiter blunted Feyerabend's first argument, which was one that contemporaries were naturally quick to make.

It was true that Galileo was not able to explain how the telescope worked – this was something that concerned Paolo Sarpi, a sympathetic critic.[21] A Venetian nobleman, Agostino da Mula, set out to write a book explaining the optics of the telescope, but abandoned the task.[22] Fortunately Kepler produced a satisfactory account in the *Dioptrics* of 1611 – a work that Galileo was slow to read, and perhaps even slower to understand. Galileo lost interest in questions of optical theory because one thing was soon clear: the telescope did not manufacture illusory moons around other planets. Had he found moons around Venus and Saturn one could reasonably have argued that some defect in the telescope was responsible, but he found them only around Jupiter. Since there was no difference in the telescope there must be a real difference in the heavens.[23]

As proof of his second claim, that Galileo's illustrations were unreliable, Feyerabend reproduces an illustration from the pirated edition of *The Starry Messenger*, as if it were an illustration from the authorised edition. This second charge is thus simply a foolish mistake, as was quickly pointed out in an article in *Science* by Ewan A. Whitaker. No contemporary accused Galileo of providing misleading illustrations – no one at the time mistook the crude Frankfurt copy for the real thing. In an appendix to the second edition of *Against Method* Feyerabend tried to defend his original account, without making clear the full force of Whitaker's criticism. (He implied that Whitaker had claimed that Galileo's unpublished watercolours were accurate, which is true, but not the point that Whitaker had made – and the confusion is puzzling, because Feyerabend had understood the real point at issue when he first replied to Whitaker in the pages of *Science*.) In the third edition, the appendix, and with it all reference to Whitaker, disappeared. In all three editions (and in the posthumous fourth edition) the illustration is the same one, from the Frankfurt edition.[24] It is difficult to imagine any defence for Feyerabend's decision to publish revised editions in which he brazenly repeated his original mistake, but had he replaced the Frankfurt illustration with one taken from the first edition it would have immediately been clear to every reader that his argument was mistaken.

Galileo had no doubt of the scientific importance of his discovery of the moons of Jupiter: here was the first direct evidence that the cosmos did not consist of an array of heavenly bodies all of which orbited around a common centre, the earth. But he also saw that it represented a quite different sort of opportunity. Because the moons were previously unknown he could claim the right to name them, and in return he could hope to obtain a reward. The Venetians, who had increased Galileo's salary at the University of Padua and given him an appointment for life in return for his improvements to the telescope, had insisted there were to be no further rewards, so there was nothing to be gained by offering them the new stars.[25] But on 12 February Galileo received a letter from Belisario Vinta, the secretary to the grand duke of Florence (and Galileo's patron at the Medici court), reporting that the grand duke was 'stupefied' by news of Galileo's discoveries.[26] Galileo replied the very next day. Would the grand duke, who was called Cosimo (he had succeeded his father a year before), like these to be named the Cosmic stars, after him? Or would he prefer them to be named the Medicean stars, given that there were four of them, and he had three brothers.[27]

Quickly the answer came back: it was written on 20 February and will have reached Galileo a few days later (Florence is 240 km from Padua, or at least two days on horseback). 'Cosmic' was too ambiguous; it would not automatically make people think of Cosimo. 'Medicean' was fine.[28] Galileo had made his arrangements for publication at the end of January. He had, in effect, booked time on the press and could not wait for a reply. When Vinta's letter relaying the grand duke's choice arrived, an erratum slip was pasted over the word 'cosmica', replacing it with 'medicea'.[29] It will have been after he received Vinta's letter that Galileo chose the final title for the book (the title page refers to the Medicean stars) and wrote the prefatory letter in praise of Cosimo.

Galileo's printer divided the original manuscript in two and printed the two parts separately, leaving Galileo free to make revisions up to the last moment (when the preface was added) not only at the end, where we would expect to find them, but also in the middle. In these additions Galileo moved from reporting his observations (the original idea encapsulated in his early titles) to engaging in mathematical arguments. It is only this section of the book which is accompanied by geometrical drawings. In Galileo's mind this must have significantly enhanced the book's status – the 'Avviso' thus became a 'Nuncius', or (for the word 'nuncius' can additionally mean an ambassador) the journalist became an ambassador.

There is another, more significant, discovery to be made from an examination of the surviving manuscript. There are three passages in the book in which Galileo declares his support for Copernicanism. The first is in the preface, the second in the discussion of earthshine (in both cases the reference to Copernicus is implicit but unambiguous), while the third is at the very end of the book, where Copernicus's name appears for the first and last time.[30] All of these represent late additions to the text.

It is thus clear that when Galileo received Vinta's letter on 12 February he had a virtually complete draft manuscript which contained not a single reference to Copernicanism.[31] The first and third sections of the book open with references to Venus and Mars orbiting the sun, but a knowledgeable contemporary would have been more likely to interpret this as a reference to Tycho Brahe than to Copernicus.[32] (This is also true of the equivalent passage in the letter he had written announcing his discoveries.)[33] In mid-February Galileo made two bold decisions. He decided to put a reference to the 'sidera cosmica' into the title of his book and into its conclusion (had Cosimo refused to allow Galileo to dedicate the book to him two whole sheets would have had to be reprinted). And he decided to commit himself to an explicitly Copernican argument, and to include within his book an implicit attack upon Tycho Brahe.

Galileo began by writing a book that was open to a Tychonic interpretation, but he ended by writing one which was clearly directed as much against Tycho as against Ptolemy. It was not Galileo's telescopic discoveries which most impressed the English ambassador Henry Wotton, but the pages in which Galileo explains the phenomenon of earthshine, and thus shows that the earth lights the moon just as the moon lights the earth.[34] As Galileo says, earthshine is powerful evidence in favour of the Copernican claim that there is no difference between the earth and the heavenly bodies. Of the discoveries reported in *The Starry Messenger* this is one of two which tell against the cosmology of Tycho Brahe: the implications of the other, the discovery that planets can have moons, so that the earth could in principle be a planet, were only drawn out in the penultimate page of the book.[35]

Had Galileo made a major new discovery just as the book went to press? Surely not: the crucial new passage on earthshine refers to a discovery made, whether by Galileo or by Sarpi, many years before, a discovery which did not require the use of a telescope. Nothing had happened to make Galileo more convinced than he had been a few weeks earlier that Copernicanism was right, so something must have happened to make Galileo more confident, less frightened of being subjected to the ridicule he had referred to in his letter to Kepler of 1597.[36] It is easy to see what had changed – or at least what Galileo hoped was about to change. When Galileo entered into a contract with his printer he was an insignificant Paduan professor; after 12 February he felt confident he would receive Medici support; and as the book was being printed he received official confirmation that it could indeed appear under Medici patronage. In return for having the moons of Jupiter named after his family, Cosimo was to become an unwitting sponsor of Copernicanism.

It is worth pausing to see what Galileo was doing in mid-February. He was on the threshold of publishing a book that would 'stupefy' the scholarly world. Success was assured. And so he decided to raise the stakes and make his success uncertain. He committed himself to dedicating the book to Cosimo before he had permission to do so. Even more boldly, he decided to maximise the opposition to his book and minimise the support for it by making it not simply anti-Ptolemaic but explicitly Copernican. Of course, since he believed in Copernicanism, this had the great advantage of allowing him to speak his mind. But it was rash in the extreme.

On 19 March Galileo told Vinta that he had written most of the book as it went through the press.[37] This was a pardonable exaggeration. More than half of the book may indeed have been written after he had contracted with Tommaso Baglioni at the end of January. The title, the central section, and the ending of the book had been significantly revised in mid-February, turning a Tychonic text into a Copernican text, and announcing the proposed masterwork on the true system of the universe. Finally, a major response to criticism – why, a critic, probably Paolo Sarpi, had asked, if there were mountains on the moon did the moon not look like a cogwheel when seen from the earth? – had been added as the book went through the press.[38]

This reconstruction explains a striking peculiarity of the book. Its rhetorical high point occurs not at the end (which rather peters out in discussion of the variation in the apparent size of Jupiter's moons),[39] or in the preface (where Galileo is busy praising his Medici patron), but where one would least expect it: in the middle.[40] This passage, it is now apparent, may have been the last to be written before the book went to press (although not the last written before the book was published). It is remarkable in no fewer than three respects. It is transparently Copernican. It announces Galileo's great project, the 'System of the World'. And it is one of only two places where Galileo uses the word for 'experiment' in Latin, something knowledgeable readers would have seen as an implicit reference to Gilbert's *On the Magnet*.[41] This passage thus constitutes Galileo's concluding thoughts on his new discoveries, and with these words he both announces an imminent intellectual revolution and provides a sketch of his great *Dialogue*, which was not to be published until 1632:

> Let these few things said here about this matter suffice. We will say more in our *System of the World*, where with very many arguments and experiments [*experimentis*] a very strong reflection of solar light from the Earth is demonstrated to those who claim that the Earth is to be excluded from the dance of the stars, especially because she is devoid of motion and light. For we will demonstrate

that she is movable and surpasses the Moon in brightness, and that she is not the dump heap of the filth and dregs of the universe, and we will confirm this with innumerable arguments from nature.[42]

Galileo's book was published on 13 March; within a week all 550 copies were gone. Galileo, who was supposed to get thirty free copies of the book, received only six because the book sold out so quickly: the printer could offer him only copies to which the illustrations had never been added, but these he rejected.[43] Fame had arrived at last.

18

Florence and buoyancy

Galileo's interests and those of the Medici were even more firmly tied together when, in May of 1610, Galileo agreed to move to Florence, to take up the world's first research professorship there. In the course of his life Galileo made two momentous decisions, decisions which were to determine everything that happened afterwards. This move was the first; the second came in 1632. Why would Galileo leave Venice? Not to improve his finances, for his salary in Florence was roughly the same as the salary he had been promised in Venice, and he gave up the revenue from private teaching. He escaped the need to teach, but the demands of teaching were modest – sixty half-hours of lecturing a year.[1] He escaped the demands of Venetian noblemen, who thought they could lay claim to his attention, but he exposed himself to the demands of the Medici grand duke and his court, which could easily have proved as great. His Venetian friends were puzzled: in Padua, his friend Gianfrancesco Sagredo complained, Galileo had become, in effect, his own master; now he was subject to the arbitrary will of a prince, one who might at any moment die and be replaced by someone with no sympathy for Galileo and his work. In Venice he had had security for life; in Florence he had security only for the prince's life (and indeed under Ferdinando II there was to be a campaign to deprive Galileo of his salary on the grounds that it was an inappropriate use of taxpayers' money).[2]

Why, the philosopher Cremonini asked, had Galileo given up 'Paduan liberty'?[3] Why, asked Sagredo, go somewhere where the Jesuits had so much influence?[4] Both knew Galileo extremely well. Both had spent long hours talking to him in private. Both believed that by going to Florence he was placing himself in danger – danger of falling foul of the authorities in Rome. Cremonini had frequently been pursued by the Inquisition for denying the immortality of the soul (or rather, for teaching that Aristotle denied the immortality of the soul), but the Venetian government had always protected him. He had vigorously attacked the new Jesuit College in Padua which was providing stiff competition for the university – until, that is, the Jesuits were excluded from Venetian territory during the Interdict of 1606 and the College closed.[5] Sagredo too was bitterly hostile to the Jesuits. It was well known that during the Interdict he had engaged in correspondence with a member of the order, pretending to be a pious woman, in order to expose their exploitation of the credulous.[6] They both thought that Galileo saw the world as they did; and indeed we have a letter written by Galileo, in which he describes the Jesuits leaving Venice by boat, in the dark, carrying candles and wearing crucifixes. His own sympathies seem clear. Their departure, he says, will be lamented by the many ladies who are devoted to them. This coupling of Jesuits and women, as if no man could respect the order, is the language of the Venetian anticlericals.[7]

Cremonini and Sagredo not only thought that Galileo was taking a risk; his departure looked like a betrayal. Venice had almost come to the point of war with Rome and Spain during the Interdict crisis of April 1606 to April 1607, and would have found herself under attack had she not had the support of the French. What was at stake in the Interdict crisis was the ability of a Catholic state to control the Church (and intellectual life) within its territories. Now Galileo was joining the opposing side, for Florence was firmly allied to the Habsburg rulers of Spain and the Holy Roman Empire. In what have been called the 'culture wars of late Renaissance Italy' Galileo was changing sides.[8]

Galileo had long been exploring the possibility of leaving Venice. In 1604 he had visited Mantua (where he had taught the duke the use of his sector) and had been rewarded with both more money than he was paid by the University of Padua in a year and the offer of a job. But the terms had not been good enough, and Mantua was a sleepy backwater.[9] In 1606, if Mauri was Galileo himself, he had turned to Capponi looking for patronage. Most importantly, in 1605 he had begun a series of summer visits to the Medici court, returning in 1606 and 1608. (In 1607 he was too distracted, being caught up in his conflict with Baldassar Capra, who had plagiarised his sector and the instruction manual that accompanied it.)[10]

At court he was employed to teach the young prince, Cosimo. Cosimo's education was in the hands of his mother, Christina of Lorraine, to whom Galileo would later write his famous *Letter to the Grand Duchess Christina*. The first visit went well: in the autumn of 1605 the Florentine government supported Galileo's efforts to obtain a pay increase from his Venetian employers; Galileo in turn arranged to have two sectors made from silver for princely use, and to have printed a small private edition of instructions, dedicated to the grand duke.[11] That winter Galileo felt sufficiently securely established in the world of the Florentine court to write directly to the prince himself, offering, in effect, to move from the Venetian territories to Florence.[12] A year later he was assuring Christina that Acquapendente, who held a chair of medicine in Padua, would surely jump at the chance to make the same move. In 1608 he successfully sold Sagredo's champion magnet to the grand duke. That summer Christina let Galileo know that she believed him to be 'the best and the most distinguished mathematician in the Christian world'.[13] In September, still in Florence, Galileo wrote to Christina suggesting that the celebrations for Cosimo's marriage (to a Habsburg, of course, though Christina herself was French) should include a tableau illustrating the motto 'Vim facit amor' – 'Love creates strength' – one which was perfectly appropriate for an absolute ruler, and which could be illustrated by a giant magnet – a neat example of Galileo trying to become a courtier-scientist.[14]

In February of 1609 Ferdinando I died (despite Galileo's having made an optimistic forecast of his life expectancy on the basis of his horoscope), and Galileo's pupil, now Cosimo II, became grand duke. Galileo immediately let it be known that he would welcome the offer of a position at court, making it clear that he was not after more money.[15] No offer was forthcoming, however, presumably because the grand duke was reluctant to match Galileo's vast salary (which would have made him one of the highest-paid employees of the Florentine government). Free time, Galileo said, was worth more to him than gold; with gold one might acquire celebrity, but with free time he could hope to win true fame.[16] The previous summer, the ducal secretary, Belisario Vinta, had taken Galileo under his personal protection, saying as they parted, 'Galileo, in all your activities and dealings [with the court], deal with me and with no one else.'[17]

Thus, well before Galileo discovered the telescope, well before he dedicated *The Starry Messenger* to Cosimo, he had set his heart on transferring from Padua to Florence. Again and again he had insisted that he would prefer to be employed by an absolute prince than by a republic.[18] In a sense, the generous improvement in his conditions of employment with which the Venetian government rewarded his work on the telescope simply came too late – he already associated Padua with frustration and Florence with success. The dedication of *The Starry Messenger* to Cosimo was simply the last step in a courtship which had been going on for five years. And when Cosimo finally made Galileo an offer of a position he knew exactly what he was doing. At this point no one had seen the moons of Jupiter except when looking through Galileo's own telescope. But Galileo had travelled to Florence in April of 1610 to show the moons and his other discoveries to Cosimo and his court (where he had been rewarded with a chain of gold worth four hundred scudi and a gold medal bearing Cosimo's image). Cosimo could be confident that others would eventually see what he had seen. There was something more important to Galileo than a gold chain: he insisted that he must be appointed, not just as the grand duke's mathematician, but as his philosopher.[19] Philosophy was a higher-status discipline than mathematics, and crucially, mathematics was supposed to be a purely theoretical subject, while philosophers claimed the right to declare how the world is. Mathematicians constructed theoretical models; philosophers were, in our language, scientists.

Galileo finally arrived in Florence on 12 September 1610. At first much of his time was taken up by new observations of the heavens, although it was not long before he found himself, quite unintentionally, caught up in a bitter debate with the local Aristotelian philosophers. The debate started in the summer of 1611, when Galileo happened to remark that ice floats because it is lighter than water. The Aristotelians – faithfully following Aristotle – were convinced that ice is heavier than water (ice, they reasoned, is condensed water), and that it floats because of its shape. Flat things float, they claimed, because they are unable to overcome the resistance of water, while spherical objects sink. It is so obvious to us that things float because they are lighter than water and sink because they are heavier than water that it is very difficult for us to imagine a world in which the best-educated and most sophisticated minds simply denied that this was the case.

Galileo, who viewed the subject as we do, had a number of straightforward responses. Not only had he worked on specific gravity; in 1608 he had got caught up in an argument with the grand duke's engineers who were building a pontoon on the Arno as part of the arrangements for Cosimo's wedding. They had argued that the flat structure of the pontoon's base would give it extra buoyancy, while Galileo had insisted that shape had nothing to do with buoyancy.[20] Galileo argued that an object whose specific gravity is heavier than water will always sink, although sometimes it will take a very long time to do so: a handful of mud may take hours to sink to the bottom of a large tub of water. The shape of a solid object has no bearing on whether it floats or sinks; it affects only the speed with which it moves through the water. Thus a small ice cube will float. This is obvious to us, but was not so to a seventeenth-century Florentine, who would have seen ice in two forms: in thin sheets on the surface of ponds in winter, and as large blocks brought down from the Apennines in summer to keep fish fresh. This disagreement involved a straightforward clash between the followers of Aristotle, who held that ice is condensed water, and Galileo, who was applying the principles of Archimedes, and who maintained that since ice floats it is evidently expanded water. To contemporaries, Galileo's view, which seems so straightforward to us, seemed 'paradoxical'.[21]

Matters became more interesting, as far as Galileo was concerned, when a philosopher called Lodovico delle Colombe announced that he had an experiment that would prove Galileo wrong. Delle Colombe was a strict Aristotelian; he had also written about the new star of 1604, and suspected, surely correctly, that Galileo had something to do with Mauri's pamphlet, which had been written in opposition to him. Delle Colombe's experiment was simple: he placed an ebony chip and an ebony ball gently onto the surface of a bowl of water. The chip floated, the ball sank. Aristotle was vindicated. Delle Colombe was supposed to meet with Galileo to debate his experiment, but he never turned up. Instead he went around the city, demonstrating his ability to float ebony on water in the public squares, and crying out that he had defeated Galileo. In September 1611 the whole question was discussed at the grand duke's dinner table, with two visiting cardinals joining in. Maffeo Barberini – a Florentine whose relationship to Galileo was to prove of the foremost importance – took Galileo's side, while Ferdinando Gonzaga took Aristotle's.

In this heated atmosphere, the grand duke told Galileo to stop trying to arrange a confrontation with delle Colombe and to put his arguments in writing, as this was the best way to ensure that good sense triumphed. Galileo acknowledged the grand duke's advice in a letter he drafted to him:

Many are the reasons, Most Serene Lord, for which I have set myself to the writing out at length of the controversy which in past days has led to so much debate by others. The first and most cogent of these was your hint, and your praise of the pen as the unique remedy for purging and separating clear and sequential reasoning from confused and intermittent altercations in which they especially who defend the side of error on one occasion noisily deny that which they had previously affirmed, and on the next, pressed by the force of reason, attempt with inappropriate distinctions and classifications, cavils and strained interpretations of words, to slip through one's fingers and escape by their subtleties and twistings about, not hesitating to produce a thousand chimeras and fantastic caprices little understood by themselves and not at all by their listeners. By these the mind is bewildered and confusedly bandied about from one phantasm to another, just as, in a dream, one passes from a palace to a ship and thence to a grotto or beach, and finally, when one awakes and the dream vanishes (and for the most part all memory of it also), one finds that one has been idly sleeping and has passed the hours without profit of any sort.

The second reason is that I desire that Your Highness should become fully and frankly informed of what has taken place in this affair; for the nature of contests being what it is, those who through carelessness are induced to support error will shout loudest and make themselves more heard in public places than those through whom speaks truth, which unmasks itself tranquilly and quietly, though slowly. Hence I can well believe that just as in the squares and churches and other public places, the voices of those who dissent from what I assert have been far more often heard than those of others who agree with me, so likewise at court they have tried to advance their opinion by forestalling me there with their sophisms and cavils; these I hope to disperse and send up in smoke.[22]

Galileo had thus learnt (or thought he had learnt) a valuable lesson. Debates on intellectual topics, conducted in the court, in the open air, or in the houses of leading citizens, rapidly degenerated into shouting matches. Moreover in such circumstances systematic reasoning became impossible; indeed one entered a sort of nightmare world in which the meaning of words and the subjects discussed were constantly changing. Better then to write down one's arguments, so that others could not misrepresent them; better to abandon the court and the city squares, and retire to one's study. Galileo was very good in debate: he could, his admirers believed and his opponents acknowledged, out-argue anyone. One of his most infuriating tricks was to offer, with an appearance of generosity, ways of improving his opponent's case, before comprehensively demolishing it.[23] But he no longer had confidence that out-arguing his opponents was enough. The retreat from public and court life that is already apparent in the *Discourse on Floating Bodies* was soon to be carried a stage further. By January of 1612 Galileo was effectively living in a villa in the countryside outside Florence: he had abandoned the house he had rented in the centre of Florence and had moved in with his friend Filippo Salviati.

Although Galileo at first retained a house in Florence, he was properly resident in the city again only between Salviati's death in 1614 and his signing a lease on a villa of his own (a rather magnificent property, the Villa di Bellosguardo) in 1617.[24] His opponents claimed that his retreat from court and city was proof that he had fallen out of favour.[25] But all the evidence is that Galileo retained the loyal support of Cosimo II, and then, when Cosimo died in 1621, of his son Ferdinando II (who came of age in 1628). In the summer and autumn of 1611 Galileo had a brief taste of what it would be like to be a court intellectual holding views which were at odds with those of the intellectual establishment. The experience was not a pleasant one, and he never repeated it, at least in Florence. He had shown many of the skills of a courtier in acquiring his position in Florence: he had wooed the Medici rulers, and had won their favour by his assiduous attentions. They certainly hoped that they would have opportunities to show him off to visiting dignitaries – he was, after all, universally acknowledged as a great man, and also as a brilliant conversationalist. But their major investment was in an unfinished book, the book on the two world systems that he had promised in the *The Starry Messenger*, and they had no objection to his hiding away in town or retiring to the countryside to write.

If we want to understand Galileo's relationship to the court after 1611, it helps to look forward to 1630, when his *Dialogue* was beginning to be read in manuscript. Niccolò Aggiunti found himself explaining Galileo's book to the grand duke, and while the grand duke was happy with what he heard, some of those around him were not. 'I won't bother to describe to you the discussion which followed', Aggiunti tells Galileo, 'because there's no need . . . the grand duke knows they are reading your dialogues in the house of the canon [Niccolò Cini], to the amazement and unending applause of those who hear them.'[26] Provided Galileo retained the patronage of the grand duke, it no longer mattered to him whether people at court agreed with his views or not. What mattered was that there existed a small circle of admirers, outside the court, with whom he could discuss his work; their admiration justified the grand duke's continuing to support him, and the grand duke's support made it possible for them to continue treating Galileo as a great man. Galileo was thus now neither a university teacher, nor a courtier, nor an engineer; he was something quite new – an intellectual, for want of a better word.

So Galileo, who had begun by drafting a letter to the grand duke outlining his views on floating bodies, resolved to turn the letter into a book. This was published in May of 1612, and appeared in a slightly expanded second edition in

December.[27] In his book Galileo offered a solution to delle Colombe's puzzle. He showed that it was not because of its shape that the ebony chip floated – he could, he discovered, float a needle on top of the water. How could objects that were heavier than water float? Because, Galileo claimed, they floated slightly below the surface of the surrounding water. What was really floating was not the ebony chip alone (this would always sink if you immersed it fully in the water, or even wetted its upper surface) but an open sandwich of the ebony chip and a layer of air, the average weight of the two combined being less than that of the water they displaced. In other words the ebony chip was a bit like a boat, even though there were no planks holding the water back. Galileo had thus recognised that in moving into his own territory, that of experimental evidence, delle Colombe had indeed identified an interesting puzzle, a stubborn fact. He had no idea why air would stick to the surface of an ebony chip, or why water would not flow in to cover the chip. He only knew that this was what actually happened. This fact could not be wished away, but had to be described in Archimedean terms. This Galileo did as best he could with the concepts available to him. We would now say that the needle and the ebony chip are supported by surface tension, but Galileo had no reason to think that the surface of water behaves any differently from the rest of the water. What he could see was that some sort of hidden force was at work – as in the case of magnetism.

In the same little book Galileo explored a puzzling feature of displacement. If I launch a large ship into the sea, as the ship sinks into the water it displaces water equal to its own weight – the weight of the ship and the weight of the displaced water exactly equal each other, as exactly as when two objects balance each other on a pair of scales. But suppose I take one of those silver buckets in which bottles of white wine are kept cold in smart restaurants. In the bucket I put some ice cubes; these slowly melt until they become half a litre of cold water. Into the bucket and the water I put a bottle of champagne that weighs about as much as a litre of water. If the sides of the bucket sit fairly snugly around the bottle then the bottle will float: half a litre of water is somehow counterbalancing twice its own weight. There is some sort of multiplier effect at work. Galileo argues that this multiplier effect is exactly the same as in a yardarm or a lever: in my silver bucket half a kilo of water can balance a kilo of champagne bottle because the water rises up the sides of the bucket twice as fast and twice as far as the bottle sinks down into the water. With a bit of care you can devise a hydraulic machine that will lift a very large weight a small distance; within the machine a small amount of water will have to travel a large distance. This is the principle on which hydraulic lifts and presses work.

Galileo's little book provoked considerable comment. Four books appeared attacking it, while Galileo's disciple Benedetto Castelli published a substantial work defending it (a book which was in fact jointly written by Galileo and Castelli).[28] Galileo's treatise on floating bodies represents an important step in the process whereby a physics of qualities ('light' objects that move upwards in air, 'flat' objects that float in water) was replaced by a physics of quantities (in this case, specific gravities). It reflected a new understanding of causation: where for Aristotle there were four types of cause (material, formal, final, efficient – it was the formal cause, flatness, which supposedly made ice float), for Galileo causes were identified by running similar but not identical tests until one could find an alteration to a single parameter which would change the outcome.[29] Galileo thus thought that you could only discover causes through experiments. This new way of thinking obliged him to give up his belief in astrology, which he had still been defending as recently as 1611.[30]

Galileo also thought that we often have to make do without a satisfactory causal explanation: by 1612 he knew the law of the acceleration of falling bodies, but he did not have, and never would have, a causal account of fall. For Aristotle all real knowledge was knowledge of necessary relationships, of logical deductions and causal interactions. For Galileo much knowledge consisted simply of the identification of what Hume would later call constant conjunctures, of laws rather than causes. Galileo's new science was much more parsimonious with causal explanations than Aristotelian science had been.

For all the elegance of Galileo's arguments, which his opponents could scarcely ignore, the debate over floating bodies confirmed his growing conviction that he would never persuade Aristotelian professors of philosophy to agree with his views, even if he persuaded them to adopt an approach that recognised the central significance of facts (Aristotle too had taught that one must always respect experience). In the whole of his lifetime not a single professor of philosophy in a European university supported him on any topic. Mere force of argument would not do the job. With the benefit of hindsight we may say that Galileo was too impatient, and Aristotelian philosophy was to retain a secure position in Europe's universities for more than a hundred years – it was still being taught when Newton was professor of mathematics at Cambridge. If Galileo was going to defeat the Aristotelians he needed not just new arguments, but new allies.

19

Jesuits and the new astronomy

The publication of *The Starry Messenger* in March 1610 was followed by a carefully managed campaign to win acceptance for Galileo's discoveries, which were so far at odds with everything that went before that they at first met with considerable scepticism. That campaign depended on mobilising the resources of the Medici government, for the honour of the Medici was now linked to the success of Galileo's publication.

But Sarpi, Sagredo, Cremonini and Galileo's other Venetian friends would have been astonished if they had seen the first letter (or at least the first surviving letter) that Galileo wrote on arriving in Florence in September.[1] It is addressed to the leading Jesuit mathematician Christopher Clavius, whom he had visited in Rome in 1587, and with whom he had corresponded in 1588. Clavius had written to Galileo in 1604, seeking to renew their friendship, but it seems evident that Galileo had not replied.[2] 'It is time', Galileo now writes, 'that I broke a long silence, a silence of my pen rather than my thoughts with regard to you, Most Reverend Sir. I break my silence now that I find myself repatriated to Florence by grace of the Most Serene Grand Duke, whom it has pleased to call me back to serve as his philosopher and mathematician. The cause of my lengthy silence, throughout the time, that is, that I was living in Padua, there is no need for me to specify when writing to someone as wise as yourself.'[3]

Both the Venetian political establishment and the University of Padua were hostile to the Jesuits, and for eighteen years Galileo (who needed to get his contract regularly renewed) had had no written communication with the Jesuits, although we know that he conversed on mathematical subjects with Jesuits teaching in their Paduan college. Now, he writes, he plans to come at once to Rome to show Clavius and his colleagues the moons of Jupiter in order to dispel any doubts that Clavius may have 'circa la verità del fatto' – this is one of the occasions in which Galileo uses the word 'fact' in its modern sense.[4]

The evidence provided by this letter is decisive. Galileo did not move to Florence thinking that the influence there of the Counter Reformation Church, though something of a disadvantage, was worth putting up with in order to seize other opportunities. The key advantage of moving to Florence was that it would enable him to travel to Rome as an intellectual in good standing, one in the service of a respectable Catholic ruler, and to renew contact with the Jesuits. Why leave Venice? In order to be welcomed not just in Florence, but more importantly in Rome (we have already seen Mauri thinking along precisely these lines). When we think of Venice we think of a bustling port, a cosmopolitan city – the most cosmopolitan city in Europe, just as Padua was the most cosmopolitan university. Galileo saw things differently. Venice was on the periphery, its only allies the far distant English, the French and the German Protestants. If you wanted to convince the people who counted, if you wanted to be heard throughout the educated world, it was to Rome that you must go. Why on earth did he think this? In the end there is perhaps only one answer: because he was a Florentine, and that is how Florentines thought. One might also wonder whether it was because he was a Roman Catholic and that is how Roman Catholics thought, but against this suggestion there is the simple fact that it never occurred to Sarpi, Sagredo and Cremonini that Galileo was going home because he really belonged amongst the Jesuits and their admirers. It never occurred to those who considered that Galileo had betrayed Venice for Florence that the explanation lay in his religious convictions.

But Galileo was certainly right to focus on the Jesuits, who were rapidly establishing themselves as the educators of the elite throughout Catholic Europe: there were 245 Jesuit colleges by the end of the sixteenth century. In Rome they ran what amounted to a research university, but one which trained educators who were sent out across the world. And he was right to think of the Jesuits as potential allies for the new science, for much of the early experimental science was done within the order. But there was a profound tension within the Jesuit enterprise. On the one hand, Jesuits were at the cutting edge of the new science; on the other, the order was committed to upholding the traditional learning of the Church as represented, above all, by St Thomas Aquinas. The crisis provoked by Galileo over the next few years was also to be a crisis for the Jesuit order. By an unlucky coincidence, their general, Acquaviva, reminded the order in 1611 that it was committed to the traditional teaching of the Church, and must be opposed to innovation.[5] As the Jesuit Christopher Grienberger, who was always sympathetic to Galileo, put it in a letter of 1613, 'I do not have the same freedom that you have.'[6] By 1632 the order as a whole was clearly opposed to Galileo, but individual Jesuits continued to convey, in private conversations, their sympathy with his work.[7] Galileo's strategy of alliance with the Jesuits was probably doomed from its inception, but he himself did more than anyone to make such an alliance impossible.

So Galileo returned home to Florence in September 1610, six months after the publication of *The Starry Messenger*. At this point he had shown his new discoveries to friends such as Paolo Sarpi in Venice. Others had written to him asking how they could confirm his reports.[8] His answer was that even if he supplied his own telescope, he was not sure that they would be able to see the moons of Jupiter without his guidance.[9] On 19 March he had drafted a letter to Florence, saying that he was sending a copy of his book together with the very telescope with which he had made his discoveries. For this telescope he obviously felt a particular affection: they had spent long winter nights in each other's company. It was scruffy and plain, but he insisted that it must not be improved to make it worthy of a prince. It should be preserved for posterity exactly as it was.[10] However, the letter was revised and an inferior telescope was sent instead – Galileo was unable to part with his favourite telescope and it was still in his possession in 1637 when he lost what remained of his sight.[11] Instead, Galileo made the journey himself during the Easter vacation. Jupiter, which he was observing in the evening sky, would be too close to the sun to be seen in the early summer, so time was of the essence; it would not reappear, this time in the pre-dawn sky, until late July.[12]

Galileo also had ten other telescopes capable of showing the moons of Jupiter, and these he intended to send off with copies of the book to relatives and allies of Cosimo – to the duke of Bavaria, to the elector of Cologne, to Cardinal del Monte (his old patron's powerful brother), to Spain, Poland, Austria, Urbino and even France.[13] We know that some of these were sent, but it looks as though in the spring of 1610 only one person had seen the moons of Jupiter through a telescope of their own – Antonio Santini, a Venetian merchant, who had made an excellent telescope for himself. Everyone else who had seen the moons had done so when looking through Galileo's telescope, in Galileo's presence.[14]

Thus readers who were unable to meet with Galileo had to make their minds up simply on the basis of his published report. Some responded with scepticism: they thought Galileo was not to be trusted because they understood that he was claiming to have invented the telescope, which they knew not to be true.[15] Others were prepared to credit Galileo's reports, while at the same time doubting their significance. What he was offering were new facts, but the knowledge of nature came from philosophy, and philosophy had little to do with facts. One great philosopher, Galileo's good friend Cesare Cremonini, was so sure of this that he refused to look through the telescope, even when Galileo offered to bring it round to his house, saying that it would give him a headache (although he was happy to use one of the first microscopes – Aristotle had said nothing about the microscopic world, so the microscope was irrelevant to Cremonini's professional concerns as a philosopher).[16] Cremonini went on to publish a big book on the heavens (a book in due course banned by the Church), in which he mentioned neither Galileo nor his telescope. Others, however, said that astronomy was a subject founded on sensory perception, and therefore might indeed be transformed by new discoveries.[17] For the most part these first readers were content to wait to see if Galileo's discoveries were confirmed by others.

There was one important exception. Johannes Kepler was the emperor's mathematician, despite being a Protestant and that remarkable rarity a Copernican. A copy of *The Starry Messenger* (but no telescope) was sent to the emperor and passed on to Kepler. Kepler immediately (19 April) wrote Galileo a long letter that was rapidly published in Prague, and then reprinted in Florence and Frankfurt.[18] Kepler sought to put Galileo's discoveries in an intellectual context: some of his own previous work was relevant, he pointed out, as was the work of the Neapolitan Gianbattista della Porta, who had published on lenses. Reading Galileo, he made clear, reminded him of Giordano Bruno and the question of whether there were other inhabited worlds apart from ours.[19] Some, reading Kepler, thought that he was putting Galileo in his place, by gently pointing out that there was nothing very new in *The Starry Messenger*. Galileo, they said, reminded them of the crow in Aesop's fables, who dressed himself in the feathers of other birds, the story ending with the crow being left ugly and naked.[20] But this was not Kepler's point at all.[21] He was prepared to believe everything Galileo had described, and was convinced it was of the foremost importance, because he recognised that only a Copernican could happily accept Galileo's new astronomy. He was prepared to make clear his support for Galileo even before he had seen the moons of Jupiter with his own eyes – which he was not to do until September.

This was fortunate for Galileo, because meanwhile the story was going around that others could not in fact see what he had seen. In late April Galileo had gone to Bologna and had tried to show his discoveries to twenty mathematicians there. Martin Horky quickly sent out letters claiming that no one had been able to see the moons of Jupiter (although Galileo had claimed that he could see them clearly), and that Galileo had gone away crestfallen.[22] Much the same report was circulated by Galileo's host and Horky's employer, the mathematician Giovanni Magini, although Magini obstructed Horky's efforts to publish against Galileo, and finally dismissed him from his service – at which point Horky did indeed publish.[23] Galileo never denied that his visit to Bologna had been a fiasco. Even Cosimo began to have doubts.[24]

Galileo was now in something of a pickle. In *The Starry Messenger* he had provided only the most basic information about the telescope he had used – not enough to make it easy for other astronomers to build replicas.[25] He had, for example, withheld (presumably deliberately) information about the diaphragm that greatly improved the clarity of the image. His motive will have been straightforward: he wanted to maintain his lead in telescope construction for as long as possible, in the hope of making further discoveries – and this was a sound strategy, as soon became apparent. But he had assumed that he would have no difficulty in convincing people that the moons of Jupiter really existed. He would show them to them by letting them look through his telescope. His mistake here was not to have tried doing this before publishing. As far we can tell he showed the moons to no one before *The Starry Messenger* was published, although rumours of his discoveries began to spread in Florence as the book went through the press. Now, with the book out, he had some important people, such as Cremonini, refusing to look; and others, such as Magini, looking, but claiming they could not see what Galileo said was there for all to see. He was already committed to giving his ten good telescopes to princes and cardinals, and this had seemed like a sensible idea, as it raised the status of the telescope and of the discoveries he had made with it. But few princes and cardinals were qualified to tell the world if Jupiter had moons or not, and as more and more of them asked him for telescopes the day on which he would have a good telescope to spare for a trusted colleague was put further and further back.[26]

The obvious thing for Galileo to do would have been to send a good telescope to the emperor, who would immediately have passed it on to Kepler. But Galileo did not do this; indeed the emperor complained that more and more cardinals seemed to be getting telescopes while he had yet to receive one.[27] I think we can assume that Galileo (or the grand duke, whose reputation was on the line, and whose views could not be ignored) was trying to get his discovery confirmed by a Catholic scientist, not a Protestant one. He could have sent a telescope to Clavius, but what if Clavius then claimed he could not see any moons? Galileo's reputation would have been destroyed overnight, and might never have recovered. He had plenty of second-rate telescopes – he had, he said, ninety second-rate pairs of lenses – and plenty of people asking for them; but he kept them to himself, for any telescope he supplied that was not good enough to resolve the moons of Jupiter would have been taken as providing evidence against his great discovery.

Galileo was thus caught in a double bind. He urgently needed a respectable astronomer to confirm that Jupiter had moons, but there was no astronomer he dared trust with a first-rate telescope of his own manufacture. What he needed was for other astronomers to make their own telescopes. Kepler tried, but his was not very good.[28] By the end of 1610 only two people other than Galileo – as far as Galileo knew – had made telescopes of the required quality. One (it was claimed) was a Jesuit in Rome, while the other was Galileo's friend in Venice, Antonio Santini. Santini was surprised at how slow others were, and Galileo probably shared his puzzlement.[29]

By July Magini was working with his own telescope, studying the moon's surface,[30] but still there was no confirmation that others could see Jupiter's satellites, now back above the horizon, although Galileo was able to observe them from the twenty-fifth.[31] It was only in September that Santini in Venice reported that they had reappeared. Santini's observations were enough to convince Magini, even though he had not yet seen the moons himself; eventually Santini supplied him with the objective lens he had used for observing the moons.[32] Further confirmation came from Kepler, who now had the use of the telescope sent by Galileo to the elector of Cologne (which certainly had not been intended for Kepler): he had observed the satellites on 5 September, in the company of Thomas Seget, who had evidently made his way from Venice to Prague.[33] Thus there were now two independent observers who claimed to have replicated Galileo's observations. They were counterbalanced by reports of failure: from Paris, and worst of all from Rome, where it was said that Clavius had tried and failed to see the moons of Jupiter.[34]

On hearing in September of the failure of Clavius's efforts to see the moon, Galileo had immediately written to him (and here we return to his first letter from Florence), saying that he was about to come to Rome and would show him the planets himself. The technique of using the telescope was not straightforward, he acknowledged; even one's pulse could make it shake, rendering the planets invisible, and it was all too easy to fur up the lens by breathing. But instead of leaving for Rome Galileo became anxious and depressed, and then he became ill.[35] One can well imagine his anxiety that a visit to Rome might turn out as badly as his visit to Bologna. By late November, however, Clavius – who had by now received two telescopes from Santini – had seen four small stars moving around Jupiter.[36] He was not immediately convinced that they were planets, but he expressed no reservations when he wrote to Galileo on 17 December praising him for his wonderful discoveries.[37] Santini had thus solved Galileo's replication problem for him: he had supplied first-rate telescopes to the two key Catholic scientists, and had done so without increasing the risk that Galileo was running. Since Galileo and Santini were on very good terms we may guess that he was acting as Galileo's agent; certainly he does not seem to have been acting on his own behalf, as he demanded neither money nor fame in recompense for his efforts.

Galileo was immensely cheered by Clavius's letter – he sprang out of bed, full of renewed confidence, and wrote straight back to him. Two months later, in February 1611, he happily declared that there was no longer any serious dispute about the existence of Jupiter's moons, although many of the most eminent mathematicians had held out against them for a long time.[38] Now people had had plenty of opportunity to see them for themselves, and were prepared to believe their own eyes – which only showed, Galileo complained, that they had previously suspected him of being a liar. This was true enough, though the more charitable had wondered if he might be deceiving himself in some fashion.[39]

Thus in April he could arrive in Rome without the least anxiety, for he could now be sure of having the support of the Jesuits. At the Roman College he was received with applause – it was, a contemporary said, insisting he was choosing his words with care, 'a triumph'.[40] In other words, Galileo was a conquering hero. While he was there Cardinal Robert Bellarmine, himself a Jesuit, and acknowledged as the greatest authority of the day on all matters of doctrine, looked through Galileo's telescope and saw strange sights. He wrote to the mathematicians of the Roman College asking if they could confirm the telescopic discoveries reported in *The Starry Messenger* – the multitude of previously invisible fixed stars, the lunar landscape, the moons of Jupiter.[41] Four mathematicians replied, confirming all of Galileo's claims, although Clavius had some new doubts regarding the lunar landscape, thinking that it might be some sort of effect of the light on a smooth surface.[42] The battle, to all intents and purposes, was over. Galileo could take heart from being invited to become the sixth member of the Lincean Academy, founded by Federico Cesi. He now had a small group of allies in Rome, as well as Cesi's financial resources to support him (Cesi spent like a prince, although he ran out of money before he died).

By this point Galileo had announced two new discoveries, which Bellarmine also wanted confirmed. Galileo had first sent a mysterious anagram to Kepler (via the Medici ambassador in Prague) in July 1610, so that he could authenticate his claim to priority should anyone forestall him. He was saving his discovery, he told the grand duke, for a projected second edition of *The Starry Messenger*.[43] He had noticed that there appeared to be two small circles attached to the body of Saturn – he had seen, though could not interpret, the rings of Saturn. Kepler proceeded to publish Galileo's mysterious communication, and in November Galileo (who was rapidly losing interest in the idea of a second edition of *The Starry Messenger*) wrote to him decoding the cipher, and giving him permission to make his discovery public, which Kepler promptly did.[44]

A month later, on 11 December, he announced, again in code to Kepler, a much more important discovery. He was so excited, however, by his receipt on 30 December of Clavius's letter confirming that he had seen the moons of Jupiter that he immediately informed Clavius of both of his new discoveries. On the same day a letter went to Castelli, who had predicted this latest discovery, and two days later one went (via an intermediary) to Kepler, decoding his anagram, which stated that the shape of Venus varies like that of the moon.[45] Kepler proceeded to publish Galileo's letters – as Galileo would have expected.

This second discovery, of the phases of Venus, was of fundamental importance. It owed something, perhaps, to chance. Between 12 September and All Saints' Day (1 November) Galileo was living in temporary rented accommodation. There he could not mount the grinding wheels he needed to make new lenses for telescopes and his view to the east (where Jupiter was to be seen in the morning sky) was poor.[46] So he naturally turned his telescope to the west, where Venus was to be seen in the evening sky. When he moved to the house which he intended to make his home, which had good views in every direction, he continued his programme of observation. Thus Galileo had seen both that Venus had phases, like the moon, and that the planet's apparent size varied greatly over time.

No astronomer before 1610 had firmly predicted that Venus would have phases, because it was first necessary to be sure that the planets shone only by reflected light from the sun. This idea was one that had previously been discussed as a possibility, but the evidence seemed to be against it: the light coming from Venus, as seen by the naked eye, did not seem to vary as much as one might expect if it had phases. Kepler, for example, took it for granted that Venus shone by its own light.[47] Still, Galileo had taken a major step towards this prediction in *The Starry Messenger* by arguing that the light of the moon was entirely reflected light, and that the earth too shone by reflected light. The implication was that none of the bodies that orbited the sun shone by their own light. Given this assumption, it became possible to predict that – if Copernicus and Tycho were right and Venus orbited the sun – Venus would have phases which, like the moon's, would extend all the way from crescent to full, while if Ptolemy was right the phases would be limited either to between full and half (if Venus was further from the earth than the sun), or to between half and crescent (if Venus was closer to the earth than the sun). The discovery of the phases of Venus, which did indeed extend all the way from crescent to full, could thus correctly be presented as evidence that Venus orbited the sun, not the earth.[48] This was the first observation that was directly incompatible with the Ptolemaic cosmology, but it was perfectly compatible with Tycho Brahe's geo-heliocentric system.

Also important was the fact that the Tychonic and Copernican systems predicted a significant variation in Venus's size, since at times it would be relatively close to the earth, and at other times very far away. Naked-eye observations of Venus showed only a modest variation in Venus's brightness, and this had been a fundamental problem for both systems. Galileo could now claim that telescopic observations showed variation on the scale predicted: crescent Venus, when Venus most nearly approached the earth, was more than six times the diameter of Venus at full, when furthest from the earth. This raised some difficulties as to why naked-eye observations and telescopic observations did not agree, but here at last was visual evidence that was impossible to reconcile with the Ptolemaic system, and which suggested that the scale of the universe might, as Copernicus claimed, have to be rethought.[49] Both these discoveries were confirmed to Bellarmine by the Jesuits in April 1611.

On 5 December Galileo's pupil Benedetto Castelli had written to him from Brescia, 180 miles from Florence, asking him if he had conducted any careful observations of Venus. If he did, Castelli suggested, he would surely find that it had phases. On 11 December (the very day, in all likelihood, on which Galileo received Castelli's letter) Galileo wrote to Giuliano de' Medici, the Medici ambassador in Prague, who was in communication with Kepler, announcing in code the discovery of the phases of Venus (based, he said, on three months of observations). This has led to the charge that Galileo had not in fact observed the phases of Venus. Prompted by Castelli, he was laying claim to observations that he believed he would be able to make in the future, and he was claiming priority for himself, when really he should have been acknowledging Castelli as the person who had done the theoretical work.[50] This is a charge of intellectual theft, and it is not the only one made against Galileo. It is not, however, one that Galileo ever had to face himself, as it was not made until the nineteenth century.

Is Galileo guilty? Paolo Palmieri has used mathematical models to reconstruct what he would have seen through his telescope in the autumn of 1610.[51]

This corresponds exactly to the description he gave on 30 December, and in a key respect it differs from any prediction he might have made. (Because Galileo did not have an accurate grasp of the planetary orbits he could not have predicted that for a month Venus would appear stuck at the half-moon position.) It follows that Galileo had indeed made the observations he claimed to have made over the previous three months. Palmieri also clarifies Galileo's delay in announcing his discovery: the key change in Venus's appearance, the beginnings of the appearance of a crescent, which, given that Galileo had already seen Venus as a full disc, disconfirmed the Ptolemaic hypothesis, took place only after 20 December. In his letters Galileo says it is taking place 'now'. Galileo was thus prompted by Castelli's letter to lay claim to his discovery, although he knew there was crucial information still to be collected. He had seen some of the phases of Venus, but he had yet to see anything that was clearly incompatible with the Ptolemaic system. By the time he received Clavius's letter he had all the information he needed and was in a position to go public. Although Galileo had nothing to learn from Castelli, he was, in prematurely laying claim to his discovery on 11 December, gambling that he could successfully predict what future observations would show.

Galileo and Castelli were both sure what they would find, because they were both already convinced that Copernicus was right. What puzzled them was the difficulty they were having in persuading others to agree with them.[52] Castelli thought that the phases of Venus might prove decisive, while Galileo, who was still not sure that the moons of Jupiter would be generally acknowledged, thought that nothing would convince the obstinate.

It was Kepler to whom Galileo had reported, first in code and then in clear, his discovery of the strange appearance of Saturn, Kepler to whom he had sent his cipher announcing his discovery of the phases of Venus, and Kepler whom he had allowed to publish these two new discoveries. But it was to Clavius that he first *revealed* his discovery of the phases of Venus, and it was his correspondence with Clavius which determined the timing of its announcement. Kepler, the fellow Copernican who had immediately recognised the importance of *The Starry Messenger*, was the expert whose testimony Galileo wanted to be able to rely on if anyone else were to claim these discoveries as their own, and it was Kepler whom he trusted to publish on his behalf; but it was Clavius whom he wanted to win over, and it was the opportunity of communicating with Clavius that was the major advantage of moving to Florence. Galileo's move to Florence exposed him to the attentions of the Inquisition; his preoccupation with a Roman audience was to lead to his downfall. His two letters to Clavius in 1610 reflect deliberate choices whose consequences were to become clear only many years later.

It is worth pausing here to note two things. First, Galileo was preoccupied with establishing his priority in making telescopic discoveries; this was why he had rushed *The Starry Messenger* through the press, and why he had announced his discoveries in cipher to Kepler. Science had not previously been a competitive activity in quite this way. Galileo had invented the scientist as the person who made discoveries, and since all you needed in order to make these discoveries was a good telescope, and Galileo was himself distributing good telescopes around the world, he knew very well that he had only a small window in which to claim priority. (Simon Marius in Germany had in fact first observed the moons of Jupiter the day after Galileo, although he did not publish until 1614.)

The model here was the discoverer of new lands, and indeed we find Galileo's Florentine friends calling him 'another Amerigo', in reference to his fellow Florentine Amerigo Vespucci. (We can also compare the naming of the Medicean stars with the naming of new lands after rulers, such as Carolina, named after Charles I.)[53] Priority was not only the means by which to attain immortality; it also gave one the power to shape interpretation. Galileo was in no great hurry to announce the triform shape of Saturn because he had no interpretation to offer; he was impatient to announce the phases of Venus because he had an interpretation ready to hand.

Secondly, we should note a peculiar feature of Galileo's behaviour. He was releasing his discoveries to the world in dribs and drabs, using Kepler to publish his findings as if he were running a scientific journal. In doing so Galileo was stepping away from an alternative model of scientific activity. As soon as *The Starry Messenger* was published he had started talking of an expanded, elegant version to be published in Tuscan with accompanying dedicatory poems – a book of which a prince such as Cosimo and a philosopher such as Galileo could properly be proud. He even hoped to publish twenty-eight engravings of the moon, showing its changing appearance day by day through the lunar month. Cosimo had advanced him the money to fund such a publication, and Galileo still had it in mind in the summer of 1611.[54]

When he accepted Cosimo's offer of a job he announced that he had a number of books that were nearing completion – a statement in which we can recognise the future *Dialogue* and the *Two New Sciences*, the works that were to secure his reputation but that were to be delayed for more than twenty years. He had said that he intended to earn his living by his books. Instead he was now getting caught up in hand to hand combat with his opponents; each new discovery was being rushed out in the hope of putting them at a disadvantage. We can see how, having been accused either of lying and cheating, or of being unable to interpret what he saw with his own eyes, Galileo must have found this irresistible. It is harder for us to understand how eager he was to grasp every excuse not to write the books he had promised. The explanation lies in a trick that Galileo constantly played on himself. We have seen that in September 1610 he wanted to go to Rome to show Clavius the moons of Jupiter, but that he put off going until April 1611. He simply could not make the journey until he knew that it would be successful, and had learnt from his experience in Bologna that it was rash to trust others to see things that he could see. So, too, Galileo could not bring himself to write and publish his great books until he was sure that they would be greeted with applause; he constantly put off the task of writing them. In the spring and through the long summer of 1610, when there was no independent confirmation of the moons of Jupiter, Galileo had learnt what failure tasted like. Better, far better, not to publish than to publish and be mocked. This was exactly what he had told Kepler in 1597. Experience, he must have felt, had proved him right. A different man would have thought the triumphant visit to Rome in the spring of 1611 a just reward; to Galileo it must have felt as though he had barely escaped disaster. Every setback loomed large in his eyes; every success seemed insufficient.

Sunspots

Sunspots can sometimes be seen with the naked eye: in 1590, for example, an English ship's captain, off the coast of West Africa, saw a spot 'about the bignesse of a shilling' on the sun. Johannes Kepler saw a spot in 1607, when looking at the sun through cracks between the shingles in the loft of his house in order to observe a transit of Mercury; at the time he thought it was Mercury, but with hindsight he realised it was a sunspot.[1] A similar spot had been seen in Florence in 1604.[2]

Galileo was probably the first to see sunspots with a telescope. In July 1610, after the publication of *The Starry Messenger* but before he left for Florence, he showed them to a number of people in Venice. In the spring of 1611 he showed them to several people in Rome, and a friend of his wrote to him in the autumn of that year reporting what someone else had seen.[3] In England Thomas Harriot independently observed sunspots in December of 1610, as did Johannes Fabricius and David Fabricius in East Frisia around the spring of 1611. At about the same time Christopher Scheiner observed them in Ingoldstadt – probably as a result of hearing of Galileo's observations.[4] Galileo may have been the first to observe sunspots, but Johannes Fabricius was the first to publish (1611) and Scheiner the second (1612).[5] As for Harriot, whose scientific discoveries were in many ways comparable to those of Galileo, he never published anything.[6] Having been sent a copy of Scheiner's very brief tract, Galileo began observing sunspots again in February 1612 and mentioned them briefly in the *Discourse on Floating Bodies*, which he finished in April. His first serious discussion of the subject was a letter to Marc Welser on 4 May 1612.

This chronology is easy to reconstruct now, but it was not so clear at the time. Fifty years ago Arthur Koestler made a number of astonishing claims in a sustained attack on Galileo – that Galileo had never read Copernicus, for example.[7] Galileo, Koestler insisted, was simply lying when he claimed to have seen sunspots before reading Scheiner's tract. He claimed to have shown the sunspots to others, but he named no names; he claimed witnesses in order to make his account seem authentic, but this was a sleight of hand. Therefore his account could safely be dismissed. It would perhaps be too much to expect Koestler to have read the seventy pages that Favaro had devoted in 1887 to the priority debate over the discovery of sunspots; but one might at least assume that he was familiar with the standard edition of Galileo's works, where Favaro provided a summary of the issue in a couple of paragraphs. There we find, for example, Fulgenzio Micanzio writing to Galileo after reading the *Dialogue*, saying that he can remember Galileo showing him sunspots through his telescope in the summer of 1610 'as clearly as if it were yesterday'.[8]

On closer examination, however, it becomes apparent that Koestler rarely consulted the National Edition. His book is written as if Favaro's labours had never taken place. He begins to be trustworthy only when he discusses the trial, where he depends not on Favaro's edition of the documents but on German and French editions from the 1870s – which were and remain (with the exception of the discovery of a handful of new documents) reliable. One example of how misleading Koestler can be is his claim that Galileo wrote only two letters (and a note introducing a traveller) to Kepler. This is strictly true. But after Galileo left Padua he could not safely correspond with a Protestant: he had entered the world of the Roman Inquisition. So, as we have seen, he communicated with Kepler through the Medicean ambassador to Prague. This was a convenient method of communication, as he was able to use the diplomatic postal service. But more importantly it was a safe method, since no one could accuse him of secret dealings with a Protestant. (In exactly the same way, the Jesuit Christopher Grienberger was to try to use Galileo as an intermediary for communicating with Kepler.)[9] It enabled Kepler to announce Galileo's latest discoveries to the world by publishing Galileo's letters within his own publications, at the same time enabling Galileo to disclaim responsibility.

Between the summer of 1610 and that of 1616 Galileo was carefully reinventing himself, a process which had begun with his departure from Padua in the early autumn of 1610 and which picked up speed after his triumphant visit to Rome in the late spring of 1611. After he left Venice and Padua he never returned, despite repeated invitations.[10] He wrote only once (if we can trust the surviving evidence) to his old friend Paolo Sarpi, who was universally acknowledged to be an enemy of the papacy, and who lived until 1623. The letter is dated 12 February 1611, and it contains a revealing slip: Galileo says that in view of his long-standing friendship with Sarpi he stands ready to spring to *his or my* defence – the 'my' has somehow intruded itself where it has no place.[11] Galileo is not interested in defending Sarpi; he is interested only in defending himself. Before Galileo's visit to Rome, Kepler was the first to hear of his new discoveries. Afterwards Galileo writes to inform him of the discoveries relating to sunspots, but only after he has communicated them to others. And there the relationship, in effect, ends, with Galileo expressing regret that he plans to write in Italian, not Latin, so Kepler will be unable to follow him.

Galileo's decision to write in Italian did not mean that he was preoccupied with entrenching his position in Florence. The revised *Starry Messenger*, which would of necessity have been dedicated to the Medici, no longer interested him. The audience he wanted to reach now was in Rome, as is clear from his strategy for disseminating his discoveries about sunspots. Galileo chose to develop his thinking about sunspots by writing a series of letters to Marc Welser. Welser was bilingual, and had studied in Italy; he was also powerful and wealthy, being a member of the ruling council of the Catholic city of Augsburg and a banker to the emperor. Galileo had been specifically warned against him by a friend, who saw Welser, perfectly accurately, as pro-Spanish and anti-Venetian.[12] Although Welser had made contact with Galileo soon after the publication of *The Starry Messenger*, he had waited until Clavius approved Galileo's new discoveries before committing himself to them.[13] But he had already contacted Galileo again, in January 1611, conveying doubts about the mountains on the moon. Galileo did not shy away from this connection. On the contrary, he embraced it. And, through Welser, Galileo soon found himself embarked on a debate with Scheiner.

In the course of 1611, Scheiner tried to make the best sense of sunspots that he could. He took it as axiomatic that the spots could not be, as they appeared to be, on the surface of the sun. If they were, this would mean that the sun was imperfect and changeable, which was contrary to a fundamental doctrine of Aristotelian philosophy: that all change was confined to the sublunary world. With the model of Galileo's discovery of the Medicean stars (or moons of Jupiter) in mind, Scheiner argued that the appearance of spots on the sun could be explained if there were numerous objects orbiting the sun at a low level. These objects could travel in perfectly regular orbits, but at times they would seem to clump together into large spots, while at other times, being small, they might be entirely invisible.[14] Galileo had not been particularly interested in sunspots before reading Scheiner's brief tract (perhaps because he knew they were sometimes visible to the naked eye, but probably also because he found them anomalous and did not know what to make of them), and in his first printed reference to them, in his *Discourse on Floating Bodies*, which was ready for the press at the beginning of April 1612, he said that it was a moot point whether the spots were on or above the sun. In this reference he made no mention of Scheiner, which was ill mannered.

In his three letters to Welser (4 May, 14 August and 1 December 1612, printed in Rome by Prince Cesi on behalf of the Lincean Academy) Galileo made three important advances. First, all recent observations of sunspots had involved looking straight at the sun through a telescope. At dawn or dusk, or when there was a light haze, one could stare at the sun for short periods of time, particularly if using a coloured-glass filter, but one still ended up temporarily blinded, and it was hard to record what one had seen. Galileo's friend Castelli, however, had discovered that one could project the image of the sun onto a piece of paper held behind the telescope's lens. The resulting image had top and bottom reversed, but it was large and clear. You could now study sunspots under conditions of comfort. Moreover you could record exactly what you had seen. If you drew a circle on the paper onto which the image of the sun was projected, you could adjust the distance of the paper from the lens until the image of the sun exactly filled the circle. You could then mark the sunspots on the page simply by painting over the spots as they appeared on the page. The process was a little bit trickier than it sounds, as the sun moves fairly rapidly through the sky, so you have to re-aim the telescope periodically. But the result was that Galileo could now make beautifully exact records of his sunspot observations. Scheiner had seen that the spots moved across the sun's face; Galileo could now follow in detail changes in their shape and calculate exactly their rate of movement. He first announced this breakthrough to Cardinal Maffeo Barberini in Rome – who had supported him in his debate at the Medici court over things that float in water – and to Kepler. To Barberini he sent a series of beautifully exact recordings of the sun's appearance day by day from 3 to 11 May 1612. A similar series of thirty-seven illustrations, for June, July and August, accompanied his second letter to Welser.

Galileo's second breakthrough evidently occurred as a result of studying these images, because on 12 May he wrote to another correspondent in Rome, Federico Cesi, saying that he could rigorously demonstrate his opinion that the spots were on and not above the surface of the sun. The argument is breath-taking as one follows it on the page. Scheiner had argued (and Galileo had at first agreed) that one could not distinguish visually between spots on the surface of the sun and satellites just above the surface. But looking at his series of images from 3 to 11 May Galileo could see three crucial facts about the spots: as they advanced from or drew near to the margin of the sun's circular image the spots became narrower; they became closer together; and they moved more slowly. These were all the effects of foreshortening, as the spots were being seen edge-on rather than from directly overhead. Given that it was easy to show that the sunspots were essentially two-dimensional phenomena, it then became a simple project to measure the effects of foreshortening and compare the measurements obtained with what one would see if the spots were in fact significantly above the surface. Galileo had no difficulty in showing that if the spots were orbiting around the centre of the sun on a trajectory whose radius was 5 per cent greater than the sun's radius the appearances would be completely different from what they actually were: the foreshortening would occur much closer to the sun's edge and would be much more rapid in its effects. Galileo was thus using basic geometry to demonstrate (the word is his, and it is entirely appropriate) that the spots must be extremely close to the sun's surface, if not actually on it. They might, for example, be hovering above the sun's surface in the way that clouds hover above the earth's surface; but they could not possibly be orbiting the sun at a significant distance from the surface. We can now understand the title under which Galileo's letters to Welser were printed: *Istorie e dimostrazioni intorno alle macchie solari.* A history is a factual record: in this case this consisted primarily of the reproductions of solar images. A demonstration is a logical deduction: in this case this consisted of the proof that the sunspots could not be orbiting objects.

In the opening pages of the *Discourse on Floating Bodies* (published in May 1612) Galileo also announced a technical advance that was to lead to his third breakthrough. In *The Starry Messenger* he had provided fairly crude images showing the relationship between Jupiter and its moons day by day. The images were based on simple estimates: Galileo looked through his telescope and made a judgement as to how far each moon was from Jupiter, using as his unit of distance Jupiter's own diameter. In January of 1612 he had begun to employ a new method. He had attached a grid of lines to a panel which stuck out from the side of his telescope – or so we believe, because the object itself does not survive. With a little adjustment he could arrange this grid so that, if he opened both eyes when one was applied to the telescope, it appeared to float over the image of Jupiter, and its centre exactly coincided with Jupiter itself. He had constructed a crude micrometer, and could dismiss the rough estimates of others as mere hallucinations.[15] In modern astronomic telescopes micrometers are mounted within the telescope itself. This is possible in the type of telescope invented by Kepler (which provides a reversed image of the heavens), but not in a Galilean telescope, so Galileo had done the best he could. He now simply had to count the lines on the grid between Jupiter and any one of its satellites to obtain an exact measurement of the distance, which he could then express in terms of diameters of Jupiter.

Galileo could now solve a problem that had been bothering him for the past two years: he could accurately calculate the periods of Jupiter's satellites. And having done this, he could predict their location in the future. He had already discovered that the relationship between the earth and Jupiter was constantly changing (which would not be the case if, as the Ptolemaics claimed, Jupiter orbited the earth), and that his calculations had to allow for this. In the summer of 1612 he made a further discovery. The moons disappeared from view when, relative to an observer on the earth, they were 'behind' Jupiter. But they also disappeared from view when Jupiter lay between them and the sun – they had been eclipsed by Jupiter. When they entered Jupiter's shadow no light from the sun reached them, and so no light could be reflected from them into space. They disappeared as if a light had been switched off.

In March 1613, as his letters on sunspots went through the press, Galileo supplied tables predicting the positions of the moons of Jupiter over the course of the next few months. He thus put the accuracy of his measurements and calculations to the test. Publication was unfortunately somewhat delayed, so by the time the book was in the bookshops half of the predictions had been overtaken by the passage of time, but it was still possible to confirm that Galileo's work was almost entirely reliable.

The discovery of the eclipses of the moons of Jupiter set Galileo thinking. An eclipse occurred because of the relationship between Jupiter, the moon and the sun – the location of the earth, or of an observer on the earth, was irrelevant. It would thus occur at exactly the same time for all possible observers on the earth. What Galileo had discovered was a universal clock. Consequently he had also discovered a way of solving the greatest problem facing sailors on long voyages, that of calculating longitude.

It is easy to calculate latitude. If the pole star is immediately overhead then you are at the north pole. If the pole star is at an angle of 30° from the vertical, then you are at 60° north. Longitude, though, is much harder to measure, since all the points on the same line of latitude see exactly the same stars in exactly the same position – they just see them at different times. Thus if you are 45° west of Florence you will see a star rise three hours later than someone in Florence. In order to measure how far west of Florence you are, you need to be able to compare the time where you are with the time in Florence. In principle the easiest way to do this is to carry a clock from Florence to where you are (Rio de Janeiro, for example). In practice a reliable clock that would tell accurate time during a sea journey was incredibly difficult to make (Harrison first solved the problem in 1761). Galileo proposed an alternative. If you saw an eclipse of one of Jupiter's moons, and if you had a table which told you when such eclipses would occur in Florence, then all you had to do was compare local time (easy to establish if you set your timepiece to solar noon every day) with Florentine time, and then you could immediately calculate your longitude.

Galileo's method was certainly feasible for anyone on dry land, on a clear day, with a good telescope. Whether it was feasible for someone at sea, on a boat moving to the waves and the wind, with clouds overhead, was a quite different question, and it is worth remembering that Galileo had only ever seen the Mediterranean, which is often comparatively calm; he had never seen the open ocean. He argued that people learn to do all sorts of things that at first seem incredibly difficult – to read and write, for example – and that with a bit of training this new task could also come to seem straightforward. Castelli was sent out on a galley in the Mediterranean to prove that it could be done – and became horribly seasick. But a real test would require an ocean-going boat, and Galileo made repeated but always unsuccessful attempts to persuade the Spanish government to engage in proper trials. Later he turned to the Dutch government, who showed significant interest, although their attempts to understand his proposal were handicapped by the fact that in the whole of the Netherlands not a single telescope could be found capable of resolving the moons of Jupiter – proof, if proof were needed, that there was nothing inevitable about Galileo's improvements upon the original Dutch telescopes.[16] Galileo continued to work on ways of overcoming all obstacles to his proposal up to the time of his death – he devised a way of clamping a telescope to your upper body, so that it remained in a fixed relationship to the eye, and of placing the observer on a seat floating in a large tub of water (a sort of vast binnacle) so that he remained still even as the ship pitched and swayed. He remained firmly convinced that his proposal was practical, although he never had much success in persuading others that he was right.[17] In his view there was nothing impractical about turning ships into astronomical observatories, and ships' captains into astronomers. In pursuit of his theoretical technology he even devised a clock that would (in principle) tell perfect time. So caught up was he in the idea of the observatory afloat on the waves that he failed to realise that if he could make a clock that would tell perfect time then with that alone he could establish longitude. He was so obsessed with solving each and every local problem that he had lost sight of the overall task.

This imaginary technology is related to a 'fantasy' that Galileo describes in the *Dialogue*.[18] Imagine you are in the cabin of a ship, writing or drawing on a piece of paper. The tip of your quill pen traces a complex pattern of lines on the paper, but if we zoom out and view what is going on from space, then the ship, the cabin, you and the tip of your pen are tracing a perfect arc across the surface of the globe. And since the ship is a metaphor for a moving earth, as Galileo is writing his pen is tracing cycles and epicycles in space as the earth turns and orbits the sun.

What happens when Galileo turns his own writing into a perfect arc in the cosmos? Galileo himself – with his fine wines, his lovingly tended garden, his memories of his distressed brother and of his rabid mother – Galileo himself simply disappears. He came from a family which specialised in disappearing acts. As we shall see, his brother went to Munich in order to disappear. His nephew went to Poland and did disappear. His grandson, Galileo junior, went to the wars and never returned. Galileo stayed close to home, but he had his own disappearing act. He picked up his pen or his telescope, and disappeared into deep space, a traveller amongst the stars.

Galileo complained that he was dependent on others to bring to fruition his method of measuring longitude, 'for in my study there are no seas, no Indies, no islands, no ports, no sandbanks, no ships'.[19] Of course not. But in his study there were new and previously unimagined worlds, so why not an ocean or two? Galileo seems to have been more than a little surprised to realise that there were limits to his mental voyaging.

The Catholic scientist

From the summer of 1611 to the winter of 1612–13 Galileo was pursuing a strategy. He compares his opponents to soldiers defending a fortress who rush to repulse an attack, leaving the rest of the castle undefended – and we must assume that he was every bit as careful in planning his attacks as he thought his adversaries were careless in their responses.[1] Historians have lost sight of that strategy because it is directly at odds with the strategy he was to pursue from 1613 to the end of his life. Galileo is famous for arguing that religion and science must be kept separate, but in 1611–12 his goal was to bring religion and science into a new alliance. In later years he had to insist that he had always been consistent, and consequently he never acknowledged the radical change in his approach to the intellectual conflict in which he was ceaselessly engaged. His silence, however, should not mislead us.

In order to understand Galileo's strategy we have to grasp that at this point his only adversaries were the Aristotelian natural philosophers. As far as Galileo was concerned the followers of Aristotle were slaves, pathetic creatures who refused to think for themselves. True philosophy required intellectual liberty. Instead of studying nature directly, Aristotelians looked only at the paper world constructed by Aristotle, assuming that it was a faithful representation of the real world, when it was no such thing.[2] This was the view not just of Galileo, but also of Federico Cesi and of the Lincean Academy – an academy, not for slaves of Aristotle, but for free minds.

Every educated person knew that Aristotle was not easy to reconcile with Christianity. According to most reputable authorities, Aristotle regarded the universe as having always existed (in other words he denied creation); he denied the immortality of the soul; and he left no conceptual space for miracles.[3] Galileo would have been well aware of the deep suspicion with which the Church regarded the teaching of strict Aristotelian philosophers such as Cremonini, philosophers who repudiated the Thomist synthesis of Greek philosophy and Christian theology, and sought to eliminate all traces of Christianity from their exposition of Aristotle.

It was thus natural for any opponent of Aristotelian philosophy to wonder if an alternative philosophy might be easier to reconcile with Christianity.[4] Galileo not only thought about this, but also took advice, and the advice was that he should emphasise the creation of the world, and the mutability of the heavens.[5] Both were denied by Aristotle and asserted by the Scriptures. (The Copernican doctrine of a moving earth, on the other hand, was, he was warned, scarcely compatible with the Bible.)

Thus Galileo was not only interested in forming a close alliance with the Jesuits, in building ties with cardinals such as Maffeo Barberini and with prominent pious laymen such as Welser, and in publishing in Rome. He was also committed to claiming that the new astronomy was more Christian than the old astronomy. Federico Cesi produced a substantial theological argument to this effect, one that he hoped would be published alongside Galileo's letters on sunspots. And the letters on sunspots contained an important passage where Galileo insisted that the sunspots demonstrated, against Aristotle, that the heavens were mutable, and that this corresponded precisely to the teaching of Holy Scripture.

When the text was submitted to the censors, however, this was one of three passages that were struck out. (In another Galileo said that he was grateful to God for having made him a human being rather than a worm – forgetting that, as a creature with an immortal soul, he could never have been a worm; he was allowed to say that he was grateful to God for having made him a human being rather than leaving him in non-existence.[6] The third passage, somewhat ironically, was a remark made by the pious Welser.)[7] The censors adamantly refused to allow any theological discussion in the book on sunspots, and insisted that the question whether the Bible was compatible with Aristotelian teaching on the immutability of the heavens was far from straightforward, and was in any case a matter for theologians rather than mathematicians. Galileo tried repeatedly to retain his core argument while revising his text (the printer had to skip over the offending pages so that the whole project was not held up while efforts went on to find an acceptable form of words), but no compromise could be reached.[8] This was, in a sense, lucky for Galileo; it meant that when he later claimed that he was interested only in discussing science, not religion, there was no obvious evidence to contradict him. And the result is that the logic of his activities in these years has become invisible to us.

As Galileo began work on his sunspot letters, in May 1612, he was full of confidence. His proof that the sunspots were on the surface of the sun, and that consequently the Aristotelian doctrine of the immutability of the heavens was simply false, would, he believed, prove decisive. This would be the funeral, as he put it – he meant the *coup de grâce* – for the 'pseudophilosophy' of his adversaries. By the time he finished his third letter, on 1 December, he felt the battle was over. The appearance and then the totally unexpected disappearance of the bodies accompanying Saturn (the rings become invisible if the viewer is in the same plane as they are) served to confirm not only that the heavens were constantly changing, but that Copernicus was right. Galileo had the wind in his sails. He had reinforcements arriving to join the conflict. Victory was now at hand, and there was little left to fear.[9] In the postscript he referred explicitly to the annual rotation of the earth around the sun – which one had to take into account in calculating where Jupiter's moons would appear to be located when seen from the earth. Of course we now know that Galileo had totally misjudged the situation. The new troops arriving to join battle were those not of his allies but of his enemies. This was the high point of his fortunes.

By December of 1612 Galileo had virtually completed his career as a prac- tising scientist. Every important discovery that he was going to make he had now made. And one great discovery had slipped through his grasp. He had noticed one night that one small star seemed slightly further from its nearest neighbour than it had been the night before. What he had seen was a new planet, Neptune, a planet which was not to be properly discovered until 1846.[10] But the idea of a new planet had not occurred to Galileo, so he did not linger over this minute anomaly. He continued to work on the tables of Jupiter's moons, but he stopped exploring the heavens. In the *Discourse on Floating Bodies* he had published a major illustration of the experimental method. He went on experimenting, now and again, but all his important experiments were in the past. He was about to turn forty-nine, and he already had a sense of himself as being old, and of time as being short. What he needed to do now was publish the work he already had in hand. This was not to prove straightforward.

PART THREE
THE EAGLE AND THE ARROW

An eagle was soaring through the air when suddenly it heard the whiz of an arrow, and felt itself wounded to death. Slowly it fluttered down to the earth, with its life-blood pouring out of it. Looking down upon the arrow with which it had been pierced, it found that the haft of the arrow had been feathered with one of its own plumes. 'Alas!' it cried, as it died, 'We often give our enemies the means for our own destruction.'

Aesop's fables

22

Copernicus condemned

The first indication that Galileo's position might be deteriorating rather than improving came on the morning of 12 December 1613.[1] Benedetto Castelli had breakfast with the grand duke, the duchess, and the grand duke's mother, Christina of Lorraine, who were on one of their periodic visits to Pisa. Also present, amongst others, was a Pisan Aristotelian philosopher, Cosmo Boscaglia. Castelli, as one might expect, had spoken in praise of Galileo's latest scientific discoveries; Boscaglia had conceded that Galileo's telescopic discoveries were indeed to be trusted, but he had rejected the idea of a moving earth, which he said was contrary to Scripture. Castelli finished his meal and left, but just as he emerged from the palace a porter came chasing after him to summon him back to Christina's room. There he found the group from breakfast had gathered, and Christina set about proving that Copernicanism was contrary to the Bible. Castelli in reply defended Copernicanism, and had the support of everyone present, apart from Christina and Boscaglia. Galileo, having received a detailed account of the discussion from two sources, wrote a substantial letter back to Castelli on the twenty-first, and copies of this letter were soon in circulation.

There was nothing particularly alarming about the debate in the ducal palace on 12 December 1613. But a year after Galileo had written to Castelli, on 21 December 1614, a Dominican preacher in Florence, Tommaso Caccini, whose text for the day was the passage in the Old Testament about the sun standing still during Joshua's battle against the Amorites (Joshua 10.12–14), publicly attacked Copernicanism. Punning on Galileo's name, he quoted Acts 1.11: 'Men of Galilee, why do you stand looking up into the sky?'[2] A number of weeks later, on 7 February 1615, another Dominican, Niccolò Lorini, sent a copy of Galileo's letter to Castelli to the Inquisition in Rome, which promptly opened a formal enquiry against Galileo.[3] (Note that Lorini complained to the head office of the Inquisition in Rome, not, as one might have expected, to the inquisitor general in Florence.) Galileo evidently got word of Lorini's intentions: on 16 February he sent a letter in his own defence to Piero Dini, a sympathetic Roman clergyman, suggesting he might pass it on to the leading Jesuit mathematician, Christopher Grienberger, and to the leading theologian on the Inquisition (and a Jesuit), Cardinal Bellarmine, which Dini promptly did – Galileo was clearly hoping that the Jesuits would take his side against his Dominican opponents.[4] With it he included his own version of the letter to Castelli.

The letter to Castelli thus survives in two different forms: the wording of the text sent to the Holy Office, or Inquisition, is less cautious than that of all but one of the other surviving copies. Thus the Lorini version says that the Bible 'perverts' key doctrines of the Christian faith, while Galileo's own version says that the Bible 'obscures' key doctrines.[5] One possible explanation is that Lorini falsified Galileo's text. But the evidence points in a different direction. As soon as it received Lorini's version of Galileo's letter the Inquisition set about trying to obtain the original from Castelli, using the archbishop of Pisa as an intermediary so that Castelli would not know what was going on (although, given the air of solemnity with which the archbishop warned Castelli against Copernicanism, Castelli would have been slow indeed if he did not see through the pretence). Castelli told the archbishop that he had returned the letter to Galileo, but that he would ask him for a copy. Castelli promptly wrote to Galileo, urging him to take this opportunity to 'make the final revisions to the text'.[6] In fact Galileo already suspected that he had been denounced to the Inquisition in Rome (Lorini was announcing to all and sundry that he had a copy of Galileo's letter to Castelli and that it was full of heresies, and rumours were clearly flying about), and so on 16 February he had sent Dini a copy of the letter, drawing attention to the possibility that this text might differ from that being circulated by his opponents – perhaps as a result of some 'inadvertent' error by a copyist.[7] This preoccupation on the part of both Castelli and Galileo with establishing the official version of the text suggests that it was Galileo himself who revised his text in order to make it rather less offensive.[8] This proved to be unnecessary: the Inquisition found that, even in the version supplied by Lorini, though the wording was sometimes unfortunate, the arguments were unexceptionable.[9]

Galileo's fundamental premise in the letter to Castelli was straightforward. The Bible is the word of God, but it is adapted to human capacities. So there are plenty of statements in the Bible that are not literally true (that God has hands, or that he gets angry, for example), but that are God's way of communicating not with philosophers but with ordinary people. Nature is, like the Bible, a book in which we can trace God's doings. But nature is not adapted or modified in order to be understood by us. So, when it comes to scientific questions, our direct knowledge of nature must always take priority over whatever the Bible may have to say on the subject because there are always grounds for uncertainty as to quite how the Bible is to be interpreted. Galileo had earlier used a 'two books' argument to claim that the book of nature must take priority over the text of Aristotle.[10] Now, with only a minor modification, he was using it to argue that the book of nature must take priority over the Bible. Having laid this foundation, he went on to argue (rather at odds with the main thrust of his argument) that the text of Joshua made little sense from a Ptolemaic point of view (according to

Ptolemaic astronomy the sun does not move independently through the sky, but moves because it is attached to the sphere of the fixed stars), and could as easily be interpreted in Copernican terms.

There is, arguably, a fundamental problem with the 'two books' argument, one which escaped the sharp eye of the anonymous Roman theologian who drew up the report on the letter to Castelli (but then his eye was being directed by the underlinings with which Lorini had carefully marked out those passages in Galileo's text which he thought most shocking). Galileo's central claim is that the Scriptures are adapted to our understanding, while nature is not. Nature, he says, is 'inexorable and immutable'.[11] But of course it is a fundamental teaching of Christianity that nature is occasionally adapted in order to communicate with us – this is what a miracle is. It is very hard to see how miracles can be made compatible with Galileo's text. This may not have worried either Lorini or the theologian reading Galileo's text, for the simple reason that Aristotelian philosophers were equally incapable of leaving space for miracles.

In the letter to Dini which accompanied the revised text of the letter to Castelli, Galileo defended Copernicanism: there had never been any previous suggestion that Copernicus's teachings were heretical, and Copernicus was a pious Catholic. But he went further. In a move which was to characterise his arguments over the course of the subsequent year (and again in 1633), he insisted that he himself was motivated by zeal for the true religion, while his opponents were relying on a simulated piety.[12] He was someone who would tear his own eyes out rather than endanger his soul – and who would unquestioningly accept his superiors' judgement in all matters relating to salvation.[13] Over and over again Galileo insisted that he was motivated by religious devotion – though there is no evidence whatsoever of religious devotion motivating his words or his deeds at any previous point in his life. Looking back after Copernicanism had been condemned he would claim that 'a saint could not have shown more reverence or more zeal towards the Church'.[14] One would have to be naïve not to suspect that this was a convenient rhetoric to adopt when facing accusations of heresy.

Matters might have ended there: Galileo had been denounced; he had been cleared; and in addition he had protested his personal piety and willingness to obey. But the Inquisition was now interested in two questions. First, what else could be discovered about Galileo, for Lorini had claimed that Galileo and his followers (whom he called 'galileisti') 'speak disrespectfully of the ancient Holy Fathers and St Thomas . . . and spread a thousand impertinences around'?[15] And second, was Copernicanism now to be regarded as heretical?

We last heard of Tommaso Caccini preaching against Galileo in Florence. In March 1615 he appeared, conveniently, in Rome, and on the advice of a Dominican cardinal presented himself to give evidence against Galileo – one is bound to suspect that he had been summoned to provide reinforcements when Lorini's original accusation had failed to stick.[16] In his evidence he first provided a summary of his December sermon. He then revealed that he had complained to the inquisitor general of Florence, implying that no action had been taken. Clearly Lorini had known this when he addressed himself to the Inquisition in Rome. Caccini then got down to business. He said that Galileo's followers had been heard to make three claims: God is not a substance but an accident; God is a sensible entity because he has senses; and the miracles attributed to the saints are not real miracles. He pointed out that Galileo was a friend of Paolo Sarpi's and that he had defended Copernicanism in print. And he cited two possible witnesses, who would be able to confirm what the 'galileisti', as they called themselves, were saying: yet another Dominican, Ferdinando Ximenes, and a member of the Attavanti family, a follower of Galileo's.[17] But he offered no new direct evidence against Galileo himself, and word began to spread in Rome that Galileo had been cleared.[18]

The Florentine inquisitor, however, was promptly instructed to call in the witnesses cited by Caccini and question them. But Ximenes had gone first to Milan and then to Bologna, and by the time he returned to Florence a new inquisitor general, who had not taken the trouble to read back through the files (or so at least he claimed), was in office, so that it was not until November that Ximenes was finally brought in for questioning.[19] He added a new accusation against the 'galileisti', that of atomism and belief in a vacuum, but he denied having heard any of the incriminating statements from Galileo himself. He had never heard the miracles of the saints disputed, and the other claims (that God is an accident and that he is material) had been put forward as part of a scholastic disputation to test the arguments of St Thomas.[20]

Ximenes had identified the follower of Galileo's referred to by Caccini as Giannozzo Attavanti and so he was summoned the next day. Attavanti insisted that he had never heard Galileo say anything heretical, and that the conversations of his students had been entirely innocent, though perhaps open to misinterpretation. He was asked what he thought of Galileo's religion: 'I regard him as a very good Catholic; otherwise he would not be so close to these Most Serene Princes [the Medici family].'[21] These words are carefully underlined in the record of his deposition.

Thus this whole line of enquiry ran into the sand. Indeed in the eyes of the Florentine inquisitor it was convenient that it should do so. The archbishop of Pisa had vouched for Castelli, insisting that there was nothing suspicious in his claim that he no longer possessed Galileo's letter; the enquiry had been delayed; and everyone had been gently reminded that Galileo was under the protection of the Medici family. In Rome they could see that there was no point in banging their heads against this particular brick wall.[22] But we need not be as quick as they were to let go of these questions, for the conversations of Galileo's disciples are of interest to us. The charge of atomism and belief in a vacuum was certainly well founded – both Galileo and his closest disciples were atomists.[23] Soon after his return to Florence in 1610, for example, Galileo converted the young Giovanni Ciampoli to atomism – a doctrine Ciampoli remained firmly attached to for the rest of his life. And Galileo's own account of his position seemed to leave no room for belief in miracles, which suggests that rumours that his disciples questioned their reality may have been accurate.

Moreover, we have advantages that the Inquisition did not have. We know much more about Galileo than they did. It really is a striking fact that in the hundreds of letters written by Galileo that survive there are absolutely no spontaneous expressions of piety – such expressions occur only when he has himself been charged with impiety. And there are no attempts to invoke the ordinary, almost routine, miracles that formed a part of Counter Reformation spirituality. It would be easy – too easy – to contrast Galileo with his daughter Virginia, who had taken the name in religion of Maria Celeste. She prays for Galileo; she supplies him with water blessed by a living saint in the hope of restoring his health; she sells a couple of chickens to pay for a Mass to be said for him.[24] Let us rather compare him with his neighbours, who believed that a statue of the Virgin paraded through the streets could save Florence from the plague. Or with his close friend and pupil Benedetto Castelli. Castelli was a scientist – indeed both a Copernican and an atomist – and a monk. When he is seasick and fears drowning, he asks that his abbot should be told, so that he can pray for him.[25] When he is ill, he prays to St Philip Neri, and the saint cures him.[26] In church he sometimes hears sermons that feed his soul.[27] Castelli is proof that Galileo could have been a pious Christian and still been responsible for a revolution in science.

But Galileo is not Castelli. Galileo never praises preachers, prays to saints or purchases indulgences, and only rarely does he ask that others pray for him (never, to the best of my knowledge, before he loses his sight). Now and again there is a glimpse of him going to church, even listening to a sermon – but his thoughts are elsewhere. All this, you may say, is argument from silence. I agree. It would be wrong to base any conclusion on a profoundly cynical letter sent to Galileo describing a girl who had the stigmata.[28] Equally, no conclusions can be drawn from the fact that Galileo once visited Loreto, where there was a famous shrine: he was following in the footsteps of Cosimo II, who had made the pilgrimage fashionable.[29] But in the end Galileo's silence becomes so remarkable that we simply cannot ignore it. For 'Galileo never spoke of Jesus. There is simply no direct testimony to what he thought of Christ.'[30]

Direct evidence once existed, but unfortunately it has disappeared. For Galileo did write a treatise on miracles, in 1627 or shortly before that time: a treatise providing a naturalistic explanation for the miracles recorded in the Old Testament. We have two references to it: one from Giovanni Battista Baliani, who had not read it but wanted to, particularly if it dealt with the book of Genesis; and one from Fabio Colonna, who wrote to Prince Cesi, urging him to warn Galileo that this was no time to be providing a scientist's views on

Shadrach, Meshach, Abednego and the fiery furnace.[31] It seems clear that this was a treatise which turned supernatural events into natural events – and if so, it suggests that Galileo was profoundly reluctant to acknowledge the possibility of miracles. It sounds, to make a wild guess, like Bishop Burnet's Cartesian account of the flood, which seeks a naturalistic explanation of where all the water came from. We are reduced to guessing, because no copy of Galileo's treatise is known to have survived. There are copies of everything else that Galileo wrote after he moved to Florence, but not of this. Perhaps a copy of the text does survive, but has been unnoticed or unrecognised; no one seems to have remarked that this crucial document is missing, so it is too soon to say that it will never be found. Or perhaps the last copy went into the fire before Galileo set out for Rome early in 1633. Later, his friends went to his house to move his papers out of harm's way, but it seems likely that this was in order to protect his working papers rather than to remove anything clearly incriminating.

The inquistors were, I feel sure, on the right track when they inquired about the miracles of saints. For Galileo seems to have been unwilling to concede the idea of a God who might intervene in historical time. He was prepared to think of a divine architect who had constructed the universe and set it going according to inexorable laws, but he was profoundly unwilling to think of a God who intervened in his own creation. There is a telling marginal comment that Galileo wrote in 1619 on a letter from Baliani. Baliani had suggested that the explanation for comets might be that they were newly created by God, and set in motion by him. Galileo's response is worded with exquisite care: 'I have absolutely no objection to these statements; and if anything I have said was at odds with them, then it would be necessary to hold what I said to be not only false but heretical. I say, however, that not only can all these things be said, but this is the easiest, simplest and quickest method of resolving these and indeed any other problems, no matter how difficult.'[32] Galileo does two things simultaneously here: he acknowledges that it would be heretical to deny that God intervenes in the world, and he implicitly rejects any argument from divine intervention as being altogether too easy, simple and quick – as being non-scientific.

There is another example of this way of thinking in Galileo's marginalia. In 1612 Julius Caesar La Galla had argued that if it were left to nature water would cover the entire surface of the globe; but God, by his peculiar providence, had arranged for there to be dry land. Galileo's comment was this: 'If water does not cover the entire surface of the globe because of the providence of the Divine Architect, then you should not say that it would be more in accord with nature if water covered the entire surface of the globe; for this providence has imposed much better laws on the parts of the universe than that nature. Assuming, that is, that it is appropriate to distinguish providence from nature in this way.' Galileo seems to be saying two things here. First, that providence only operates through general laws: there is no such thing as the *particular* providence of which La

Galla writes. This is certainly heretical, so we should not be surprised if it is implicit rather than explicit in Galileo's comment. Second, if providence only operates through general laws, then there is no meaningful distinction between providence and nature – between divine decree and natural regularity.[33] In other words the idea of 'God' is superfluous.

Galileo's enemies were right to suspect not only his views on miracles. We have seen that he was indeed interested in the Old Testament account of God as having limbs and feelings.[34] And so we may even suspect that there is something to the strange claim that God is not a substance but an accident. The strangeness of the claim does not lie in the suggestion that God might have accidents: having two hands is an accidental feature of a human being (if one is amputated you remain a human being), while having a rational soul is an essential or substantial feature, so discussion of divine accidents followed inevitably from a discussion of whether God has limbs or feelings. The strangeness of the claim lies rather in the suggestion that God might be *only* an accident. From an Aristotelian point of view this makes no sense, because every accident is also, from another point of view, a substance: a beard is an accident on the face of a man, but constitutes the very essence of facial hair. For an atomist, however, there are no such things as substances or essences, because all beings are simply arrangements of matter in space. From this it follows, of course, that God too, if he exists, must be a material entity (a view that Hobbes, for example, seems to have been prepared to adopt). It so happens that Paolo Sarpi held the view that materialism rescues one from belief in substances or essences (substances are deeply problematic, because they seem to have a sort of ghostly existence), and we can therefore assume he had discussed it with Galileo (indeed, he may well have acquired it from Galileo).[35] When that great and irrepressible rebel Tommaso Campanella met Sarpi and Galileo in 1592 he came away with a clear understanding that they were both atomists.[36] The peculiar claim that God is only an accident is simply a restatement of Sarpi's materialism in the language of Aristotelian scholasticism.

So at every point the charges made by Caccini and Ximenes against Galileo's disciples have a distinct air of plausibility. The fact that the case was dropped for lack of evidence does not mean that the charges were mistaken (and we will find very similar charges being repeated towards the end of Galileo's life, so that we can postpone for the moment any decision regarding his religion or lack of it). But the Roman Inquisition, unable to proceed further, simply returned to Caccini's claim that Copernicanism should be regarded as a heresy. According to Caccini, Galileo had avowed his Copernicanism in his *Letters on Sunspots*, so the Inquisition ordered that a copy of the text be supplied to a panel of theologians. On 24 February 1616, these theologians proceeded to condemn two propositions: that the sun does not move, which they held to be 'foolish and absurd in philosophy, and formally heretical since it explicitly contradicts in many places the sense of Holy Scripture, according to the literal meaning of the words and according to the common interpretation and understanding of the Holy Fathers and the doctors of theology', and that the earth moves (around the sun) and rotates once a day, which was held to be 'at least erroneous in faith'.[37]

This condemnation was, it is fair to say, inevitable. In January of 1615, Bellarmine had told Prince Cesi that he thought Copernicanism was heretical, and Bellarmine's views were surely known to the panel of theologians who condemned Copernicanism a year later.[38] They were certainly known to Galileo, who had heard about them not only from Cesi, but also from Dini and from another ally, Ciampoli.[39] Galileo had also received a copy of a letter that Bellarmine wrote on 12 April 1615 to a Carmelite friar, Paolo Foscarini.[40] Foscarini had published in Italian a pamphlet arguing that Copernicanism was compatible with the text of the Bible. The pamphlet was important: Foscarini was the first Italian to support Galileo in print, and his argument addressed directly the now central question whether Copernicanism was compatible with the Scriptures. He had then written in Latin a brief defence of his pamphlet, defending it against the charge that it was 'contrary to the common explication of the Holy Fathers'.[41] Both these texts he had sent to Cardinal Bellarmine, appealing for his support. Bellarmine was, however, unconvinced: what Foscarini was asking, he said, was that the Church should 'tolerate giving Scripture a meaning contrary to the Holy Fathers and to all the Greek and Latin commentators', whereas the Council of Trent had specifically forbidden 'interpreting Scripture against the common consensus of the Holy Fathers'. On this there was no room for prevarication. One could not say, for example, that 'this is not a matter of faith, since if it is not a matter of faith "as regards the topic", it is a matter of faith "as regards the speaker"; and so it would be heretical to say that Abraham did not have two children and Jacob twelve, as well as to say that Christ was not born of a virgin, because both are said by the Holy Spirit through the prophets and the apostles'.[42] (We know that Galileo had a reply to this argument: it is commonplace for people to have two children or twelve, so there is no reason for us to doubt the accuracy of the Bible on such matters; it is not commonplace, however, for children to be born of virgins.)[43]

Yet Bellarmine was not advocating a ban on the discussion of Copernicanism. He was perfectly happy for mathematicians to argue that Copernicanism provided a mathematical model that was superior to any other for calculating the positions of the heavenly bodies – one could do this without claiming that it was true. And he was prepared to concede that in principle it might be possible to demonstrate that Copernicanism was true: 'I say that if there were a true demonstration that the sun is at the centre of the world and . . . the earth circles the sun, then one would have to proceed with great care in explaining the Scriptures that appear contrary, and say rather that we do not understand them than that what is demonstrated is false. But I will not believe that there is such a demonstration, until it is shown me.'[44]

It is important to understand that Bellarmine was arguing as a theologian, not as a scientist (although he had, it seems, consulted with the leading Jesuit scientist, Christopher Grienberger).[45] As a theologian, he knew one could trust the Bible. Indeed in lectures he had given as a young man he had placed more trust in the Bible than in the Aristotelian philosophers: he had rejected the crystalline spheres of Ptolemaic astronomy, and accepted the possibility of change in the heavens, because this was what the Bible appeared to say.[46] We can be reasonably confident that, like the Aristotelian philosopher at the grand duke's breakfast table, he was prepared to accept all Galileo's telescopic discoveries. What he was not prepared to accept was that one could conclude from them that Copernicanism was true.

Thus by 12 April 1615 Bellarmine had read Galileo's letters to Castelli and Dini; he had read the original pamphlet by Foscarini and Foscarini's defence of it; he had talked at length with Piero Dini and had consulted with Father Grienberger; and he had made up his mind. The only thing preventing a formal decision at this point was that there was an ongoing heresy investigation against Galileo, an investigation that was supposed to be secret, although Galileo correctly suspected its existence: until the result of that was known it would be premature for the Church to announce anything. To an intelligent bystander, however, it looked very much as though the affair had reached a conclusion, and the conclusion, it seemed (although we can now see that this inference was mistaken), was that no decision need be taken.[47]

All the information that Galileo was receiving from Rome conveyed the same message – that there was no longer any real danger. He had written at length to Dini on 23 March 1615. He was evidently replying to Dini's account of his discussions with Bellarmine, for the letter reads as if it were a reply to Bellarmine's letter to Foscarini, which had not yet been written.[48] Contra Bellarmine, he insisted that Copernicanism could not be treated as a mere mathematical model, and that the text of the Bible could be reconciled with Copernican science. But Dini, once he had been shown Bellarmine's letter to Foscarini, did not think it worth passing Galileo's letter on to its intended recipient because it seemed clear that further argument was pointless.[49]

This, unfortunately, was not how Galileo saw things. He was sick with worry as a result of the rumours that were circulating. He objected to being told by Bellarmine that he could not argue that Copernicanism was compatible with the Scriptures, while at the same time he insisted that as an astronomer he wanted to play no part in the interpretation of the Scriptures. He wanted to rise to Bellarmine's challenge, and prove beyond doubt that Copernicanism was true. And he felt that he would do far better if he could argue the question in person rather than in letters – evidently he had forgotten the advice given him by the grand duke during the dispute over buoyancy. And so he resolved to go to Rome.[50] Delayed by illness (an illness which, by his own account, was partly psychosomatic or hysterical), he arrived there in early December 1615; by chance this was a few days after the Inquisition finally returned to the question whether Copernicanism itself was heretical or not, having received the reports on the interrogations conducted in Florence.[51] Galileo carried with him letters from the grand duke instructing the Florentine ambassador to take good care of him, his secretary, his servant and his mule, and to put them up in the Villa Medici.[52]

The ambassador, Piero Guicciardini, was not pleased. Hearing that Galileo was on his way, he wrote urgently (carefully putting key phrases into cipher), warning his masters that the Inquisition, and Bellarmine in particular, had had suspicions about Galileo since 1611, and that he was not popular with the Dominicans. This was not, he said, a good time to be trying to persuade the Roman authorities to adopt new doctrines – the conservative outlook of the pope was well known.[53] Over the next three months his opinion of Galileo did not improve: 'I do not understand what he is doing here,' he wrote, 'or what he hopes to achieve by staying.'[54]

As for Galileo, as soon as he arrived in Rome, and felt that he was directly engaged in battle with his opponents, his health picked up and he began to feel positively cheerful.[55] Everywhere he went he announced that Copernicus had provided an account of the physical structure of the universe, and not a mere mathematical model (as thought by many who had been misled by a preface that Andreas Osiander had added to Copernicus's book).[56] Staging debates in which Galileo, heavily outnumbered, was attacked by his opponents became a sort of spectator sport.[57] Onlookers admired his bravura performances, while remaining completely unconvinced by his reasoning. By bringing the conflict out into the open, by refusing any compromise, by insisting that matters be settled one way or the other, Galileo hoped to obtain one thing above all else: peace of mind.[58] But being on the scene proved to be of little real help. He soon found that the people who counted were more willing to take delivery of written arguments than to enter into discussions and debates.[59] On the other side, an attack on Copernicanism was produced by Francesco Ingoli, who, once Copernicanism had been banned, was to be assigned the task of censoring Copernicus's *On the Revolutions*.

Because he put all his arguments in writing we can reconstruct what Galileo took to be the main contentions to be refuted in his campaign in support of Copernicanism.[60] Of these there were five:

First, that Copernicus had never intended to provide an account of how the universe is really constructed, only to provide an aid to calculation (which was the way in which his work was commonly used). It was fairly straightforward for Galileo to show that this was simply untrue.

Second, that Copernicanism need not be taken seriously because there were hardly any Copernicans; if Copernicanism were right it would have general support. Galileo insisted that there were rather more Copernicans than was generally recognised, particularly as many were cautious about declaring their views in public. But he also tried to turn a major weakness in his position (particularly in a society which laid great emphasis on consensus and on orthodoxy) into a strength. What was crucial was that every Copernican had once been an orthodox Aristotelian. How then to explain their conversion to Copernicanism? The only possible explanation, he claimed, was the strength of the arguments in favour of Copernicanism. Galileo thus sought (not entirely successfully, it must be said) to turn being in a minority from a disadvantage into an advantage. Where his opponents were happy to declare a bias in favour of tradition, he was prepared to argue in favour of a bias towards innovation. New arguments were more likely to be right than old arguments because new arguments implied new thinking.

Third, that because theology was the queen of the sciences, it was perfectly proper for theologians to settle disputes between astronomers. Here Galileo argued that one had to distinguish between the subject matter of a body of knowledge and the expertise of those who studied it. The subject matter of theology, which concerned itself with the salvation of souls, was certainly superior to the subject matter of astronomy, but this did not make theologians expert in astronomy. In arguing in favour of technical expertise, Galileo was presenting a claim that seems absolutely obvious to us, but that would have seemed much less obvious to his contemporaries. In the first place, Renaissance culture was much more unified than ours is. Every theologian had studied philosophy, including astronomy, before studying theology. It was therefore plausible to insist that every theologian was capable of understanding and evaluating the arguments of astronomers. Second, in a society which was sharply stratified between those who performed manual labour and those who did not, it was perfectly normal for gentlemen to give orders to builders, or armourers, or estate managers, even though they lacked any specialist expertise. In arguing that there were some choices that were properly the preserve of experts, Galileo was defending a new type of authority.

Galileo then had to overcome the most important of all arguments: the argument that the Bible could not be wrong when it said that the sun moves. This was the main purpose of his (undated) letter to the grand duchess Christina of Lorraine. It is worth stressing that there is no evidence that this letter was ever sent to Christina, or that it was actually intended for her. Galileo later wrote a letter to Ingoli which was circulated widely amongst his supporters but probably never shown to Ingoli, and his final letter on sunspots, though addressed to Welser, was on its way through the press and into print before Cesi suggested it might be polite to send Welser a copy. We know that the letter to Christina was originally addressed to someone else, because we have an early draft in which the intended recipient is a priest. In these open letters the addressee was symbolic rather than real. In this case Christina is the addressee because, as far as Galileo was concerned, the debate on the legitimacy of Copernicanism had begun in her room in the archducal palace in Pisa, and because Galileo was utterly dependent on the patronage of the Medici family.

In support of his claim that the Bible should be read in the light of current scientific knowledge, rather than science being reconstructed to fit the Bible, Galileo cites Augustine eight times (five quotations are from the commentary on Genesis, including 'as for the structure of the heavens, the [Bible's] authors knew the true state of affairs, but the Holy Spirit, speaking through them, has not wanted to teach men something that has nothing to do with their salvation'; others are from a text mistakenly attributed to Augustine), St Jerome three times (including 'as if, in the Sacred Texts, many things are not stated in line with the popular opinion at the time the texts were written, and not in line with the truth of the matter'), Denys the Areopagite (or rather pseudo-Denys) twice, Alfonso Tostado (d. 1455) twice, St Thomas (who said that the fact that the Bible states that the earth is suspended in a void should not be taken as an argument in favour of a vacuum), Tertullian ('Our view is that God should first be known through nature, and then recognised through religious doctrine: through nature by studying his works, and through religious doctrine by listening to preachers'), the Jesuit theologian Pereira (d. 1610) ('We must be very careful, when discussing the teachings of Moses, not to contradict manifest experience and the arguments of philosophy and the other disciplines'), Cardinal Baronius (d. 1610) (who said to Galileo that the intention of the Holy Spirit was to teach us how to go to heaven, not how the heavens go), Cajetan (d. 1534), Cosmo Magaglianes (d. 1624), and Diego de Zuñiga, whose commentary on Job (1591) was, together with Copernicus and Foscarini's letter, to be banned in 1616 (until revised), because it explicitly claimed that the Bible was compatible with Copernicanism.

It would be very easy to form the impression from this regiment of quotations that Galileo was widely read in matters of theology – indeed this is certainly the impression he wants to leave with his reader. But it would be a false impression, for Galileo probably had at least four sources to hand when assembling his paper army. First, there was (or probably was) Foscarini's Latin *Defence*, in which reference is made to many of the same passages. Second, Castelli had passed on to him a collection of useful quotations, particularly from Augustine, compiled by a sympathetic theologian.[61] And third, he had probably been sent a copy of a text written by Cesi, entitled *Watcher of the Skies*, in which Cesi argues that the Bible, the Holy Fathers and the new philosophy were compatible. (This was the text that was originally intended to accompany the *Letters on Sunspots*.)[62] Moreover Galileo – or one of his assistants – evidently used the commentary on Joshua by Magaglianes which had been published in 1612: this one book provided him with seven of his references.[63] These four sources may well have furnished him with each and every one of the quotations he used.

If Galileo actually read Augustine's commentary on Genesis, the work he quotes most frequently, then it must have been in a borrowed copy, for there is

1 The frontispiece of Niccolò Tartaglia's *New Science* (1537) shows Euclid guarding the outer gate to the fortress of philosophy; within are the mathematical sciences (including music and astronomy). Plato guards the inner gate, insisting that no one can enter without a knowledge of geometry – a view Galileo shared. On the left two cannon are firing; the paths of their projectiles are evidently not symmetrical.

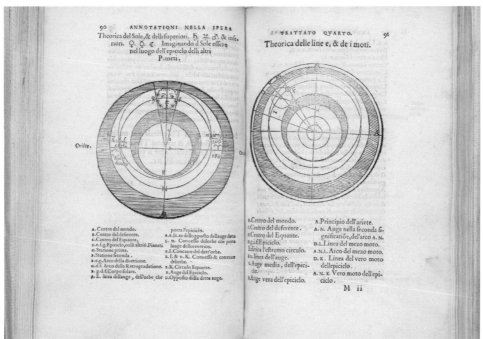

2 The Ptolemaic universe as portrayed in the commentary on the *Sphere of Sacrobosco* published by Mattei Mauro in Venice in 1550. Galileo lectured on Sacrobosco's *Sphere* as part of his duties at the University of Padua. Sacrobosco is (it is generally thought) the italianized name of John of Holywood, an Englishman (c.1195– c.1256). The images here show the movements of the planets and of the sun – the two are the same, we are told, if one thinks of the sun as occupying the epicycle of a planet.

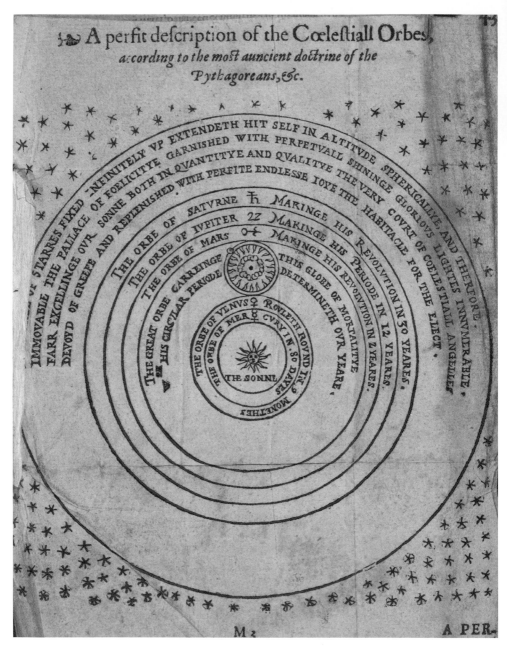

3 The Englishman Thomas Digges, in his *Perfit Description of the Cœlestiall Orbes* (1576, here from the edition of 1596), added as an appendix to a reprint of his father Leonard Digges's *Prognostication*, was the first to portray the Copernican system within an infinite universe. The image seems to have been an afterthought, and binders were unsure whether to bind it as a foldout plate (as here) or as a double-page spread. Galileo is unlikely to have known this work, but he will have been familiar with the idea of an infinite Copernican universe from the work of Giordano Bruno, and may well have read Digges's Copernican *The Wings of Mathematics* (on the comet of 1572), written in Latin, a copy of which was in Pinelli's library – and of course it is conceivable that one of Galileo's English-speaking friends showed him this image in the *Prognostication*.

4 Galileo's sector (invented in 1597; an instruction manual was privately printed in 1606). The sector (or *compasso geometrico et militare*) has three attachments, which are here seen fitted to it: the sighting device (at the end of the left arm), the bob, and the quadrant which would have been used for surveying and for measuring angles of elevation.

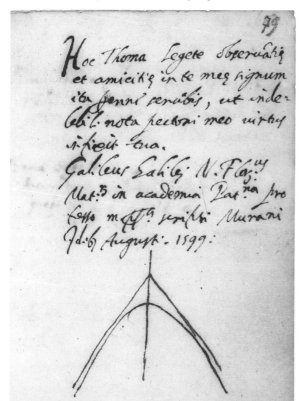

5 This 1599 entry by Galileo in the friendship book of Thomas Seget, with a sketch that is almost identical to the illustration of the parabolic path of projectiles in *Two New Sciences* (1638), conclusively shows that Galileo had already discovered the parabolic path of projectiles – and it presumably indicates a concern to record his discovery in case he should find himself caught up in a priority dispute in the future.

6 In 1600 Peter Paul Rubens visited Venice. There he surely met Thomas Seget, for Rubens's brother
Philip and Seget were both pupils of Lipsius; and perhaps, through Seget, Galileo. In March 1604
Rubens, his brother, and Galileo were all in Mantua, where Galileo was teaching the duke the use of his
sector. Frances Huemer has proposed that this mysterious painting by Rubens (now in the Wallraf-
Richartz-Museum, Cologne), which shows Rubens, his brother, and Lipsius in Mantua (Lipsius never
went to Mantua, and so can have been present only in spirit), also portrays Galileo. Eileen Reeves
has further suggested that the strange lighting in the picture shows the aurora borealis or Northern
Lights – a subject Galileo discussed in lectures given in January 1605. They view the painting as a
'friendship portrait' which records the shared commitment to neo-stoicism of those portrayed in it,
like a later painting (c.1615) which shows Lipsius, Philip (by now both dead), Rubens, Jan Wowerius
and Lipsius's dog Mopulus, all transported to Rome.

    The figure on the left is Galileo, but the painting is not a 'friendship portrait'. Rather it reflects a
philosophical disagreement. Galileo claimed – mistakenly – that the aurora borealis was a reflection

7 One of only two telescopes made by Galileo that survive. This one has a magnification of x 14 and is 1,273mm long. It is made of wood and covered with paper. Exactly when it was made is unclear, and it certainly was not the telescope with which he made his astronomical discoveries.

8 The objective lens of Galileo's telescope; it has been heavily stopped down in order to reduce distortion.

or refraction of the sun's rays; the classical view was that it was a fire in the heavens. Galileo's explanation for the aurora borealis was linked to his view that the moon and the earth shine only by reflected light, and this was part of his argument for Copernicanism. Lipsius, on the other hand, dismissed Copernicanism in 1604 as 'madness', and there is no evidence that Rubens disagreed. So Rubens presents his relationship with Galileo as one of amicable disagreement – turning to the audience, he invites us to take his side, invoking the aurora borealis as a refutation of Galileo's claims, while Galileo reaches out a hand to restrain him, but is silenced. On the right we see Lipsius and two of his disciples, opponents of Copernicanism, and on the left Galileo and two of his disciples (Seget, perhaps, being one), supporters of Copernicanism. The painting, recording a dispute at the Mantuan court, will have been made for Vincenzo Gonzaga (1562–1612), who employed Rubens and Galileo. (Reeves, *Painting the Heavens*, 57–90; Frances Huemer, 'Il dipinto di Colonia', in *Il Cannochiale e il Pennello*, ed. Lucia Tongiorgi Tomasi and Alessandro Tosi (Florence: Giunti, 2009), 60–70).

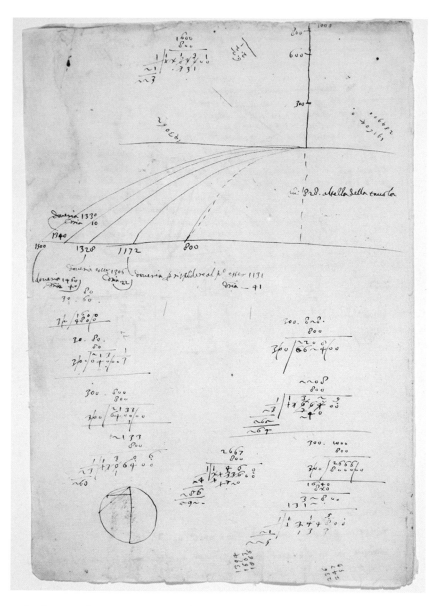

9 This notebook page, whose importance was first recognized by Stillman Drake, provides decisive proof that Galileo carried out experiments (probably in 1604) to test his theory of the acceleration of falling bodies. Balls were allowed to roll down an inclined plane, and then their speed at the end of their descent measured indirectly by seeing how far they were projected off a tabletop. Galileo's manuscript records both the actual distances travelled by his projectiles, when 'dropped' from different heights, and his theoretical predictions – the two correspond nearly exactly. A related experiment is described in *Two New Sciences* (1638).

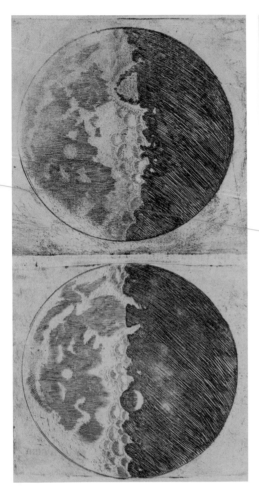
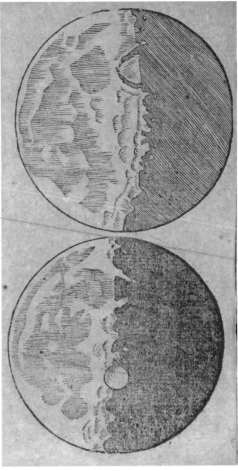

10 and 11 On the left is a page of illustrations of the moon from the first edition of *The Starry Messenger* (1610), and on the right the equivalent page from the pirated second edition. The pages of the first edition are larger than those of the pirated edition (23cm tall against 17cm tall). The first edition uses a combination of etching and engraving; the pirated edition uses a much cruder woodcut. The printer of the pirated edition rotated the original image through 180° – his image is here shown upside down for ease of comparison. In both editions the 'boss' the size of Bohemia, now thought to be the crater Albategnius, is exaggerated (bottom image). But in the pirated edition shadows are not cut so that they can be seen as having been cast by a single light source, which means that the features of the moon, such as the 'boss', can no longer be securely read as two-dimensional images of three-dimensional objects. Feyerabend made the mistake of reproducing an image from the pirated edition in support of his claim that Galileo's observations of the moon were unreliable and inaccurate.

12 Lodovico Cardi, *The Immaculate Virgin* (1610–12), dome of the Borghese Chapel of Santa Maria Maggiore, Rome. Paintings of the Immaculate Virgin standing on the moon (itself an image of purity) were a recognized genre. Galileo's friend Cardi was the first to paint the moon as revealed by the telescope, imperfect and irregular. See Eileen Reeves, *Painting the Heavens*, 167-76.

13 This is the engraving of Galileo by Francesco Villamena (1564–1624) which appears as the frontispiece to Galileo's *Sunspots* (1613) and here as the frontispiece to his *Assayer* (1623). Villamena may have based this engraving on a drawing or painting of Galileo that he had made when Galileo was in Rome in 1611, or he may have been working from a lost portrait which we know was painted in 1609 by Galileo's friend Cardi. At any rate, it shows Galileo at around the age of 46. Above Galileo, the putto on the left holds Galileo's sector; the putto on the right holds a rather fanciful telescope. On Villamena, Francha Trinchieri Camiz, 'The Roman "Studio" of Francesco Villamena', *The Burlington Magazine* 136 (1994), 506–16.

14 This engraving of Cardinal Bellarmine, by Francesco Villamena, was produced in 1604. It shows Bellarmine, aged 62, seated in his study in Rome, a devotional painting of the Annunciation on his desk, and a portrait of Ignatius Loyola on the wall behind him. Through the window one can see the Jesuit church of the Gesù, and the Casa Professa, the headquarters of the Jesuit order. Trajan's column can be seen in the distance. This image will have been produced so that it could be framed and hung on the wall of a Jesuit's study, just as Bellarmine has framed and hung an engraving of Loyola on the wall of his own study.

15 A portrait of Prince Federico Cesi (1585–1630), founder of the Accademia dei Lincei, by Pietro Facchetti (1539–1613).

16 This painting, attributed to Leandro Bassano (1557–1622), of a Venetian nobleman, has recently been identified by Nicholas Wilding as a portrait of Galileo's close friend Gianfrancesco Sagredo, who was a patron of Bassano's. Wilding argues that this is the portrait given to Galileo by Sagredo in 1619, in which case it was painted not by Leandro but his brother Gerolamo. The harbour of Aleppo is in the background – Sagredo was consul there from 1608 to 1611. ('Galileo's Idol: Gianfrancesco Sagredo Unveiled,' *Galilæana* 3 (2006), 229–45).

17 Attributed to Francesco Furini (1600–46) or to Cecco Bravo (1601–61), this fresco of Astronomy showing the moons of Jupiter to Cosimo II, was painted in 1621–3 in a Medici pleasure house, the Casino di San Marco, and was probably commissioned by Cardinal Carlo de' Medici (1595–1666). It shows that the family continued to take pride in Galileo's discoveries after the condemnation of Copernicanism in 1616, but its private location suggests a certain caution in publicizing the family tie to Galileo.

18 Justus Sustermans (1597–1681), a court painter to the Medici, painted *c*.1640 this group portrait of the Medici family. It shows Cosimo II de' Medici (1590–1621) and his wife Maria Magdalena (1589–1631) as they might have looked in 1620, with their son Ferdinand II (1610–70), presumably as he looked at the time the painting was produced.

19 This recently discovered fresco by Jacopo Zucchi (*c*.1541–*c*.1590) shows the gardens of the Villa Medici in Rome as they were in 1576–7. It was here that Galileo lived a riotous life in 1616, and here that he was briefly imprisoned after his condemnation in 1633. Since Galileo loved his own garden he must have delighted in walking and talking here.

20 Orazio Grassi went white when he picked up a copy of Galileo's newly-published *Assayer* (1623) in a Roman bookshop – presumably because he saw that it bore on its title page the arms of the pope (St Peter's keys and the Barberini bees at top centre). It also bears the symbol of Cesi's Academy of the Lynx–Eyed below the title.

21 This portrait of Maffeo Barberini, Urban VIII, comes from the first edition of his poems, published in 1631. Urban, who had a connoisseur's taste, was also portrayed by Caravaggio and Bernini. Campanella's success at Urban's court was based in part on his unstinting praise for Urban's poetry, of which Urban was inordinately proud.

22 The fresco of Divine Wisdom (1629–33) by Andrea Sacchi (1599–1661) in the Palazzo Barberini. The painting will have been commissioned by Taddeo Barberini, one of the pope's nephews. This is the best evidence we have for the cosmological views of the Barberini family prior to the condemnation of Galileo, although its interpretation is far from straightforward (see below, no. 24, for an interesting comparison, which suggests that there is nothing Copernican about the earth being portrayed as a globe floating in space). (John Beldon Scott, 'Galileo and Urban VIII: Science and Allegory at Palazzo Barberini', in Lorenza Mochi Onori, Sebastian Schütze, Francesco Solinas eds, *I Barberini e la Cultura Europea del Seicento* (Rome: De Luca Editori, 2007), 127–36).

23 The frontispiece to Galileo's *Dialogue* (1632) invites the reader to think of the book as a performance in which Galileo plays the part of Copernicus. In the Latin edition the frontispiece was changed, so that Copernicus looked like Copernicus and not like Galileo, and the little model of the Copernican system that Copernicus is holding, instead of pointing downwards in symbolic defeat, points upwards in symbolic victory. The setting clearly refers to Galileo's new physics, based on arguments about ships and projectiles, and to his explanation of the tides.

24 The title page of Melchior Inchofer's *Summary Treatise* (1633) shows the Barberini bees holding the earth fixed in space – a clear indication of the pope's personal commitment to the condemnation of Galileo. This text was published to justify the decision, taken during the trial of Galileo, to regard Copernicanism as a heresy.

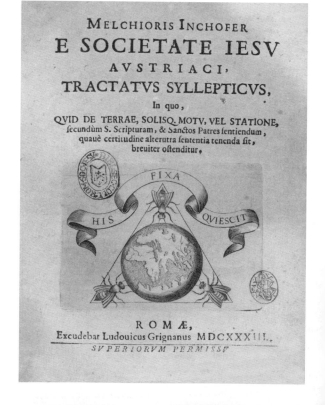

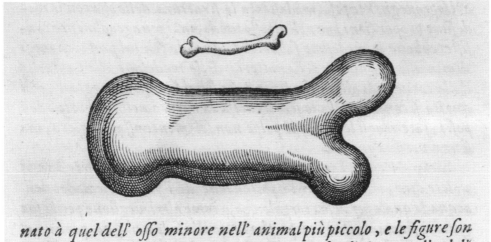

nato à quel dell' osso minore nell' animal più piccolo, e le figure son queste: doue vedete sproporzionata figura, che diuiene quella dell' osso ingrandito. Dal che è manifesto, che chi volesse mantener in

25 This illustration from Galileo's *Two New Sciences* (1638) shows how much you would have to increase the thickness of a bone if you increased the height of the animal it belonged to three times; the bone has to be scaled up to carry a weight which is twenty-seven times as great.

26 This illustration from *Two New Sciences* (1638) brings home the fact that Galileo's first new science is concerned with structural failure; if his treatise on mechanics was concerned with idealized machines, without flaw or fault, he is now concerned with their exact opposite, machines on the very point of failure.

del muro, da quella che è dentro; e per le cose dich to della forza posta in c al momento della resist

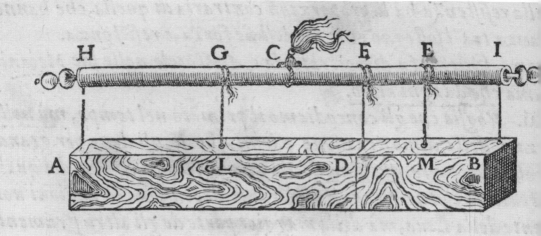

27 Using this illustration, Galileo 'proves' in the *Two New Sciences* (1638) that two weights are in equilibrium at distances from a point of balance that are inversely proportional to their heaviness. The proof is not mathematical but visual and depends on a thought experiment: first imagine a beam hung from points H and I on a yardarm which is itself suspended from point C; the system will balance because C is midway between A and B. Then imagine cutting the beam at D and suspending the two ends at D from a single thread at E – evidently nothing will have changed and the system will still be in balance. Now imagine suspending A–D from its mid-point L and D–B from its midpoint, M. Again, the system will balance. So A–D counterbalances D–B, or any weights equivalent to them. If one imagines making the cut D at various different places, it soon becomes apparent that the distances of L and M from C are inversely proportional to the heaviness of the weights suspended from them – one sees this is true even if a mathematical proof is beyond one. Since this is the principle of the lever, and, in Galileo's view, all machines work on the principle of the lever, no demonstration could be more important. This example is similar to Galileo's early work on centres of gravity, which he submitted to Guidobaldo del Monte and Clavius in 1588.

28 A nineteenth-century construction of Galileo's design for a pendulum clock, based on Viviani's drawing. No working model of Galileo's clock was built before Christiaan Huygens came up with his own design in 1656 (and Viviani's drawing was not made until 1659), but it is clear that Galileo, had he not been blind, could have produced a working pendulum clock (Silvio A. Bedini, *The Pulse of Time: Galileo Galilei, the Determination of Longitude, and the Pendulum Clock* (Florence: Olschki, 1991)).

no evidence that he ever owned any of the theological works that he relies on.[64] Rather the evidence suggests that his theological knowledge was secondhand, a fact somewhat at odds with his claim to be acting out of piety. Indeed we can trace the movement of a key passage from hand to hand: in 1612 Carlo Conti told Galileo about Diego de Zuñiga's commentary on Job – at that time the only instance of a theologian defending Copernicus; Galileo used the passage in the *Letter to the Grand Duchess Christina* and passed it on to Foscarini.[65]

Finally, Bellarmine had admitted that in principle it might be possible to demonstrate the truth of Copernicanism. Ideally what was needed was proof that the earth was moving. As it happened, Galileo had had such a proof (or at least what he took to be such a proof) to hand for more than twenty years, for he believed that the tides could be explained only as by-products of the earth's double movement through space, both orbiting the sun and rotating on its own axis. He had evidently been reserving this proof (which he had described to Sarpi in 1594) for publication in his planned book on the Ptolemaic and Copernican cosmologies. Now he needed to produce his secret weapon. He chose to confide it to the most sympathetic of the cardinals with whom he had held meetings, Cardinal Orsini, in the last days of December, and on 8 January 1616 he supplied Orsini with a written text.[66] Was the body of the text written specially for the occasion? We cannot tell. The substance of it reappears in the *Dialogue* of 1632, and for all we know much of it already existed when Galileo told Kepler in 1597 that he had explained many natural phenomena in the light of Copernicanism.

Galileo's theory of the tides was straightforward. The speed at which different parts of the surface of the earth move through space varies because the earth is both orbiting the sun and rotating on its axis: two constant movements produce a combined movement which involves acceleration and deceleration. The result, on long bodies of water that run from east to west, is that water piles up at one or other end, creating tides. Some (Koestler, for example) have claimed that there is a simple logical error involved in Galileo's argument. They maintain that while the acceleration and deceleration may be apparent to someone looking at the earth from an external fixed point, they will not be apparent to someone on the surface.[67] This is just wrong, and the physical principle Galileo is invoking does exist; it only happens to play no noticeable role in the formation of tides on this planet, where they are caused by the gravitational pull of the moon and the sun. One can, however, construct theoretical models for imaginary planets in which such forces do produce tides, and there was nothing hopeless in Galileo's aspiration to construct a mechanical model to illustrate how such forces could generate tides.

By the time Galileo published the *Dialogue* he knew that tides vary in size according to a lunar and a solar cycle and he sought to develop his model to account for these variations. He also knew that there are two tides a day, not (as his model implied) one. It would seem that he originally thought that having two tides was a peculiarity of small bodies of water, like the Mediterranean, where the second tide was an 'echo' of the first (Galileo thought tidal waters oscillated like a pendulum). But an Englishman, Richard White, set him straight in 1619.[68] Galileo's argument was much harder to apply to an ocean, and he really needed there to be a different number of tides in different bodies of water. What he seems not to have known, even in 1632, is that the time of high and low tide varies in a regular fashion each day – a variation that his model also fails to explain. His friend Baliani wrote to him as soon as he had read the *Dialogue* saying that this seemed to be a fundamental flaw in Galileo's argument, and indeed it was.[69] Galileo had gone to some lengths to collect evidence about tides in different parts of the world, but this key bit of evidence had somehow eluded him.

Galileo's theory of the tides was thus defective. It was at odds with the facts, and it was, in its final formulation, internally inconsistent. But in 1616 he offered a second, supplementary argument in support of the earth's rotation: the fact that the trade winds blow from east to west.[70] Now it so happens that the rotation of the earth is the true cause of the directionality of the trade winds. Bacon, in a text surely unknown to Galileo (although its arguments had probably been reported to him), had tried to explain this phenomenon in Ptolemaic terms, as a result of the rotating heavens rubbing against the earth, and the possibility of such an explanation may have prevented Galileo from laying nearly as much weight on the argument from trade winds as on the argument from tides. Moreover nearly everyone has some experience of tides (although over much of the Mediterranean they are very small), while Galileo, like most Italians, will have had no direct experience of the trade winds.

There is another reason why Galileo was reluctant to acknowledge the superiority of the argument from trade winds over the argument from tides: he had invented the latter himself, while he had borrowed the former from someone else. Galileo was presented with the trade wind argument by Paolo Foscarini in a long letter written to him at some date between Bellarmine's letter of 12 April 1615 and Galileo's letter to Orsini on tides. Foscarini (in an attempt to get round the restrictions imposed by Bellarmine's letter) was working on a dialogue in which the case for and against Copernicanism would be argued hypothetically (a dialogue, in fact, like Galileo's later *Dialogue*), and having come across a discussion of trade winds, he wrote to Galileo to ask if this physical phenomenon could be used as evidence in favour of Copernicanism.[71]

Galileo may have felt no need to acknowledge Foscarini in a letter to Orsini. By the time the *Dialogue* was published Foscarini's work had been banned and he himself was dead, so that no one would have benefited from such an acknowledgement. Still, Galileo's silence is disturbing, and a little splinter of evidence adds to our unease. Both in 1616 and in 1632 Galileo claimed that a careful study of the arrival and departure dates of Venetian ships travelling the length of the Mediterranean showed that it took 25 per cent less time to make the journey from east to west than vice versa.[72] Since everyone knew that Galileo had not been to Venice since 1610, this implied that his theory had been conceived by that date. Except that Galileo is wrong – the journeys from west to east were the shorter ones. Is it possible that he invented this convenient piece of information in order to create the false impression that he had long been familiar with the trade wind argument?

So Galileo marshalled his arguments in favour of Copernicanism. He knew very little, however, about what was really going on. It took him quite some time to establish that there was no prospect of his being found guilty of heresy – and when he did he promptly attributed it to his own presence in Rome and the skill with which he had presented his case, when in fact he had been in no danger from the day of his arrival.[73] His greatest success was with Cardinal Orsini, who, armed with Galileo's argument about the tides, approached the pope on his behalf on 24 February, the very day on which the panel of theologians ruled that Copernicanism was heretical.[74] It was already too late – indeed it had already been too late in May 1615 when Galileo first talked of coming to Rome.

The decision of the theologians, however, had now to be turned into a programme of action. On 5 March the Congregation of the Index issued a decree banning Foscarini's book; requiring Diego de Zuñiga's commentary on Job and Copernicus's On the Revolutions to be amended (it was not until 1620 that the emendations required were published, but Galileo duly entered them into his copies of Copernicus – or at least into the one that survives – although in fine enough ink for the original text to be still clearly visible); and banning all books that taught Copernicanism as true. There was no explicit mention of Galileo's Letters on Sunspots, presumably in deference to the grand duke, who had made clear his support for his court mathematician. Even more surprising is that the word 'heresy', which had been central to the judgement of the panel of theologians, appears nowhere in the Congregation's decree, Cardinal Barberini later claiming credit for its omission. The result was that the exact status of Copernicanism was far from clear. The decree was the only public document to emerge from the Inquisition's enquiry, so there was no public record of the theologians' view that Copernicanism was heretical.[75]

Something had to be done about Galileo, who was still actively campaigning on behalf of Copernicanism – indeed he was now pestering the pope via Orsini. Just what was done is not entirely clear, for we have four official documents, and they tell conflicting stories. On 25 February the pope instructed Cardinal Bellarmine to speak to Galileo and warn him to abandon Copernicanism. If he demurred an officer of the Inquisition, a Dominican as it happened (the Jesuits and the Dominicans were famously at odds, having been long caught up in a bitter conflict over the nature of free will and predestination), was to read out to Galileo a formal warning in front of witnesses, advising him that he would be imprisoned if he taught or defended Copernicanism, or even if he discussed it.[76] On 3 March Bellarmine reported to a Congregation of the Inquisition that he had warned Galileo and that he had acquiesced.[77] And on 26 May, in response to rumours (rumours that had reached Florence, Pisa and Venice) that Galileo had been found guilty of heresy, had abjured and had been ordered to do penance, Bellarmine issued Galileo with a certificate that he could take with him back to Florence stating that he had not abjured or been ordered to do penance: 'On the contrary, he has only been notified of the declaration made by the Holy Father and [later] published by the Sacred Congregation of the Index, whose content is that the doctrine attributed to Copernicus . . . is contrary to Holy Scripture and therefore cannot be defended or held.'[78]

The obvious conclusion to be drawn from these three texts is that Galileo had been given a friendly warning, but that he was bound only by the public declaration of the Congregation of the Index. In 1633, however, when the Inquisition reopened its files on Galileo, a fourth document was discovered: the text of a formal warning read 'successive ac incontenti' (thereafter, indeed immediately) to Galileo by Commissary-General Michelangelo Seghezzi on 26 February after Bellarmine had given his informal warning.[79] This text forbad Galileo from holding, teaching or defending Copernicanism 'in any way whatever, either orally or in writing' – a form of words which might well be taken to preclude even a hypothetical or disputative defence. In 1633 it was regarded as a document of fundamental importance, since, it was claimed, Galileo would never have been licensed to publish on Copernicanism had it been known that the document existed. What is most peculiar about this document, however, is that it makes no mention of Galileo's having demurred when issued with Bellarmine's informal warning. This peculiarity must have struck the officials who studied it in 1633, but there was no way for them to obtain clarification. Bellarmine was dead, and so was Seghezzi.

When, in the 1860s, the record of the Inquisition's dealings with Galileo was made public, this document became the object of heated debate. Some said it was a forgery, introduced into the file in 1633 in order to make it easy to convict Galileo; modern scholars, however, are nearly all agreed that the document is genuine, and consequently that this warning was read to Galileo at his meeting with Bellarmine.[80] There would then appear to be three possibilities: (a) Galileo did not demur, in which case Seghezzi exceeded his instructions (he was, after all, a Dominican, and it was the Dominicans who had from the beginning set out to have Galileo condemned, while Bellarmine was a Jesuit, and the Jesuits were widely believed to be sympathetic to Galileo): the document should not have been read to Galileo – which is why Bellarmine made no later reference to it; (b) Galileo did demur, in which case Bellarmine's report was somewhat misleading; and (c) there was some sort of muddle – Galileo perhaps opened his mouth to speak but never got a word out – so that it was unclear, even to the participants, whether Galileo had demurred or not, in which case Seghezzi jumped the gun.[81] In any case, Galileo was never given the text of this formal warning.[82] Reading Bellarmine's certificate, he was entitled to assume that the formal warning simply repeated the terms of the public decree of the Congregation of the Index, which were available in print a few days later. As the years went by he either forgot or pretended to forget that Seghezzi had played any role in the proceedings.

We will probably never know what really went on in Bellarmine's room on 26 February 1616. Seventeen years later, when Galileo (by then an old man of seventy with a fading memory) was questioned about these events, he too may have been unclear as to what exactly had taken place, and over the years he may have put a good deal of effort into reconstructing the precise sequence of events in such a way that he could, more or less honestly, say that he knew nothing about any formal warning.[83] Even if we could read Galileo's thoughts in 1633, then, we would not necessarily know what really happened. But two things are perhaps worth noting. The proceedings of the Inquisition were secret. Bellarmine will have felt no obligation to give a full account of the Inquisition's decisions in the certificate he presented to Galileo. On 7 March 1615 he solemnly assured Piero Dini that he had never heard anything about charges made against Galileo, and yet he had been present, indeed he had been in the chair, on 25 February, when Lorini's charges against Galileo had been discussed by the Holy Office.[84] And he may even have felt no obligation to give a full account of the events of 26 February 1616 to the Congregation at its meeting a few days later. A proper record was on file, which was all, arguably, that mattered. Second, we may note that in the years immediately after the condemnation of Copernicanism, Galileo carefully avoided mentioning it in print, suggesting that he may well have understood that he was prohibited from discussing it, at least in public. He did send out a few copies of his discourse on tides, but they were sent to heads of state and to Lord Chancellor Bacon rather than to other scientists, and were carefully labelled as documents of historical rather than scientific interest. (A copy was also sent to Sagredo, who could be trusted.)[85]

My guess, for what it is worth, is that Galileo did indeed demur when told to abandon Copernicanism. I think this, because it would have been entirely in character. As Guicciardini put it, 'He gets fired up with his own opinions, he becomes extremely passionate in his commitment to them, and he shows little strength and prudence in knowing when to get control of them: so that being here in Rome, especially as things are at the moment, is very dangerous for him . . . he will put himself in harm's way, and with him all those who give him their support or allow themselves to be persuaded by him to do as he wants them to.'[86] The man Guicciardini had observed so carefully would not have meekly accepted some friendly advice, even from Bellarmine. When he had recently met Caccini, he had been unable to hold back, and had lost his temper. Now he was face to face with Bellarmine: he had come all the way to Rome to convince him that he was wrong, and this was his first and last chance to do so. Can we really imagine Galileo remaining silent in such circumstances?

Did Galileo's going to Rome have any effect on the outcome? Probably not. Copernicus, Foscarini and Diego de Zuñiga would have been condemned anyway, and the charges against Galileo would have been dropped. The one thing that would certainly have been different if he had stayed in Tuscany and awaited the outcome of events is that there would have been no memorandum from Seghezzi in the file, which would have significantly improved his situation during the trial of 1633, and this might have resulted in a milder sentence. Hindsight is a wonderful thing, but with hindsight we can say that Galileo did himself more harm than good in visiting Rome in 1615–16.

In all Galileo made six visits to Rome (in 1587, to introduce himself to Clavius; in 1611, to demonstrate the telescope; in 1615–16, to defend Copernicanism; in 1624, to celebrate the election of Barberini as pope; in 1630, to prepare the way for the publication of the *Dialogue*; and in 1633, for his trial). The visit of 1615–16 was the longest, and it gives us a valuable opportunity to judge his skills as a courtier, although he was admittedly working under unfavourable conditions. In an important book, Mario Biagioli has portrayed Galileo as someone who fashioned himself in order to excel within the culture of the absolutist court. Strangely, Biagioli's book has nothing to say about these crucial six months in Rome, during which Galileo's performance was being carefully monitored by an experienced and intelligent courtier, Piero Guicciardini.[87] Guicciardini's assessment of Galileo as a courtier was harsh: Galileo simply did not understand the rules of the game, and he was not interested in learning them. He took his own advice, not that of his friends. He did not understand that, whatever your private beliefs, you must shape your avowed opinions to match those of the ruler – in this case the profoundly conservative Paul V. Galileo was a danger not only to himself but to all those associated with him.[88]

Guicciardini's assessment, made on the ground in Rome, was somewhat at odds with those of his masters in Florence. He wanted Galileo out of the way as soon as possible; Galileo wanted to hang on after the decree of 5 March, and did hang on for three months, until 4 June. For him it was a matter of saving face: he needed to put himself on public display in order to show that the Inquisition had cleared him and that he still enjoyed the protection and support of the grand duke. This had to be shown in ways that experienced observers would have no difficulty in interpreting: in the provision, for example, of a ducal litter to carry Galileo home when he eventually returned to Florence.[89] (A litter, in Renaissance Italy, was a small cabin, capable of holding two people, which was carried by two mules or horses, one in front and one behind, while the driver rode alongside.[90] It must have been an uncomfortable experience: since the horses did not keep step with each other, the passengers were jostled and jounced about. No one travelled in a litter if they were capable of riding.)

Moreover a few days before Copernicanism was condemned, the grand duke's secretary had told Galileo that the grand duke would like him to be in

Rome for the celebrations arranged for the arrival of Carlo de' Medici, recently elected cardinal, the idea evidently being that Galileo's celebrity status (and Galileo was certainly a celebrity) would make the great and the good eager to dine with the twenty-one-year-old cardinal.[91] Galileo had naturally told all and sundry that he was regarded as indispensable, and he insisted that he could not now leave before the cardinal arrived without a serious loss of face.[92]

On 11 March he managed to obtain a lengthy interview with the pope, in which he complained about the attacks of his enemies. The pope reassured him, telling him (according to Galileo's account) that 'I could feel safe as long as he was alive';[93] and on 26 May, as we have seen, he obtained Bellarmine's invaluable certificate. He could now retreat in reasonably good order.

Guicciardini, in any case, was now ready to kick him out. He had prepared the ground by calling in the accounts of Annibale Primi, the administrator at the Villa Medici. Galileo, it became apparent, had more than made himself at home: he had run up huge expenses, which Primi was promptly fired for having permitted. There had been, the ambassador complained to Florence, 'strange and scandalous . . . goings on'; Galileo had led 'a riotous life'. He also, Guicciardini complained, appeared determined to 'castrate' the friars who had opposed him.[94] A letter from the grand duke's secretary addressed to Galileo came by return of post. He was to return to Florence. There was no mention of the bills he had run up, but he was told it would be dangerous for him to stay, since there was concern that he was poking sleeping dogs with a stick.[95] The young cardinal came to inspect the Villa Medici and met Galileo. Perhaps he invited Galileo to dinner: if so, Galileo had strict instructions to avoid any reference to Copernicanism.[96] As for the ducal litter, Galileo had been told politely but firmly that none was available.[97] We must assume that he left Rome as he had come, on foot (for the mule, grown fat in the Medici gardens, will surely have been sufficiently burdened with his clothes, his books and his papers).

The one lesson that Galileo carried away from his six months in Rome was that courts are dangerous places. Before he left Rome he began discussions with Spanish diplomats about his methods for determining longitude. He would, he said, be happy to come to Spain to show how to put his proposal into effect, on one condition – that he was guaranteed against ill treatment: 'It often happens in courts that someone finds himself being judged by people who don't understand the matter he is dealing with, something which through ample experience I have found to be the most painful that commonly occurs to those whom blessed God has permitted to raise themselves, with some uncommon invention, above the capacities of the mob.'[98]

Galileo was to retain this distrust of courts.[99] He was next in Rome in 1624, and finding himself caught up in the usual round of visits to potential patrons, he wrote, 'These make me all too aware of the fact that I am old, and that playing the courtier [il corteggiare] is an activity for the young, who, because they are physically robust and feed on their aspirations, are capable of withstanding the effort involved. I, lacking these, want to return to my peace and quiet, and will do so as soon as I can.'[100] Galileo had been a courtier once; but those days were over.[101]

In the early modern period every gentleman was involved in a network of patronage relationships that extended both above him, to those from whom he sought patronage, and below him, to those who sought patronage from him. Galileo's withdrawal from court life also manifested itself in a noticeable indifference towards the opportunities and responsibilities represented by such relationships. In 1617 he rented a villa outside Florence, and spent almost all his time there. He passed over to Castelli the task of teaching the prince, although his own position at court had been founded on teaching first Cosimo and then Ferdinando.[102] On one occasion we find him apologising for not having replied to a letter because he has not been into Florence to collect his post for two months.[103] In 1621 he was elected consul of the Florentine Academy (an organisation to which he had read, in 1588, two lectures on the topography of Dante's hell), a prestigious position that should have given him an opportunity to reward those close to him with marks of esteem. But he stayed in his villa, did absolutely nothing, and eventually appointed a deputy.[104]

Galileo had simply become careless about the most elementary social obligations. He was genuinely well intentioned towards a young mathematician called Bonaventura Cavalieri, a self-professed Galilean; indeed he thought he would prove to be a great mathematician. At one point Cavalieri wrote to Galileo to ask if he could put in a word on his behalf with Cardinal Borghese, the pope's nephew and so the fount of all patronage in the Papal States. Cavalieri received no reply. Later he met Borghese, who seemed strangely friendly towards him. Was this because Galileo had spoken up for him? He could not tell. Like every young man on the make, Cavalieri lived in a world in which he was expected to produce a constant flow of requests for assistance, accompanied by promises of obedience and service. In response he expected to receive a counter flow of declarations of assistance delivered and of promises of assistance to come. These should be met in turn by a flood of expressions of gratitude, and of further promises of future service. Everyone knew the rules governing these exchanges. Now Cavalieri was at a loss. How could he thank Galileo if he did not know if he had spoken up for him? How could he risk not thanking Galileo if he was indeed acting as his patron? Cavalieri wrote to Galileo outlining his predicament: he was trying to play by the rules, but Galileo was not playing his part.[105] Later he sent Galileo some of his mathematical demonstrations. Despite repeated enquiries, he could not learn what Galileo thought of them. Was he, he wondered, expected to understand that silence itself was the response? Did silence imply that his work was not good enough to deserve any comment?[106] Neither courtier nor patron nor colleague, Galileo stood on the brink of social limbo.[107] The ties that bound him to others were being cut one by one. In social terms he was ceasing to exist – he was fading away.

23

Comets

In 1617 Galileo made a last attempt to find a proof for Copernicanism that no one could question: he tried, yet again, to measure the parallax of fixed stars to see if their relative position changed over time.[1] But he was no more successful than those who had tried before him. He was also preoccupied with his schemes for determining longitude by reference to the eclipses of Jupiter's satellites. Then between August and November 1618, three comets appeared in the sky. All three were visible to the naked eye, the last to appear being 'of such brightness that all eyes and minds were immediately turned toward it, and great throngs gathered on mountains and other very high places, with no thought for sleep and no fear of the cold north wind. Suddenly, men had no greater concern than that of observing the sky.'[2] Even the laziest slugabeds, we are told, were prepared to give up a night's sleep to see such a sight.[3]

Galileo's friends wrote to him alerting him to these extraordinary events.[4] Of course they asked him for his views. Virginio Cesarini cautiously added that if he had an opinion that ought not to be confided to paper, then he could pass it on via Ciampoli: he was thinking that Galileo might have something to say about comets and Copernicanism.[5] With the same thought in mind, Galileo was assured that a printing press in Paris was at his disposal – the printers did a better job there, he was told.[6] It would have been tactless, and perhaps even dangerous, to mention the fact that there was no French Inquisition. Yet although there was a period of four months in which anybody who wanted to could look at a comet, there was one (perhaps only one) mathematician in Italy who never turned his eyes to the sky, and that, amazingly enough, was Galileo.[7] He was, apparently, ill throughout this period, although even a sick man could have had himself carried outdoors to look at such a remarkable sight – and Galileo was certainly not too ill to receive visitors (including Grand Duke Leopold of Austria) and to discuss astronomy with them. He was too ill, he tells us, to write, although again he was not too ill, according to his own account, to reply to at least some of the letters he received.[8] Galileo's illness was assuredly genuine, but it was also a diplomatic illness. He was officially hors de combat, under no obligation to express any view on what comets might tell us about the structure of the universe.

According to traditional astronomy, which held that change was impossible in the heavens, the comets had to be terrestrial phenomena – they had to be located somewhere in the upper atmosphere. According to the new astronomy new stars (novas) and comets were obviously located in the heavens: they were, until the discovery of sunspots, the crucial evidence in favour of change in the heavens (evidence which it was easy to discount, since they only appeared at lengthy intervals). The most important work on both these phenomena had been done by Tycho, who had published on the new star of 1572 and the comet of 1577.[9] In both cases the main argument was from parallax: if novas or comets were close to the earth, then observers looking at them from different places on the earth's surface ought to see them located differently against the backdrop of the fixed stars. (Galileo had used precisely this argument in relation to the nova of 1604.) You can see how this works if you hold a finger up before your eyes and look at it first with one eye and then with the other: the finger will appear to move against the background of whatever is in front of you. The closer the finger is to you, the greater the movement. Such measurements could be carried out with the moon, and could be used to calculate the moon's distance from the earth; since no parallax could be measured in the case of novas and comets, they must be much further away.

The comets of 1618 were thus important events for astronomers. But they were particularly important for Catholics, for they came hard on the heels of the condemnation of Copernicanism. If Galileo's Copernicanism was false, then there were two alternatives available, both of which placed the earth unmoving at the centre of a universe of fixed size: Aristotelian-Ptolemaic astronomy, which denied the possibility of change in the heavens, and held that all heavenly objects circled around the earth, and the astronomy of Tycho, which recognised the existence of change in the heavens, and held that the planets circled around the sun while the sun circled around the earth. Every one of Galileo's *astronomical* discoveries – which were the only relevant discoveries that he had published – was at odds with the teaching of the Aristotelians; every one of them was compatible with the new astronomy of Tycho.[10] For the Jesuit scientists in Rome, who had followed Galileo's discoveries closely and who had carried out their own observations with telescopes, the conclusion was straightforward: Ptolemy was wrong. Their acknowledged leader, Clavius, had, before he died in 1612, expressed in print the view that the old cosmology would have to be abandoned and replaced by a new one. But now Copernicus was forbidden, and so, by elimination, Tycho must be right. This reasoning was so straightforward that the Jesuit scientists will have been convinced that even Galileo would adopt it.

It was in these circumstances that a Jesuit scientist in Rome, Orazio Grassi, published in 1619 a little book entitled *On the Three Comets of the Year 1618*. Grassi's 'chief aim', as Galileo later put it, 'was to attack Aristotle's opinion about comets', and his core argument was that the comets were in the heavens: in Grassi's mind this was, surely, the third major attack on Ptolemaic astronomy to be published in Italy, the first two being Galileo's *Starry Messenger* and *Letters on Sunspots*.[11] From Grassi's perspective he and Galileo were involved in a common enterprise. Thus the second paragraph of the book summarises the discoveries of Galileo (admittedly without mentioning Galileo by name, though in this respect Galileo is not singled out, for not a single scientist is named in the whole pamphlet), while the third begins, 'Thus, only comets have remained aloof from those lynx eyes' – a transparent reference to Galileo, who always signed himself 'Lynceus'.[12] Grassi must surely have expected that Galileo would read his book with approval.[13]

Grassi will not have been surprised when, that same year, a friend and pupil of Galileo's, Mario Guiducci, published a lecture given to the Florentine Academy entitled *Discourse on the Comets* – a lecture which Grassi will immediately have suspected was nothing other than a report of Galileo's views, and which we now know was indeed co-written by Galileo (so that I will call its author GG). The pamphlet was printed in Florence by Pietro Cecconcelli, who had named his printing works after the Medicean stars discovered by Galileo. The contents of GG's lecture will, however, have astonished Grassi, for it amounted to a defence of the Aristotelian account of comets as terrestrial objects! Nothing had prepared him for this.

One of the central objections to Copernicanism had always been that it required the universe to be too big – so big that, seen from one of the stars, the sun would be no bigger than a star. Yet GG's first objection to Grassi's book was that if comets were in the heavens then the universe was far larger than anyone had imagined possible. The longest-lasting of the comets of 1618 had traversed one quarter of the heavens in ninety days, but (since there had been no similar comet since 1577) must have an orbital year of at least forty earth years: 'Now how many worlds and how many universes will have to be assigned to give it space enough for an entire revolution when one four-hundredth part of its orbit exceeds half of our universe?'[14] This is, to say the least, odd, as we know that Galileo was prepared to consider the possibility of an infinite universe.

Another of the objections to the Copernican universe and to Galileo's discoveries was that they implied that God had created a universe much of which served no conceivable purpose. What, for example, were the mountains on the moon, or Jupiter's moons, actually for?[15] GG turns exactly this argument against the idea that every comet might have an orbit of it own: 'This multiplicity of spheres . . . I cannot reconcile with the extreme neatness which nature maintains in all her other works by retaining nothing which is superfluous or idle.'[16] Something odd is going on when we find Galileo relying on the arguments of his opponents.

Instead of simply defending Aristotle, GG propounded a quite new theory: comets were, as Aristotle had claimed, in the upper atmosphere, but they were not fires, as Aristotle had assumed (an argument that immediately ran into difficulties when one tried to work out exactly what was burning), but visual distortions, like rainbows, triple suns, or haloes around the moon. They belonged to the category of 'mere appearances, reflections of light, images, and wandering simulacra which are so dependent for their existence upon the vision of the observer that not only do they change position when he does, but I believe they would vanish entirely if his vision were taken away'.[17] What they would require for their appearance would be some distorting material. Just as a rainbow requires raindrops, so a comet would require some vapour extending into the upper atmosphere. This theory had one great advantage: one cannot measure how far away a rainbow is by measuring parallax, so that the core argument on which Tycho and Grassi had relied was undermined. But it also had an obvious and absolutely fatal disadvantage (one that was immediately apparent to intelligent bystanders): comets do not change their position when the observer changes his, while rainbows, for example, move when we do. The logical flaw is so obvious that we are bound to ask if we are really supposed to take this seriously.[18] One astute reader decided that we are not, and that Galileo was simply being provocative.[19] It is evident that, although he never published any sort of retraction, Galileo later abandoned this line of argument, and it is striking that his friends thought of his contributions to the debate as a form of theatrical performance ('questo theatro literario').[20] They do not use such language about any of Galileo's other publications.

GG had one powerful argument against Grassi. Grassi had said that telescopes magnify things that are far away less than they magnify things that are nearby; this is why stars still look like points of light when seen through a telescope. Since comets still seem tiny when seen through a telescope they must be far away. Galileo made short work of this argument, which was generally accepted at the time, while doing his best to explain why sources of radiant light (candles, stars) can seem larger than they ought to when viewed with the naked eye (an effect he put down to reflection in the surface of the eye, and that we would explain in terms of diffraction).[21]

Historians of ideas are always, and properly, reluctant to solve an interpretative crux by claiming that an author does not mean what he says, but certainly we are bound at this point to wonder quite what Galileo is up to. One thing is absolutely clear: he has correctly recognised that Grassi's account of comets is based upon Tycho, and in attacking Grassi he is seeking to demolish Tycho. There can be no doubt about his strategy: it is to forestall the claim that Tycho's cosmology is the only viable one. This overall strategy (and it is worth remembering that Galileo himself compares intellectual debate to warfare) is far more important to Galileo than the question of how comets are to be explained. Galileo knows Tycho had hoped, and that the Jesuits continued to hope, that comets would provide evidence to refute Copernicanism.[22] Rejecting their fundamental contention, that comets are in the heavens, is a useful precaution should they later claim to have found such evidence. It is also worth remembering that Galileo enjoyed improving the arguments of his opponents – and here we find him taking up arguments that had been used against him and turning them against Grassi. In Galileo's mind Grassi is a representative of the Jesuit order, and the Jesuits, who should have defended him in 1616 against the Dominicans, had betrayed him. Galileo is giving them a taste of their own medicine. Rather foolishly (as Ciampoli pointed out at the time), he made clear that his enmity was not merely towards Grassi, but towards the whole of the Roman College – a fact which was immediately noted by the Jesuits.[23] Indeed Galileo's behaviour should remind us of a story he himself tells, though he uses it against Grassi: 'He reminds me of a teacher of philosophy in my time at the University of Padua, who, being angry at one of his colleagues (as frequently happened), said that if the latter did not mend his ways he would send someone to spy secretly on the opinions he voiced in his lectures, and always maintain the contrary.'[24]

Battle was joined. Grassi replied, still in 1619, with a publication entitled *The Astronomical Balance*, writing under the double disguise of a pseudonym, Lothario Sarsi Sigensani, an anagram of Orazio Grassi Savonensi, and a pretence to be one of his own pupils. Guiducci replied to Sarsi in a *Letter to Tarquinio Galluzzi*, published in 1620. Matters might have been left there, with two publications on each side, and the last word going to Guiducci. Galileo, however, egged on by his fellow Linceans, replied to Sarsi in a substantial book entitled *The Assayer* (1623), published, like the *Letters on Sunspots*, in Rome by the Linceans. Both Galileo and Guiducci took great offence at Grassi's perfectly correct claim that Guiducci was simply Galileo's mouthpiece. By this point Galileo was in favour with the pope, so Grassi's final response, *The Reckoning of Weights*, had to appear in Paris, in 1626. There is no need to follow this dispute through every twist and turn, or (in the metaphor to which the disputants became attached) to test (assay) and weigh every argument. What Grassi had to say about comets was not a significant advance on Tycho. What Galileo had to say about comets convinced no one; the interesting question is whether it convinced Galileo himself.

But the debate is significant for two reasons. First, because it tells us something important about Galileo. And second, because it provoked him to publish *The Assayer*, which is his most extended discussion of scientific methodology. Let us begin with what it tells us about Galileo. We have seen that there was nothing in Grassi's *Astronomical Balance* that could have been expected to offend Galileo – Grassi probably expected him to agree with almost every word. Sarsi begins by insisting that Galileo 'knows the great benevolence of my teacher toward him, immoderate in his praise both in private conversations and in public discussions'.[25] Grassi has been in touch with Galileo: 'My master undertook that it be made known to him through friends that it was farthest from his thoughts to harm him in speech or writing; and although Galileo indicated to those from whom he had received this information that his mind was at peace and that he accepted their words, yet afterward he much preferred to lose a friend than an argument.'[26] It was Galileo not Grassi who had opened hostilities; it was Grassi not Galileo who had sought to be a peacemaker; and yet Galileo rewrote the history of their relationship, insisting that it was Grassi, like so many before, who had picked a quarrel with him, while he, all innocence, was simply trying to advance the cause of knowledge.[27] His friends accepted this version of events: Grassi had been 'the first to attack, opposing himself like a mad dog to the truth'.[28]

We have plenty of evidence of Grassi's attempts to restore good relations with Galileo. In 1623 (admittedly after the election of Maffeo Barberini as Pope Urban VIII) he was reported to have hopes of establishing an 'intimate friendship' with Galileo should he come to Rome.[29] Mario Guiducci, Grassi's chosen interlocutor in Rome, thought in the autumn of 1624 that there was a real possibility that Grassi would adopt all of Galileo's views, including Copernicanism.[30] He was certainly engaged in careful research: he had read Bellarmine's letter to Foscarini and Galileo's discourse on the tides, and he had talked at length to a former pupil of Galileo's, who had explained how Galileo understood vertical fall on a moving earth. In February 1626 Grassi was reported as saying that during the previous year he had tried to make peace with Galileo, but that Galileo had not been willing.[31] And this is clearly true. Grassi had made approach after approach after approach – and Galileo never sent back a single word of encouragement, continuing to lambast him as violently as ever in print. Indeed we know that he had directly refused any reconciliation unless Grassi gave up his right of reply.[32]

This is not just a story of Galileo's relations with Grassi, instructive as they may be about his bloody-mindedness. We have seen that when Galileo moved to Florence he had a key objective – to win over the Jesuits to the new astronomy. Now the Jesuits were offering him a central role in constructing a new anti-Aristotelian astronomy, one which would be explicitly founded on Tycho Brahe, but which might be compatible with a covert sympathy with Copernicanism. The price of this alliance would be that Galileo would have to become a team-player. Galileo understood this perfectly well, and insisted that he was uninterested in making alliances; he was interested only in truth and falsehood.[33] Where the Jesuits thought they were offering him an alliance between equals, he insisted they were offering nothing but slavery, that they were demanding that he never disagree with them.[34] Galileo was not interested in collaborating with the Jesuits on their terms, and since he could no longer openly defend Copernicanism, he was no longer interested in seeking their approval.

Every Jesuit knew of the hostility Galileo had shown, first to Scheiner in the debate on sunspots, and now to Grassi, and when Galileo found himself on trial in 1633, the Jesuits – particularly Scheiner – worked actively against him.[35] Christopher Grienberger would later say that if Galileo had not fallen out with the Jesuits he could have written freely on any subject he chose – including Copernicanism.[36] His strategy in 1611 had been an intelligent one; his proud rejection of Grassi's overtures in the years after 1618 was foolish and self-destructive, for he had nothing to gain by it, and he placed his own security at risk.

Indeed Viviani blamed the controversy with Grassi for all Galileo's later difficulties.[37] But it is easy to understand that the condemnation of Copernicus had left Galileo very angry, and Grassi must have seemed infuriating because he was behaving as if nothing had happened, as if the truth had not been suppressed.

If we turn to *The Assayer* we find that Galileo presents in it a self-portrait which is designed to justify this sort of dog-in-the-manger behaviour:

> Sarsi perhaps believes that all the hosts of good philosophers may be enclosed within walls of some sort. I believe, Sarsi, that they fly, and that they fly alone like eagles, and not like starlings. It is true that because eagles are scarce they are little seen and less heard, whereas birds that fly in flocks fill the sky with shrieks and cries wherever they settle, and befoul the earth beneath them. But if true philosophers are like eagles, and not like the phoenix instead, Sig. Sarsi, the crowd of fools who know nothing is infinite.[38]

Galileo thought of himself as a phoenix, alone of all his kind. Certainly his views could not possibly be held by large numbers of people. Popular philosophy, he argues, is bound (like scholasticism) to claim that it knows all the answers; real philosophy begins with a recognition of the limits of our knowledge. If an eagle finds itself surrounded by starlings, it inevitably thinks it is being mobbed and flies away. This is how Galileo responded to Grassi's overtures, beginning with the favourable reference to him in *On the Three Comets*.

In part this attitude derives from Galileo's background as a mathematician: a good mathematician inevitably finds that his less able colleagues cannot keep up with him. For Galileo, mathematical demonstration involves a sort of conquest, and conquerors are of necessity different from the rest of mankind: 'One must with but few words and at the first assault become Caesar or nobody.'[39] Reasoning, he says, is not 'like carrying burdens, where several horses will carry more sacks of grain than one alone ... reasoning is like running and not like carrying, and one Arab steed will outrun a hundred pack horses'.[40] But this view is in tension with Galileo's egalitarianism (his hostility to social hierarchy), and his insistence on writing in Italian (Grassi, of course, writes in Latin). It is hard to think that he writes to be read only by another phoenix! Indeed Galileo had claimed that he wrote in Italian because he believed that there were many intelligent people who had been denied a formal education but who were entirely capable of following his arguments – it was their support he hoped for, rather than that of the university philosophers.[41] (Even Italian, of course, was, outside Florence, a language for the educated. The uneducated spoke the sort of dialect in which Cecco di Ronchitti had written, and this varied greatly from region to region.) Rather the view of himself as an eagle mobbed by starlings tells us exactly how he felt about the condemnation of Copernicanism. First the Aristotelians and then the Church had tried to cage him up, but he was determined to fly free, even if he flew alone; he was determined to run ahead of the pack horses, even if he hoped that one day the pack horses would arrive at the right destination.[42]

Galileo's *Assayer* belongs to a literary genre that has now (thank goodness) disappeared: it is a paragraph-by-paragraph, line-by-line refutation of Sarsi's *Astronomical Balance*. Indeed Galileo reprints every sentence of Grassi's text in the course of his refutation. As a consequence, of course, Galileo's book lacks any structure of its own. It is, in effect, a collection of miniature essays on a range of subjects that has been generated almost randomly. But three amongst these many essays are of great importance (and we will come later to a fourth, which was decisive in shaping Galileo's relations with the authorities).

First, *The Assayer* contains the clearest exposition of Galileo's understanding of the role of facts in scientific argument (even though he does not use the word 'fatto' in any of the key passages). Grassi believed that as objects fly through the air they become hot, so that metal cannonballs often melt before they arrive at their target. Under pressure from GG, Grassi produces more and more authorities, including poets ('I begin with the poets, satisfied with those whose authority is considered great and is usually presented in very serious matters, because they were very well trained in knowledge of natural phenomena. Ovid, skilled not only in poetry but also in mathematics and philosophy, testifies that not only arrows but leaden balls hurled from Balearic slings have often been kindled in their course') and historians (Suidas reports that the Babylonians cooked eggs by whirling them around in slings).[43] Guiducci replied that it seemed 'a rather unreasonable and injudicious thing to try to prove by testimony the effects of nature'.[44] Galileo devoted ten pages to making the same point. Testimony, he insisted, was appropriate only for events which were not suitable for replication. Could one replicate the cooking of eggs by whirling them in a sling? Grassi's response was that one could do so only under exceptional conditions, conditions so rare that the event was almost a miracle. Galileo, protesting all the time that he is not deriding Grassi, proceeds to do just that:

> If we do not achieve an effect which others formerly achieved, it must be that in our operations we lack something which was the cause of this effect succeeding, and if we lack but one single thing, then this alone can be the cause. Now we do not lack eggs, or slings, or sturdy fellows to whirl them; and still they do not cook, but rather they cool down faster if hot. And since nothing is lacking to us except being Babylonians, then being Babylonians is the cause of the eggs hardening . . . Is it possible that Sarsi in riding post [i.e. galloping] has not observed the coolness brought to his face by the continual change of air?[45]

It is easy for us, I think, to miss the impact of this exchange, because we all live in a world in which facts are thought of as 'hard', and, in a sort of paper/

scissors/rock game, always defeat testimonies and authorities, which are soft and vulnerable. But Galileo did not live in such a world. For centuries a good collection of authorities – and Grassi had a very good collection of authorities – was seen as sufficient to establish a truth. In rejecting all the authorities and relying instead on his own experience of galloping, and in insisting that he would pay no attention to Grassi until he claimed to have actually cooked an egg by whirling it around in a sling (clearly Galileo had never hired a sturdy fellow and supplied him with an egg and a sling – as far as he was concerned this was a thought experiment), Galileo was redefining the nature of what we now call evidence.

It should be stressed that Grassi was not a conservative thinker. He had accepted with enthusiasm Galileo's telescopic discoveries. He and Galileo were engaged in a lively debate about experimental evidence: if you rotated water and air in a half-sphere, did the air and water rotate with the sphere, being carried along by it, or did they stand still while it slipped past them? He was perfectly happy to put this question to an experimental test.[46] So why did he not realise that the question of slingshot cookery was also a practical one? The answer, I think, is that the telescope was new, and one could therefore acknowledge that it represented a new beginning. So was the question about the rotating sphere – it could not be settled by authority because the authorities had never spoken on it. In his work on floating bodies Galileo might have shown that the Aristotelians could not give a coherent account of flotation, but he had been forced to admit that heavier-than-water bodies could, under certain circumstances, float – Aristotle's theory might not be right, but his facts arguably were. Thus in the mathematical sciences the facts had never defeated the authorities in a fair fight.

Grassi lived in a cultural world founded on consensus. As a Jesuit, he accepted that theology and philosophy could be harmoniously integrated and that the teaching of the Church had remained the same through the ages. And so he simply had not grasped that Galileo's method left no space at all for arguments from authority. He was in the position of the Aristotelian philosopher Galileo describes in the *Dialogue*, who could not accept that what he had seen with his own eyes in an anatomical dissection must take precedence over the text of Aristotle.

The Assayer also contains Galileo's third reworking of the two books theme. We have seen that he first used this trope to contrast the book of nature and the books of Aristotle; then to contrast the book of nature and the Bible. Now he argues that philosophy is not a fiction like *The Iliad* or *Orlando Furioso*, something constructed (as we would now say); it is something we discover by reading the book of nature:

> Philosophy is written in this very great book which always lies open before our eyes (I mean the universe), but one cannot understand it unless one first learns to understand the language and recognize the characters in which it is written. It is written in mathematical language and the characters are triangles,

circles and other geometrical figures; without these means it is humanly impossible to understand a word of it; without these there is only clueless scrabbling around in a dark labyrinth.[47]

This is perhaps the most famous passage in all of Galileo. It represents a firm commitment to a realist theory of knowledge. It makes clear his view that mathematics is the language of science. It may puzzle us that Galileo equates mathematics with geometry, but it should not: algebra was entering general use only towards the end of Galileo's life, and it was Galileo's pupils who were to lay the foundations for calculus. For Galileo geometry was the most powerful tool at the mathematician's disposal. But the power of the passage derives, I think, from a systematic ambiguity that runs through it. What does Galileo mean by 'book', 'characters' and 'written'? He starts with *The Iliad* and *Orlando Furioso*. For a seventeenth-century reader these would, of course, have been printed books. So the book is a printed book, and the characters (and the word 'character' normally implies a letter that is impressed or incised) are the printed letters produced by movable type. If this is the case the universe is, like a printed book, an object which has been mass-produced, an object which is internally uniform because every 'a' is like every other 'a', every 'b' like every other 'b'. Its legibility derives in large part from its uniformity. I think this is certainly part of Galileo's meaning: for him every falling body obeys exactly the same law as every other falling body, and every projectile traces the same path – always a parabola (the parabola must be a strong candidate for being one of the 'other geometrical figures' he does not name). Wherever you go in the universe you see the same characters and they make the same sense; hidden in this thought is the idea of the repetition which makes every copy of an edition like every other copy. In such a book we cannot get lost, as in a labyrinth: we have a table of contents and an index, and everywhere we go we meet a classic text that has often been reproduced. And of course this is a book which everyone can read, in that copies are widely available.

On the other hand, Galileo says this book is *written*. *The Iliad* can be written in Greek and still be a printed book, but if the characters are *written* then this is a manuscript, a unique object of which God is the author. The characters then become not the metal letters in a printer's font, but the letters of the alphabet, and the alphabet out of which this book is composed consists of circles, triangles, parabolas. This universe has not been mass-produced; it has been drawn by God's own hand. And if we can all read it, it is because we are all gathered around the one unique copy. This is still the medieval 'book of nature' – an illuminated manuscript, with all the implied associations of beauty and unreproducibility.

The power of Galileo's text derives from the fact that it flickers between these two meanings, between the unique object and the mass-produced object, between manuscript and printed book, between ancient technology and modern technology, and as a consequence it seems to mean more than it actually says – more than if the wording were unambiguous, and committed Galileo to one meaning or the other. It declares both a continuity between Galileo's understanding of the world and that of Pythagoras or Plato, and a discontinuity – only now, in the age of mass production, can we imagine a nature which is truly 'inexorable and immutable', only now can we begin to think in terms of the uniformity of nature. Galileo's text generates a gestalt-shifting image, such as the duck/rabbit described by Wittgenstein.

The irony, of course, is that at the very moment at which Galileo is insisting that the meaning of the universe is unambiguous for those who can understand it, he is relying on ambiguity to convey his meaning.[48] He tells us we can escape from the labyrinth, but the very text in which he tells us this is itself a labyrinth. No wonder this passage is constantly quoted, because it permits us to think of the universe simultaneously as a work of art and as a blueprint, a technical drawing which can be reproduced over and over again, and which has none of the sensory qualities (touch, smell, colour) that one associates with a fine manuscript. Galileo is describing a disenchanted universe (in Max Weber's phrase), but he describes it in a language which is itself enchanted.

In the final pages of *The Assayer*, Galileo also took the opportunity to provide what we might call a disenchanted account of matter: matter, he argued, is made up of tiny, indivisible particles whose only certain characteristic is that they have shape. Through touch, the sense with which we perceive shape, we thus perceive reality (touch here, unusually in Galileo, takes precedence over sight). Our other senses do not tell us how the world really is, but how it is for us. On this theory, smells are merely particles of a particular shape which enter our noses. The smell is not in the particles but in our noses. (Galileo makes this distinction vivid by using the example of tickling: if someone tickles me with a feather the tickle is in me, not in the feather.) Sounds are our way of interpreting some sort of movement in space. Colour is something our eyes read into the world – if there were no eyes, there would be no colours. And heat and cold (Galileo's original topic) are the means by which our skin interprets the impact of certain particles upon us. If reality consists of shapes, then it consists of quantities, and quantities can be measured: this is Galileo's way of eliminating wholesale the Aristotelian way of thinking about nature, which relies on arguments from *qualities* (and all arguments about essences are in the end qualitative arguments). Galileo is arguing that quantitative knowledge is objective, while qualitative knowledge is subjective.

This argument is now known as the distinction between primary and secondary qualities, and it becomes an absolutely central argument in the philosophy of the early modern period – in Descartes, Hobbes and Locke, for example. It is first formulated in the modern era by Galileo, and of all Galileo's arguments it is the one that has the greatest impact on later philosophical debate. Galileo's account of what knowledge is becomes in the course of the seventeenth century the standard philosophical account – at least for non-Aristotelian philosophers, for of course the vast majority of philosophers (particularly of university philosophers) remained Aristotelians throughout the century.

The atomism of *The Assayer* provoked a suspicion of heresy. One of the key points of disagreement between Protestants and Catholics was and is the belief of Catholics that, during Mass, the bread and wine are turned into the body and blood of Christ in their essence, even though in their accidents (their outward appearance) they remain bread and wine. By abolishing essences Galileo seemed to have declared the impossibility of transubstantiation. The book was consequently denounced to the Inquisition, and in his reply Grassi too argued that this passage was heretical.[49]

In Galileo's mind there was surely another subject that was inseparably connected to the topic of atomism (the two had already been connected in the Inquisition's enquiry of 1616): the question of the vacuum. As Galileo will have been well aware, classical Greek and Roman atomists all believed in the void. The ancient text – a fragment from Democritus – which announces the distinction between primary and secondary qualities, also declares the existence of the void: 'By convention there are sweet and bitter, hot and cold, by convention there is colour; but in truth there are atoms and the void.' It is possible to formulate an atomistic theory which denies the existence of the void – Descartes and Hobbes were to try to do just this. But no such theory can really be coherent because if atoms have defined shapes minute voids must be created whenever they move against each other. Galileo was to demonstrate in the *Two New Sciences* that infinitesimal vacuums are logically necessary.

In the *Two New Sciences* the argument about the vacuum was to be at the heart of Galileo's account of the strength of materials. There Galileo reports that if you take two smoothly polished pieces of marble and put them together they will cling to each other; you can attach a heavy weight to one of them and still it will not break free. Thirty years later Robert Boyle was to use this as a test of whether he had created a vacuum with his vacuum pump: if he had created a space in which one piece of polished marble would fall away from another, then he had created a vacuum. Galileo argued that if resistance to the creation of a vacuum could produce a solid bond, then much of the strength of materials might derive from this invisible force.

Galileo's *On Motion* had included a refutation of one of Aristotle's arguments against the possibility of a vacuum. He now knew that it is not difficult to create a vacuum. Sagredo had pointed out to him that there must be a vacuum at the top of a thermometer, and reported that he had conducted an elementary experiment to see if a vacuum would transmit sound. Galileo had devised a system for weighing the air which presses down on us; such a measurement assumed that beyond the atmosphere was empty space. In 1630 he was approached by Baliani, who was having difficulty raising water by a system of siphons.[50] Galileo replied that there was a strict limit to how far one could raise water with a siphon; if one tried to go past this limit one simply created a vacuum.[51] Baliani was astonished: he had no idea that a vacuum could be created so easily.[52]

Galileo had been interested in the vacuum for as long as he had been interested in Copernicanism. His first important work and his last deal with it. It represented a crucial weakness in Aristotelian natural philosophy: if he wanted to destroy scholasticism, this was a good point of attack. For the next generation it was also to become the crucial battleground between those who believed in a deductive science (Hobbes, Descartes) and those who believed in an experimental science (Pascal, Boyle). And whatever one might think of atomism, no one had ever said that belief in a vacuum was heretical – which was precisely why Galileo's pupils, such as Torricelli, were later to turn from astronomy to vacuum experiments, and why vacuum experiments were being conducted outdoors in the streets of Rome a few years after Galileo's death. Just as Galileo could have published his account of falling bodies at any time after 1608, so he could have turned at any time to work on vacuums.

Why, when he was forbidden to write about Copernicanism, did Galileo insist on writing about comets (a subject on which he had nothing helpful to say), rather than turning to projectiles or vacuums? The answer is that he had too much invested in Copernicanism to step away from the fight. Part of his investment came from his oppositional character: the fact that he had been forbidden to discuss Copernicanism must have made him all the more eager to do so. But part of it came from somewhere else. Projectiles and vacuums are of fundamental importance, but they do not alter our understanding of our place in the universe. One cannot imagine Milton's Adam anxiously questioning Raphael about the law of fall. Copernicanism is different: it replaces a universe made to be a home for mankind with one that mankind is lost in – we become, as one of Galileo's contemporaries complained, like ants on a football that is flying through the air. If Galileo stuck with Copernicanism as the key topic he wanted to write about, it was because he was attracted by the idea of making human beings seem insignificant. He certainly did not foresee that this would bring him into conflict with the Church – at first he had imagined he could form an alliance with the theologians against the Aristotelians. But even when it did bring him into conflict with the Church he was too attached to the idea to abandon it, just as when he discovered that the oceans had two tides a day he was too attached to his proof that the earth moved to give it up. Copernicanism mattered to Galileo, and the reasons for this were not simply scientific.

Late in 1622, after he had sent the manuscript of *The Assayer* off to Rome to be printed, Galileo began to revise his discourse on the tides: it seems he was preparing to reopen the question of Copernicanism, despite the fact that he had been forbidden ever to do so.[53] What made him think that circumstances had changed? Bellarmine had a died the year before: perhaps Galileo thought that with him had gone all memory of the injunction of 1616.

Even so, Galileo surely knew that there was no prospect of publishing or even circulating in manuscript a defence of Copernicanism in Italy under present circumstances. What could be the point, then, of updating his discourse on the tides? What we have here is a puzzle, and there has to be a piece missing. Fortunately it is not difficult to discover a piece that fits, and that enables us to make sense of Galileo's behaviour. In 1633 Leone Allacci published a book on intellectual life in Rome entitled *The Urban Bees*. While the 'urbs' in question is Rome, the title is also a punning reference to Pope Urban VIII, the bee being the symbol of the Barberini family. In it we find the following passage:

> The French have such a high opinion of Galileo that some of them have come to Italy with the sole purpose of meeting him. Indeed I have learnt from a reliable source that a person called Diodati, of noble birth, known for his science and his virtue, hurriedly came from France to Florence for this sole reason, and that, having spent the whole of thirteen days talking to Galileo about various mysteries of nature, in order to satisfy his intellectual curiosity, judging that as far as he was concerned seeing Italy at Galileo's side was all the Italy he needed to see, he returned to France, dropping all his other business, travelling in long stages.[54]

Diodati had written to Galileo in 1620 asking if he had plans for publication, and offering to help him get round any 'local obstacles' – news of the condemnation of Copernicanism was only now beginning to circulate amongst French intellectuals.[55] Galileo replied making clear that he was prevented from publishing.[56] The surviving correspondence would seem to suggest that there was no further contact between Galileo and Diodati for five years, but it seems unlikely that matters were really allowed to rest there. What Allacci reports is exactly what one would expect: a semi-clandestine visit by Diodati to discuss the possibilities of publishing abroad. If he came in the autumn of 1622 he may have brought with him a copy of Campanella's *Defence of Galileo*, written in 1616, before the condemnation of Galileo, but published in Frankfurt in the spring of 1622, with a suitable cover story absolving Campanella from all responsibility.[57] This will have provided the model for a planned publication by Galileo: all he need do was supply Diodati with a revised version of his discourse on the tides, one that could be passed off as having been written before 1616. Diodati could look after the rest (as he did in 1636 with the publication of Galileo's letter to Christina, written in 1615). Galileo was evidently intending to implement a plan of this sort when he was overtaken by events. In later years, any evidence relating to it will have been carefully destroyed, if not by Galileo and Diodati, then by Viviani, for such evidence would have been proof that Galileo had never intended to respect either the condemnation of Copernicanism or the injunction of 1616.

The death of Gianfrancesco Sagredo

Sometime in 1621 or 1622 Galileo spent thirteen days lost in conversation with Diodati. With a foreigner and a Protestant he could speak freely, as once he would have spoken to Sarpi and Sagredo in Venice. Briefly, in the garden of his Florentine villa, Galileo relived the liberty he had experienced in Padua. For now, after the death of his friend Salviati in 1614 and the condemnation of Copernicanism in 1616, Galileo was increasingly isolated. He had to be careful not only about what he wrote, but about to whom he spoke and what he said. Above all, he had lost the companionship of his closest friend, Gianfrancesco Sagredo.[1] The two of them never met after Sagredo left Venice to take up the post of Venice's consul in Aleppo in 1608; by the time Sagredo returned Galileo was in Florence. But for eight years Sagredo wrote to Galileo almost every week, and of these three or four hundred letters, seventy-nine survive. Unfortunately, we have not a single letter from Galileo to Sagredo, so trying to make sense of their relationship is a bit like listening to one half of a telephone conversation. Even if we had all Sagredo's letters to Galileo, and all Galileo's replies, our difficulties would not be over. Indeed, they would just be beginning, for what we are trying to make sense of is a relationship, and our assessment will depend on what we make of the characters and motives of the two individuals involved.

Let us be honest. The key fact about Sagredo, as he recognised himself, was that he was a failure. He belonged to one of the most powerful families in Venice, a family dedicated to the pursuit of political power. In 1615 his father came within a vote of being elected doge.[2] But Sagredo achieved only minor offices – offices that would have done credit to someone who was making his own way, but that in his case amounted to a snub by his fellow nobles. There was something about him, he knew, that they did not like. He seems to have been clear in his own mind what it was, although he did not put it into words: he probably thought Galileo would have no difficulty in filling in the gap. He was certainly arrogant, but in a hereditary ruling caste, such as that of Venice, arrogance is not necessarily a disadvantage. More to the point, perhaps, he felt a barely concealed hostility towards his fellow human beings. If people were honest with each other, he told Galileo, no one would survive for a day – if you let people know what you really thought of them, they would kill you. This tells us a good deal about Sagredo's feelings towards his fellow men. Evidently he had to hold in check murderous thoughts and feelings that the rest of us suppress so successfully that most of the time we do not realise we have them.

We get a picture of the man from a letter of 1618.[3] He had gone on a little holiday with the painter Leandro Bassano, accompanied by their mistresses and one or two others. On the first day the coach tipped over as it was rolling along, and Bassano was thrown out and hurt. Sagredo and his companions pretended to misunderstand his cries and gestures. They pretended to think he was putting on a burlesque imitation of an injured man. The more they laughed the more furious Bassano became; the more furious he became, the more they laughed. The next day, Sagredo was the one who fell out. He had in his hands a small painting by Bassano (Bassano had been sketching and painting constantly). As he fell he contrived to twist his body so that the painting was held clear: he compares himself to Caesar swimming to escape his enemies, holding his *Commentaries* clear of the waves. Bassano, who had hated him only the day before, was now delighted, for Sagredo had treated his painting as if it were more valuable than his own person.

Sagredo was a connoisseur of fine art, so of course he valued Bassano's painting. But his appropriation of the Caesar anecdote tells us that he thought of the painting not as Bassano's, but as his own. Sagredo liked collecting not only art but also living things – the pleasure house (the 'casino') in which he kept his mistress he compares to a Noah's ark, full of animals of every sort, including a tame wolf who would race around and then curl up by the fire like a cat.[4] But this was an ark which would never reach land. The creatures, including the mistress, were his to keep or to toss aside – we have a long letter in which he describes his – entirely arbitrary – decision to dismiss his mistress. His letters to Galileo are full of complaints about the difficulty of obtaining good and faithful servants; Galileo supplied him with a couple, but of course they too proved entirely unsatisfactory. Deprived of power in the state, excluded by his family even from playing a significant role in their vast commercial enterprises, Sagredo sought to make a little world of his own which he would rule over as an absolute, and sometimes cruel, despot.

Sagredo had time on his hands, and he occupied this time with his own scientific researches. He was the first to understand the paradox of what we now call the international date line: at any moment, he pointed out, there is always someone on the earth for whom it is tomorrow or yesterday, and indeed there must be two contiguous points on the earth's surface that are twenty-four hours apart. As a good anticlerical, he liked the resulting paradox that there was no moment at which it was Christmas or Easter for everybody.[5] He improved the thermometer, first invented by Galileo: brilliantly, he recognised that there was a vacuum at the top of the thermometer's tube, a vacuum through which sound could not travel. (Sagredo's thermometers were rather large, so that the space at the top of the tube was sizable).[6] He studied optics: unlike Galileo he recognised that the image of the world must be projected onto the retina upside down and be set the right way up by the brain, and unlike Galileo he recognised that nothing reaches the eye except light, just as nothing reaches the ear except vibrations.[7] It is striking to find him telling Galileo, whose knowledge of optics was clearly limited, that Kepler is the person to read.[8] (People kept telling Galileo to read Kepler, but Galileo protested that he found him virtually unreadable.)[9] He worked hard to produce good lenses for telescopes, and kept Galileo (who was constantly having to respond to demands for an authentic Galileo telescope) supplied.[10] His family had no respect for these artisanal activities (even though much of the hard work was done by skilled craftsmen whom he employed).

Sagredo died after a brief illness on 5 March 1620. He quite rightly distrusted doctors, and his brother Zaccaria thought that his failure to pay attention to the experts had been responsible for his death.[11] Zaccaria immediately cleared out the workroom in which Gianfrancesco had kept the tools for his experiments, stating that he did not want his children taking after their uncle. (As in many Venetian noble families, only one of the brothers had married in order to keep the family estate intact.)[12] It was presumably at this point that Galileo's letters to Sagredo, of which there will have been hundreds, were thrown away. When Galileo asked for some of Gianfrancesco's tools – the tools that embodied their shared intellectual life – he was told there were none left. He tried to get a picture or two, but the descriptions he was sent were useless – no artists' names were provided. It was quickly clear that it was (as Galileo wrote on the back of a letter from Zaccaria) more trouble than it was worth to try to obtain a memento of his best friend ('a loan at too high a rate of interest'). Gianfranceso had owned a fine painting of Galileo, but that too soon disappeared, along with all Galileo's letters. Indeed the family preserved no memory of Gianfrancesco at all. His name is not to be found in the family history.

Sagredo had the makings of a good scientist, but he had no intention of publishing, and he constantly advised Galileo to give up his own ambitions, and to think of science merely as a hobby or pastime.[13] If he was to publish, Sagredo insisted, he should publish demonstrations, not discourses – advice Galileo was to ignore when he turned Sagredo into a figure in his own dialogues.

So just what was the basis of their relationship? Galileo had taught Sagredo, who was eight years younger than him. Sagredo had consistently fought to have Galileo's pay increased while Galileo was teaching at Padua – although he naturally felt that this meant he virtually owned Galileo, who was once summarily ordered to accompany him on holiday. He seems very early on to have recognised Galileo as a truly great man, a great man who was under his personal protection – he had collected him, as he later collected Bassano, to be part of his menagerie.

Sagredo was away from Venice when Galileo discovered the telescope, published *The Starry Messenger* and moved to Florence. He returned in 1611 to find that many of their old friends wanted nothing more to do with Galileo, seeing his departure as a betrayal. Strangely, Sagredo did not. For a long time he persuaded his father to keep Galileo's old job in Padua unfilled, in case Galileo should decide to return. Yet in the great conflict between those who supported and those who opposed the Catholic Reformation, Sagredo was as firmly on the anticlerical side as anyone. During the Interdict he had won a certain brief notoriety by pretending to be a wealthy pious gentlewoman and entering into correspondence with a leading Jesuit after the Jesuit order was expelled from Venice: he managed to extract from his gullible interlocutor indiscreet comments on papal authority and an indecent interest in the idea of a bequest. Thus Sagredo too should have abandoned Galileo in 1610 and renewed contact with him, as other Venetians did, only after he began to get into trouble with the Church authorities.

Why did Sagredo's friendship with Galileo survive the rupture of 1610? Galileo was no longer his protégé, no longer his companion. Galileo was also not very good at respecting the unwritten rules of friendship, rules that were much more clearly understood in the seventeenth century than they are now. Sagredo, for example, sent Galileo lenses, asking for a bottle of Tuscan wine in return. He had to ask and ask before the wine was sent.[14] He asked Galileo to explain the principles of optics to him, but Galileo had better things to do with his time (and was perhaps aware of the limits of his own knowledge). There was constant tension over the terms of exchange between them. Sagredo asked Galileo to have his portrait painted by one of the finest Florentine painters of the age – he suggested Bronzino – and to send it to him; what he offered in exchange was not a portrait of himself that had been painted by Bassano, but a copy by Bassano's less able brother, touched up by the great man – an unequal exchange, as he himself acknowledged.[15] Galileo was now famous, but he had used his fame to win independence from Sagredo and the other Venetian nobles who demanded so much of his time and energy.

Part of the answer is that Sagredo was indeed a connoisseur. He had recognised Galileo as a great scientist, and nothing would make him step back from that valuation. Of all the judgements he made, this was the most astute, for his friendship with Galileo ensured that he continues to be remembered four centuries later, when his far more successful relatives have disappeared into the obscurity of Venetian political history. History's judgement has proved to be the reverse of that of his family, which remembered Zaccaria but not Gianfrancesco. Through Galileo he had found a series of occupations that helped him fill the days left empty by his failure to impress either his city or his family. To Galileo, if to no one else, he could present himself as a model of how to live one's life. When he got into an argument with the Jesuit Scheiner, for example, the contempt with which he treated him (giving him, as he said, 'a hairwash without any soap') was an example Galileo ought to follow. As for Galileo, he too lived on the margins of his world, from 1613 having effectively abandoned both

Florence and the life of the court and escaping to the seclusion of life in a villa. Rumours spread that he had been driven out of polite, pious society. Two lonely men, unable to find a place in the worlds they loved and whose admiration they longed for, turned to each other for comfort and companionship. It is not entirely surprising that Galileo came to idolise Sagredo – the word 'idol' is his own – who was free, free even of the ambition that Galileo could never shake off.[16] This was a true, if always uncomfortable, friendship.

From Sagredo's point of view, moreover, it had worked out very well. He had stuck with Galileo when Galileo had abandoned Venice and sought to establish close relations with Rome; by 1620 Galileo had fallen out with the Jesuits, and instead of engaging them in serious debate was treating their arguments with an almost aristocratic playfulness and disdain. In 1610 Galileo's preferred interlocutor had been Clavius; by the end of 1620 he was in touch with Diodati, a Genevan Protestant who had long been a close ally of Paolo Sarpi's. Galileo was safely back, or so it would have seemed to Sagredo, within the company of anticlericals and unbelievers. Back where he belonged.

Urban VIII

Just when Galileo was feeling lonely and isolated, Maffeo Barberini, a Florentine, was elected pope on 6 August 1623, taking the name Urban VIII. He and Galileo had been on friendly terms since 1611, when Barberini had supported Galileo in a debate, conducted over the grand duke's dinner table, about why bodies float or sink. He praised Galileo in a poem he wrote in 1620, a copy of which he sent to Galileo accompanied by a letter signed 'your brother'. As soon as he was elected he promoted two friends of Galileo's, fellow Linceans and fellow Florentines, Virginio Cesarini (who was sickly, and had not long to live) and Giovanni Ciampoli, to key positions in his administration, and he soon brought Galileo's closest collaborator, Castelli, to Rome to teach at the papal university.[1] In return the Linceans promptly elected the pope's favourite nephew, Francesco, soon to be a cardinal, to their society.[2] In October Galileo's *Assayer* was published – by the Linceans. It was now dedicated to the pope, and the pope's arms were prominent on the frontispiece. The first copy sold (a copy that had been given to the censor, and was sold as secondhand) was purchased by Grassi: when he opened the book and saw the pope's arms he turned white. As soon as the book was printed Ciampoli began reading it to the pope at meal-times.[3] This was, as Galileo already recognised, a 'marvellous combination of circumstances', a moment that must be seized.[4] To take full advantage of it he needed to be in Rome.

Yet Galileo delayed.[5] The weather was bad. He was ill in bed. He had to arrange for his orphaned nephews to enter monasteries. We can imagine that, after the experiences of 1616, he was reluctant to revisit the scene of his defeat. When he finally set out in early April 1624 he delayed his arrival by staying for two weeks with his friend and patron Federico Cesi in Acquasparta. It was during this visit that Galileo performed the only experiment (or rather demon-stration) of which we have an eye-witness account. He had been taken out for a jaunt on Lake Piediluco by his fellow Lincean Francesco Stelluti. They were travelling fast on still water in a boat rowed by six strong men, Galileo sitting on one side of the boat and Stelluti on the other. Galileo, apparently at random, asked if Stelluti had anything heavy in his pockets. Stelluti did – the key to his room (which, being for a seventeenth-century lock, will have been a bulky object). Galileo took it and threw it hard, straight up in the air. Stelluti immediately thought the key was lost: it would fall behind the boat in the water and would never be recovered. To his astonishment, despite the fact that the boat had moved forward six or seven metres while the key was in the air, it landed between the two of them, thus demonstrating the principle that the key would keep moving forward with the momentum acquired from the movement of the boat, even when it was no longer attached to the boat.[6]

This story is a simple refutation of the view that used to be prevalent – that Galileo never conducted any experiments. But what is really striking is that Stelluti was so convinced of the truth of Aristotelian physics that he would have been unable to throw the key in the air without having it fall in the water, for if it had been given to him to throw he would have thrown it forward and out over the bow in a misconceived attempt to prevent it from falling behind the boat. Having seen this display, however, it was easy for him, nine years later, to follow Galileo's demonstration in the *Dialogue* that the rotation of the earth would not be perceptible to anyone on its surface – birds would not be left behind in the sky as the earth turned, and arrows shot straight upwards would not fall to the west.

Galileo finally reached Rome (where he stayed with a friend – he was no longer living at the Florentine government's expense) on 21 April 1624. Eight months had slipped away. Within a week of his arrival he was eager to leave: he had decided that he was too old for a courtier's life, and that he lacked the ambition for preferment which could alone make it tolerable.[7] Galileo was interested in one thing and one thing only – persuading the pope to allow him to reopen the question of Copernicanism. In all Galileo met with the pope six times, but if they ever discussed Copernicanism we have no direct knowledge of what was said. We know that a cardinal had raised the matter on Galileo's behalf: Urban replied that Copernicanism had not been condemned as a heresy, but that there was in any case no risk of it ever being demonstrated to be true.[8] When Galileo left Rome on 16 June he had plenty of evidence of papal goodwill.[9]

Judging by what happened later, however, it must have been during these conversations that Urban laid down two conditions for any publication on Copernicanism. First, the issue must be presented as if there were good arguments for and against; Copernicanism could be defended as a useful hypothesis, but it could not be claimed to be indisputably true. Second, Galileo must present the argument that there are profound limits to human knowledge when it came to scientific questions. We might argue that the obvious cause of heat is burning, but the simple truth is that God is omnipotent, so that there are all sorts of ways, for the most part unimaginable to us, in which he might bring about heat (friction, for example). In requiring such an argument of Galileo, Urban was stressing a point which he himself believed to be of fundamental theological importance, although it was one that would have been profoundly puzzling to most contemporary philosophers, who were happy to argue that they knew the one and only true cause of heat, or light, or sound. Urban was thus trying to form an alliance with Galileo in the cause of a new, more modest, philosophy of nature.

In proposing this alliance, Urban must have thought that he was making an entirely reasonable request. Galileo had frequently claimed that what distinguished him from his opponents was his willingness to acknowledge that there was much he did not know. And he had ended *The Assayer* with a parable which had delighted the pope.[10] The parable is about a shepherd boy who tries to understand the nature of sound. Each time he thinks he has solved the problem, he discovers a new cause of sound (voice, violin, bagpipe, etc.) and realises that his previous understanding is incomplete and inadequate. The story ends with his trying to understand where the noise emitted by a cicada is coming from – and killing the harmless insect in trying to answer his question.

All Urban wanted from Galileo was a book that repeated his performance in *The Assayer*. What request could be more reasonable? Galileo evidently felt that he could meet these conditions, so he decided to test the waters. That summer, he retreated to his villa and wrote a fifty-page reply to Francesco Ingoli, who had written against Copernicanism eight years earlier, shortly before it was officially condemned, and was now secretary to the Congregation for the Propagation of the Faith.[11] Galileo's pretence was that he wanted to defend the Catholic Church against a charge which gave particular delight to heretics – that it did not know what it was doing when it condemned Copernicanism. By showing the strength of the arguments in favour of Copernicanism, Galileo claimed, he would also be demonstrating 'how little one should rely on human reason and human wisdom, and therefore how much one owes the higher sciences [i.e. theology]; they are the only ones capable of clearing up the blindness of our mind and teaching us those things which we could never learn from our experience and our reasoning'.[12]

The letter to Ingoli is a dress rehearsal for Galileo's great work, the *Dialogue*. In it Galileo seeks to destroy all Ingoli's arguments intended to prove that the earth is stationary (from the claim, for example, that objects dropped from a tower would be left behind as the earth turned) and that what we can see in the heavens would look different if the earth were not at the centre of the universe (because, for example, we would be able to see the stellar parallax resulting from the earth's orbit around the sun).

Galileo also adopts a tactic that puzzles and confuses modern readers. The letter to Ingoli, like the *Dialogue*, is about two cosmic systems, the Ptolemaic and the Copernican. As we see it, there were three world systems available at the time – the Ptolemaic, the Tychonic and the Copernican – and readers are often puzzled that Galileo appears to largely ignore Tycho. Galileo's assumption, however, is that the discovery of the phases of Venus, and the marked changes in the size of Mercury and Venus over time, and to a lesser degree in that of the other planets, had provided decisive evidence that the planets orbit the sun. He thus assumed that all intelligent Aristotelians had converted to the Tychonic system. Indeed as long ago as 1601 Kepler had written, 'Today there is practically no one who would doubt what is common to the Copernican and Tychonic hypotheses, namely that the sun is the center of the motions of the five planets.'[13] Galileo viewed the Tychonic system as an adaptation of the Ptolemaic system because, like the Ptolemaic system, it upheld the principle of a stationary earth. Tycho thus defended Aristotle's physics, while adopting the calculations of Copernicus's astronomy. It was this compromise that Galileo's arguments were directed against. When he writes,

> Aristotle was absolutely clear that he took the centre of the universe to be the centre of the planetary orbs, and at the latter he placed the earth. Now, in our time, it is as clear and glaring as the sun itself that the sun, and not the earth, is located at such a centre; indeed I believe you [Ingoli] understand this ... If any place in the world [i.e. the universe] is to be called its centre, that is the centre of celestial revolutions; and everyone who is competent in this subject knows that it is the sun rather than the earth which is found therein[14]

he is not saying that every competent person is a Copernican; he is saying that every competent person is a follower of either Tycho or Copernicus – there is no longer any scope for a defence of the traditional Ptolemaic view that all the planets orbit the earth.[15] The key question then is which is it that moves – the earth or the sun? Galileo claims that it would appear far more probable that the earth moves than the sun, for the earth, like the planets, shines by reflected light, and all the other bodies that shine by reflected light are in orbit around another body. The sun, on the other hand, appears to be indistinguishable from the stars, and the stars are all stationary.

But why, if Galileo was really arguing against the Tychonic rather than the Ptolemaic system, did he not simply say as much? His exchanges with Grassi suggest that he was furious that it was Tycho who seemed to be ending up as the beneficiary of his astronomical discoveries. Galileo was determined not to give him the recognition that would be implied in presenting the argument as one between the followers of Tycho and the followers of Copernicus. He was determined to polarise the debate as far as possible, forcing the philosophers of Tycho either to admit that they were really Aristotelians, or to acknowledge that they had no arguments against the Copernicans. Moreover he was not alone in seeking to polarise the debate in this way. In 1611, when the Jesuit mathematicians had seemed keen to accept Galileo's new discoveries, the General of the Jesuits had decreed that the Jesuits must avoid all innovation in intellectual matters. So if the Jesuit mathematicians, since the condemnation of Copernicus, were for the most part followers of Tycho, they were reluctant to say as much – rather they were keen to insist that they were orthodox Aristotelians. Galileo need scarcely fear that they would protest that he had given Tycho insufficient space. But as we shall see, there may have been a further, more pressing reason for leaving Tycho out of the debate.

In one respect the letter to Ingoli goes further than Galileo was prepared to go in the *Dialogue*. Like most of his contemporaries, Ingoli thought the Copernican system required the universe to be implausibly large. Galileo was not fazed by this issue at all: indeed, he is prepared to say that there is no way of telling whether the universe is finite or infinite.[16] This was dangerous territory. Bruno had argued that the universe was infinite, and that not only were the stars suns, but that they were circled by innumerable inhabited planets (a view which would imply innumerable Christs, for each world would need its own saviour). Moreover, it was difficult to see how an infinite universe could be the work of a creator: an infinite universe must surely have existed (as Aristotle claimed) throughout eternity.

While Galileo was writing the letter to Ingoli his portrait was painted by Domenico Crespi da Passignano, who had made the engravings of the sun for Galileo's *Letters on Sunspots*. The commission came from the brothers Marcello and Mattia Sacchetti, Roman artists and architects. When the painting arrived in Rome, having travelled with the reply to Ingoli, Galileo's close friend Mario Guiducci said that it was a very good resemblance, except that Galileo's beard was too white – he must have aged, Guiducci concluded, in the four months that had passed since he had last been in Rome.[17] It depicts Galileo at the age of sixty. He is wearing the toga of a professor. His simple linen collar will have been made for him by his daughter Maria Celeste. His face is weatherbeaten and tanned, from the many hours spent working in the gardens of his villa, on the hill of Bellosguardo. There is a look of anxiety about the eyes, his brow is furrowed, his mouth is slightly set. This is neither a happy nor a confident man.

Galileo's reply to Ingoli was sent off to his friends in Rome in October, and was soon circulating amongst them.[18] Ingoli heard of it, and naturally asked to be shown a copy.[19] The intention was for Galileo himself to come to Rome: the letter to Ingoli was to be used to test the metal of the pope.[20] But then the situation changed, and Cesi strongly advised that this was not the time. Galileo's friends in Rome learnt that *The Assayer* had been denounced to the Inquisition: they assumed that Galileo had been charged with Copernicanism, although we now know that he had been accused of drawing a distinction between primary and secondary qualities that was incompatible with belief in transubstantiation. The charge had been dismissed, but there was concern that the letter to Ingoli would provoke new accusations of heresy.[21] Moreover war had broken out in the Valtelline, the crucial pass which controlled entry across the Alps into Italy. Spanish troops were involved, and the French seemed likely to commit their forces in order to drive out the Spanish. The pope had more urgent concerns than the debates of philosophers. And so the letter to Ingoli was withdrawn from circulation; Ingoli himself was never allowed to see a copy.

Galileo was already reworking his theory of the tides – the key argument in favour of Copernicanism that had been omitted from the letter to Ingoli. And now he settled down to write his major work, promised since 1610, on the two world systems, the Ptolemaic and the Copernican. He worked slowly: when things were going well, he said, he was able to write for an hour a day.[22] He was easily distracted, wasting months improving on Gilbert's method for increasing the power of a magnet by arming it with an iron plating.[23] He found reading difficult – not just the works of Aristotelians, but also that of fellow Copernicans such as Kepler.[24] Writing was even worse.[25] The years passed and little happened – these were, or at least seem to have been, the most uneventful years of Galileo's long intellectual life.[26] It is during this period that he must have written his treatise on miracles, which circulated briefly among his close friends and was quickly judged to be imprudent and withdrawn. Had this text survived we might have a very different picture of his intellectual life during these years. The *Dialogue* was nearly finished, or so Galileo claimed, by December 1624,[27] but as time went by he seemed to be going backwards not forwards. By the autumn of 1628, with the letter to Ingoli and the treatise on miracles withdrawn, and with the *Dialogue* still unfinished, he had come to a complete standstill. Word spread that his great project had been abandoned and would never be finished.[28] Galileo had plenty of intellectual and political difficulties which might be sufficient to explain the way in which his work ground to a halt. But in these years his relations with his family became increasingly problematic; indeed, they were a constant source of anxiety. Let us now concentrate on a number of these relationships, before once more resuming the story of Galileo's intellectual life.

Family ties

Amongst professional historians, biography is not an intellectually respectable genre. This is because there is a good deal of confusion about what biography is for, and – if we are honest – just as much confusion about what history is for. As we turn to take a closer look at Galileo's private life we need to sort out some of these confusions.

My primary purpose in this book is to provide an intellectual biography of one of the world's greatest scientists – to reconstruct the development of his ideas over time. The obvious objection to such an enterprise is that people do not think alone: they think by talking to others, by sharing ideas, by engaging in common enterprises. This objection is always going to be sound, but there is less merit to it in the case of Galileo than in that of almost any other major intellectual. Galileo really did think alone. Offered opportunities to collaborate with major scientists such as Tycho Brahe and Kepler he shied away from them. The last major intellectual influence on Galileo's scientific thought was William Gilbert, whom he had read thirty years before issuing his most important publication. He had a close circle of friends with whom he shared his ideas – a private science is as conceptually problematic as a private language – but their role was primarily to validate the importance of his work, not to criticise and develop it. One can thus write an intellectual biography of Galileo which focuses narrowly on Galileo, and leaves (for example) the Linceans in the background.

In the case of Galileo, however, it is not just his ideas that matter. The condemnation of Copernicanism in 1616 and his own trial in 1633 are important events for understanding the cultural location of science in the early modern world. A satisfactory account of these events ought not to start with Galileo at all, but with attitudes to tradition, authority and innovation in the Counter Reformation Church; its focus would be as much on Jesuit scientists as on Galileo. Still, in both 1616 and 1633, Galileo had a central role to play. Understanding that role involves recognising that his approach was confrontational. He did not turn away from Copernicanism to concentrate on other topics, such as the motion of projectiles. He did not present Copernicanism in cautious and hypothetical terms. He staked everything on facing his opponents and overcoming them – either Caesar or nobody. It is evident that we cannot escape questions about Galileo's character and his private commitments here. Why did he leave Venice, where he was protected from the Inquisition? Why did it take him thirty years to write his system of the world? Why did he gratuitously clash with the Jesuits over the comets of 1618? Why did he not publish abroad? The answers to questions such as these cannot be separated from an assessment of Galileo's character. Here biography – understood as the study of a person's habitual responses and reflex reactions – becomes an inseparable part of the study of scientific progress and cultural change. If historians cannot find a place for biography, then their history will be the poorer for it.

This characterological approach to biography is sufficient to justify a much closer study of Galileo's private life than has been undertaken in the past. But let me acknowledge two further motives that underlie the discussion that follows. First, like any biographer I am interested in the story of a life, and personal relations must be central to any such story. History, as a profession, often shies away from story-telling because it seems to lack intellectual rigour, in the process underestimating its role in preserving an interest in and sense of the past. Galileo's flirtation with Alessandra Buonamici is a story – a story of true love – worth telling in its own right.

And this brings us to the third, and most complicated, justification for the study of Galileo's personal life. This is a book about Galileo because Galileo was a great scientist; and yet no one can feel comfortable with a history which has space only for the great. We need to find room for Galileo's brother and for his son, even for the daughter he never acknowledged, not because they were important, but because each and every one of us is entitled to be treated with respect. Edward Thompson famously sought to rescue the handloom weavers from the condescension of history; my intention (a more modest one, to be sure) is to rescue those close to Galileo who have been written out of the story every time they have failed to perform the roles assigned them – the roles of dutiful daughters, admiring disciples, faithful friends. Galileo's relationship with his daughter Maria Celeste has come to symbolise his intimate life, and in the process to substitute for all the other aspects of his private life, for two reasons. First – and this is a good reason – Maria Celeste wrote like an angel. Second – and this seems to me a bad reason – she was unfailingly dutiful, admiring and faithful, and he was always generous to her. She thus plays a central role in accounts of Galileo as a secular saint. But Galileo was no saint; had he been the course of his life would have been very different. If we are to get a better sense of his strengths and weaknesses we need to explore other aspects of his personal life – and, in due course, re-examine his relationship with Maria Celeste.

Sometime in 1627 Galileo encouraged his nephew Vincenzo Landucci, his sister's son, to marry Anna di Cosimo Dieciaiuti, despite the opposition of his father, Benedetto Landucci, with whom Galileo's relations had never been good.[1] We have no contemporary report of the marriage: we learn of it only because of the conflicts and disputes that sprang from it. Benedetto promptly refused to support his son. (Vincenzo later unsuccessfully sued his father for maintenance.)[2] This was a strange marriage indeed. Galileo had been paying for Anna to be brought up in a convent. He now paid for her wedding clothes, and the marriage took place from his home. He may have supplied the eight hundred ducats of dowry, since he seems to have provided all the rest of the couple's funding. Certainly he promised Vincenzo a hundred ducats a year and sixteen ducats for his rent, and he found a place for the couple to live. This agreement, apparently sealed only by a handshake, was placed in writing in 1630, with the limitation that Galileo could cease paying once he had fixed up profitable employment for Vincenzo. These arrangements naturally caused a great deal of talk, and embarrassed and angered Galileo's own son, also called Vincenzo, and his brother Michelangelo. What, everyone asked, was Anna di Cosimo Dieciaiuti to Galileo, that he should arrange a respectable marriage for her and pay out as if she were his own daughter?[3]

Of the marriage there were two children – a little boy, Benedetto, born in February 1631, and a little girl, Maria Virginia.[4] Some months before Benedetto was born Galileo reneged on his obligation to pay his nephew, and in May 1631 the two were suing each other.[5] Key details of the case are lost to us, because the court heard the parties expand on their written submissions *in camera*, but we know that Vincenzo had been reduced to such povery that he had sold all his furniture with the exception of his bed, and had been forced to take work in the hospital for plague victims, the *lazzaretto*, while Galileo was claiming that his primary obligation was now to support his deceased brother's family. The court's ruling was far more favourable to Vincenzo than to Galileo: it was that Galileo must, for as long as he lived, pay Vincenzo at a reduced rate of seventy-two ducats a year, but that he must continue to pay even if Vincenzo later obtained profitable employment. He need no longer pay Vincenzo's rent, however. The court dismissed out of hand Galileo's claim that it was Vincenzo who owed Galileo money from the proceeds of the furniture he had sold.

In July 1633 Anna died of the plague. Galileo wrote from Siena to ask Maria Celeste how Vincenzo had taken it, to which she replied, 'I cannot know how much Landucci is mourning his wife; all I know is what Giuseppe [Galileo's servant] told me the day he went there, together with Signor Rondinelli, to take him the six ducats [a month's allowance], which was on the eighteenth of this month. They left the money on the doorstep, and saw Vincenzo inside the house in the distance, and he looked very distressed, with skin the colour of a cadaver rather than a living person, and with him were the two babies, a boy and a girl, for only two have survived.'[6]

Soon after he returned to Arcetri from his trial in Rome Galileo halted his payments to Vincenzo, saying he had undertaken them only to subsidise the marriage, but the courts reinstated the contract, although they gave Galileo formal permission to bring an action to have it annulled on the grounds that the death of Anna should have been the end of his liabilities. In September of 1638 Galileo and Landucci reached a new agreement: instead of paying Vincenzo, Galileo would pay the cost of Virginia's entering a nunnery.[7] In December of 1639 he visited Virginia in her nunnery and at the same time was urgently trying to set aside seven hundred ducats in the government savings scheme, the Monte di Pietà – an action which required the permission of the grand duke, and the reasons for which were to be explained to the grand duke orally by Galileo's representative.[8] In writing Galileo said only – cryptically – that he was acting against his own wishes, but not against his own will. Around the same time he was also trying to explain to Fulgenzio Micanzio why he found himself supporting Anna's family, but Micanzio found his explanation so enigmatic as to be incomprehensible.[9] Soon the grand duke (presumably in the light of the oral explanation he had received in December) began subsidising Virginia out of his own pocket, and in August 1640 Galileo put four hundred ducats into the Monte to pay for Virginia, a sum withdrawn by the nunnery when she took her vows in December.[10] From Virginia there survives only a brief note, sent at Christmas 1639, thanking Galileo for all that he had done for her.[11]

We catch a last glimpse of Virginia. In 1641 Alessandra Bocchineri Buonamici (the sister of Galileo's daughter-in-law) happened to meet someone she calls Galileo's 'parentina', his little relative, in Prato, although it was only later that she discovered who she was. Had she known at the time, she told Galileo, she would have talked to her about her esteem for him. As it was, at least she had seen her once, and she came away with a clear impression of her as 'molto bellina e spiritosa'.[12] Just as when Maria Celeste carefully draws for Galileo a picture of the two infants, Benedetto and Maria Virginia, deprived of their mother, so here too one feels that more is being conveyed than is being said. This little relative, it is understood, is a person who matters to Galileo, and consequently to those who care for Galileo. As for her brother Benedetto, Viviani met him when he was in middle age. He was, it seems, the spitting image ('il vero ritratto') of Galileo.[13]

So – Galileo arranged a marriage for Anna as if she were his own daughter, and he took on exactly the same responsibility for Anna's daughter as he had for his own daughters. Ostensibly Galileo's payments were made out of goodwill towards Vincenzo Landucci, his nephew, and Maria Virginia was simply his great-niece. We can, though, be sure that there was some secret hidden behind this facade: Vincenzo Galileo's and Benedetto Landucci's anger, the secret evidence presented to the tribunal in 1631, the verbal explanation conveyed to the grand duke in 1639, and Micanzio's puzzled reaction to the facts as they were presented to him – each of these suggests that only some hidden motive could explain the contract with Vincenzo Landucci, a contract which Galileo had assumed would be annulled on Anna's death. Only one explanation fits all the evidence: Anna was Galileo's daughter, and Benedetto and Virginia his grandchildren. Galileo had three children by Marina Gamba, but he also had a fourth,

Anna, whose mother's name was Cassandra.[14] Anna must have been born very soon after Galileo moved from Padua to Florence in 1610, and will have been in her early twenties when she died. It is difficult to resist the impression, given the expense to which he went in order to arrange a marriage for her, and his willingness to take responsibility for her daughter, that Galileo was very fond of her, and that when she died he hoped to hear that Landucci had truly loved her. Of Galileo's relationship with her mother Cassandra no other trace survives.

THE RETURN OF MICHELANGELO GALILEI

Galileo's brother Michelangelo, who was almost twelve years younger than Galileo, had always been a disappointment. In 1593 he had gone from Padua to Poland to make his fortune as a musician; at some point he returned, but in 1600 he left for Poland again, and Galileo had spent sixty ducats kitting him out. Michelangelo failed to repay this loan (although he was earning three hundred scudi a year, and was supplied by his employer with servants and a carriage and four), as he failed to contribute to either of his sisters' dowries. In 1606 he was back in Padua, but in 1608 he was in Munich, where four ambassadors attended his marriage to Anna Chiara Bandinelli (known as Chiara); at this point the brothers were squabbling about whether the money spent on the wedding feast ought not instead to have been put towards Livia's dowry.[15] Michelangelo was extravagant in other ways – he insisted, for example, on drinking imported wine rather than the local beer.[16] The years passed; Galileo saw little of his brother (there was a brief visit in 1615), but the two exchanged occasional sentimental letters. The legitimation of Galileo's son Vincenzo, which disinherited Michelangelo, seems not to have caused any rift between them.

And then in January of 1627, after some twenty years in Munich, Michelangelo announced that he was coming home with his wife, his sister-in-law and his seven children (his wife was just about to give birth to the eighth). He had run out of money and planned to start again. Galileo was worrying about finding someone to look after him as his sight failed – he already knew that he was going blind – and had turned to Michelangelo for help. By May the scheme had changed: Michelangelo now planned to send his wife and some of his children, while he stayed behind with the others. It would be good for Galileo, he assured him, to have a woman to take care of him, and company to amuse him; he and the rest of the children might follow later. But by August they were back to the original plan: the whole family was coming, together with a servant (although in the end one daughter was left behind in the care of nuns).[17] Galileo wanted his brother to sell everything in Munich before setting out, but Michelangelo was not willing to burn his boats behind him; he counted, mistakenly, on his brother's paying the cost of the journey.[18] They presumably arrived in September. In February 1628 Michelangelo returned to Munich, leaving the rest of his family behind with Galileo (with the exception of Vincenzo, who, as we shall see, was studying in Rome). He had left, symbolically, with the keys to Galileo's house in his pocket.[19] The truth is that Galileo could have afforded to support the whole family in Florence, and indeed had offered to do so; but Michelangelo could not stand the prospect of unemployment and dependence, and was unwilling to move into accommodation in town (probably the parental residence).[20] His intention, he made clear, was never to return to Florence if he could help it.

It looks as though the two brothers had incompatible fantasies. Michelangelo dreamed of living with his brother in one, big, happy family. But Galileo wanted him to occupy their mother's home, and drive out the memory of her presence. This was too much to ask.

Galileo was thus living with Chiara and five children under the age of twelve.[21] When he fell ill in March 1628 he had himself carried to a friend's house in town for a bit of peace and quiet, since at home he found himself surrounded by a constant, intolerable hubbub.[22] But he was fond enough of Chiara to promise her that she would be well looked after in his will; if her report is to be trusted he must have intended to disinherit his son Vincenzo, who had recently been in trouble again.[23] Sometime after August 1628, Michelangelo returned to Florence and took away his wife and children. He and Galileo parted on bad terms: Michelangelo was ill, penniless and distressed, and Galileo may well have told him that his children were better off staying in Florence, and that he should stay there too.

Michelangelo Galilei died on 3 January 1631. His illness was mental before it was physical: for three years he had suffered from an uninterrupted melancholy, and it was only during this time that he fell physically ill. He left his wife and children without means of support.[24] Galileo was far from eager to come to their assistance, and pressure had to be brought to bear on him before he would promise to do so: the only payment we know of was for the modest sum of fifty scudi, and he was soon being urged to give more.[25] Further disasters befell the family. By 1636 only three of the children, Vincenzo, Alberto and Cosimo, remained alive, while the family's few possessions had been destroyed when Munich fell to the Swedes in 1632. Chiara was dead.[26] Galileo, who had been trying to obtain news of them, wept when he heard of their pitiful state. He immediately resolved to send them a hundred scudi and invited Alberto to visit him.[27] Alberto finally obtained permission to come in the autumn of 1637, and spent a year with his uncle. Galileo went to great lengths to obtain a fine violin for him and arranged for him to visit Rome.[28] In his wills of 1633 and 1638 Galileo left a thousand scudi to be divided amongst Michelangelo's surviving sons (the will of 1638 was made while Alberto was in Florence), but in a codicil dated 19 November 1638 he struck them out of his will. Alberto had left only six or seven weeks before. Apart from a small pension to support his surviving daughter Arcangela (the name Livia had taken on entering the convent), Galileo left everything to his son Vincenzo, though he did his best to protect the capital so that it would be inherited by future generations.[29] On his deathbed, Michelangelo had sent a message to his brother pleading with him to take care of his wife and children. But Munich was far away, and when Alberto left Florence in October 1638 Galileo would have had no expectation of ever seeing him again.

Alberto, for his part, continued to write faithfully for some time, despite the fact that his letters received no reply. In November of 1639, entirely unaware that he had been disinherited, he even wrote formally asking for Galileo's approval for his forthcoming marriage – for, as he eloquently put it, Galileo was the only safe haven he had in all the world.[30] Galileo certainly received the letter, but was only shamed into replying much later, when he learnt that his old friend Fulgenzio Micanzio had discovered that he had abandoned Alberto to his fate.[31] Galileo had made considerable sacrifices to provide dowries for his sisters. He had taken in his brother and his family, and had tried to keep them close to him. He had supported his own illegitimate children. He surely felt that he had done enough, but he knew very well that his behaviour would not stand up to the scrutiny of his friends and neighbours.

But there is another reason why Galileo hardened his heart against Alberto. For over a decade he had been worrying about who would look after him as he grew old; for over a decade he had known that he was going blind. He had servants living in his villa – a man and a woman – but he must have feared that once he was blind he would lose all control of his affairs, and become incapable of further intellectual work. It was probably this fear that had prompted him to encourage Michelangelo to return to Florence, and had led him to go to extraordinary lengths to keep Chiara by his side. Shortly before Alberto arrived the grand duke, alerted by the news that Galileo expected never to use his telescope again, had visited Galileo and had promised (we may gather) that, come what may, he would be cared for – a promise that he was to keep.[32] But the year of Alberto's visit was the year in which Galileo finally lost his sight and had to learn how to live with blindness. These were months of great uncertainty, and if there was any closeness between Galileo and his nephew, then Galileo must have become heavily dependent on Alberto during the course of his visit. It must have been an exquisite pleasure to a blind man to have a skilled musician play for him. We do not know if the possibility of Alberto's staying was ever discussed – he had a job to go back to, and there is no evidence that he had any competence in mathematics, which was the essential prerequisite for anyone who was to serve as Galileo's companion. But, reasonably or unreasonably, Galileo must have felt his departure as a desertion.

VINCENZO, SON OF MICHELANGELO

On 3 June 1628, a few months after Michelangelo had returned to Munich, Benedetto Castelli wrote from Rome to Galileo in Florence. Castelli had been looking after Galileo's young nephew Vincenzo, Michelangelo's son, who had been sent to Rome to study music.[33] Vincenzo, unfortunately, had been nothing but trouble. He had complained at having to sit through endless sermons during Lent – the words, he said, went in one ear and out the other. He refused to perform his devotions, even putting off Easter confession to the very last moment. He stayed out overnight. He consorted with foreigners. He was extravagant – he had tried to buy a diamond ring. On being told off, he had shouted at Castelli, 'I'm only here because my father and uncle want to get rid of me.'[34] Castelli regarded this remark – which was all too close to the truth, for there was a suggestion that Vincenzo had struck his mother – as too shocking for a reasoned response.[35] Castelli told Vincenzo that he was at the point of having to beat him with his own hands, as one would a madman.

All that was bad enough. But Vincenzo had now gone further in an argument with his landlord: 'I'm not an idiot, like you lot, who worship a bit of painted wall', he had said. 'Prudently', Castelli now reported to Galileo, 'his landlord had replied that he trusted he did not mean what he said, for if he really meant it then he was obliged to denounce him to the Holy Office, and he would be burnt alive in the Campo de' Fiori.' Castelli's advice was that Galileo must get Vincenzo back to Florence, without telling him what was in store for him, and there Galileo must himself report Vincenzo to the Inquisition, for otherwise Galileo was in danger of falling under suspicion of being the accomplice of a heretic.[36]

Vincenzo was indeed removed from Rome as fast as possible, but Galileo did not denounce him, nor have him imprisoned for insubordination as later recommended by Castelli, who seems to have had no doubt that at a word from Galileo space would be found in the prisons of Florence for a disobedient young man. Instead he was shipped off to Poland, where his father had connections, and where he was safely out of reach of the Inquisition.[37] What became of him we do not know: even his family eventually lost track of him.[38]

Castelli writes that when he was told what Vincenzo had said he felt as if his arms had been broken – he may have been imagining the standard torture used during interrogation by Italian authorities, the *strappado*. The words Vincenzo had spoken immediately invoked the image and the smell of a burning heretic; and to be in the presence of such a person was to be oneself at risk of torture, or worse.

Galileo was all too familiar with that feeling. And he understood that Castelli was alarmed, and rightly alarmed, on both his own and his friend's behalf. By 1628 Galileo had been denounced to the Inquisition five times: first in Florence, then in Padua, then again both in Florence and in Rome, and most recently in Rome. There were people who said that the only reason he had never been tried for heresy was that he had powerful protectors.[39] But just as Galileo could not have protected Vincenzo if Castelli had denounced the young man to the Inquisition, so there was a point beyond which even those who protected

Galileo would find themselves powerless. Above all, everyone chose carefully the words they entrusted to paper.[40] In advising Galileo to denounce his nephew to the Inquisition, and in carefully describing Vincenzo as mad, Castelli was constructing a defence for himself as well as giving his friend good advice. Thereafter he tried to spend as little time with Vincenzo as possible, fearing that he would say something in the presence of witnesses that would leave him no option but to call in the authorities.

Vincenzo had been born and raised in Germany, where there was no Inquisition, and where Protestants and Catholics often mixed freely. It would have been better, said Castelli, if he had been born and raised in Geneva. Then his beliefs might have been as diabolical, but at least he would have known that there were lines one must never cross. In studying Galileo, we too find ourselves in a foreign country, as Vincenzo did. We do not understand the rules that Galileo, together with his relatives and his friends, his patrons and his enemies, took entirely for granted. It takes the arrival of an outsider or a madman, someone who cannot or will not play by the rules, to bring into the open what would otherwise have remained unspoken and unwritten. In order to understand Galileo we need to understand the lessons that Vincenzo stubbornly refused to learn.

Galileo understood these rules and respected them. Because he was tried by the Inquisition, threatened with torture, condemned to house arrest, it is easy to think of him as a rebel. But rebels (and those who misunderstood the rules) came to very different ends from that of Galileo. Galileo's friend the Dominican monk Tommaso Campanella was horribly tortured, spent twenty-seven years in prison, and died in exile, in France.[41] Galileo's critic Francesco Sizzi went to France, where he published a pamphlet taking sides in a French political dispute. He was condemned to be broken upon the wheel – that is, he was tied to a wheel, and the executioner clubbed his body, breaking his bones one by one until he was dead.[42] Galileo was far more cautious than his nephew, his friend and his critic. He not only died in bed; he died at home. Did Vincenzo's close shave alarm Galileo? I doubt it. It would never have occurred to him to make the sort of rash statements that Vincenzo had made – not, at any rate, to someone like Vincenzo's landlord, someone who could not be trusted.

Permission to publish

In March of 1629 Galileo fell seriously ill. He came face to face with his own mortality. And he resolved that come what might he would at last finish the *Dialogue*.[1] His friends were delighted. It was high time, one of them had recently written to him, that he stopped murdering himself with his stubbornness, stopped betraying his intellectual gifts, and recognised his responsibility to posterity.[2] Within a few weeks he was once more well enough to work in his garden, but he was not yet ready to pick up his pen.

Finally in October 1629 he announced that he had begun writing again after an intermission of three years – three years which had seen the return of Michelangelo and his family and their rancorous departure, a crisis in his relationship with his son (the details of which escape us), and on top of that, the marriage of Galileo's unacknowledged daughter.[3] He was spurred on by the news that the Dominican Niccolò Riccardi had been elected Master of the Sacred Palace. Effectively the pope's personal theologian, the Master had the authority to prevent any book from being published; conversely his support more or less guaranteed publication. Riccardi carried with him a nickname given him by Philip III of Spain, who had dubbed him 'Father Monster' on account of his vast bulk. Riccardi had been appointed to license *The Assayer*, and instead of merely putting his imprimatur on the book had written words of praise for Galileo. He was, he did not hesitate to say, entirely at Galileo's service.[4]

Galileo hoped to be finished by the spring. His friends could not find the words to describe their relief. They awaited the manuscript, they said, like manna from heaven.[5] Work proceeded even faster than Galileo had hoped. By early January 1630 he had finished writing a book he had begun before 1597.[6] Soon chapters were being read aloud to an admiring audience. The grand duke was delighted to learn that the great work was finished at long last – although harsh words were spoken in his presence by others, who presumably feared the book would be heretical.[7]

Something other than his near-death experience in the spring of 1629, the good news of Riccardi's promotion and the departure of Michelangelo and his family speeded up Galileo's progress. He had long known that his sight was failing: cataracts were growing in both his eyes. Soon he would be unable to make out the faces of his friends.[8] And he knew that some of his most important scientific discoveries had yet to be published. He had to get a move on if he was to finish while he could still see enough to write.

In November Galileo had written to the Tuscan ambassador in Spain, seeking information about trade winds and ocean currents.[9] Giovanfrancesco Buonamici's reply contained information that Galileo could not have hoped for. Since 1616 his theory of the tides had explicitly acknowledged that there were normally two high tides and two low tides a day, although his theory appeared to imply there should be only one of each. Galileo explained the second tide as a reverberation or echo of the first. But Buonamici told him that there were reports (which he himself refused to credit) that in the Eastern Ocean there was only one tide a day. It seemed possible that there might eventually prove to be decisive empirical evidence to support Galileo's theory.[10]

In February 1630 it was decided that Galileo should go to Rome and make arrangements for publication with Father Monster. He did not actually leave until the end of April (he had pains in his legs that prevented him from travelling), by which time he had arranged to travel in one of the grand duke's litters.[11] At the same time, his trip was given official endorsement, which meant that he would be staying with the Florentine ambassador in Rome. Cosimo II de' Medici, who had come to power in 1609 and had brought Galileo, his maths tutor, to Florence the following year, had died in 1621; his successor, Ferdinando II, was now aged twenty, and had been the nominal ruler of Florence since he came of age in 1628, although much power remained with his mother. He was to prove a loyal supporter to Galileo, just as his father had been. Indeed his loyalty had recently been tested, for as soon as Ferdinando came of age an attempt was made to deprive Galileo of his salary on the grounds that he was being funded out of a tax on the clergy that was supposed to support the University of Pisa, despite the fact that Galileo gave no lectures in Pisa. The opinions of canon lawyers were sought, and Galileo's salary was continued, partly on the grounds that he was about to complete three great works which would fully justify all the money that had been spent on him.[12]

Galileo arrived in Rome on 3 May 1630 and left on 29 June. While there he met up with old friends, amongst them Campanella, with whom he had a lively debate in which Galileo, Ciampoli and others (including the Scotsman George Conn, who was later sent as the pope's representative to negotiate with Charles I for tolerance towards Catholics) defended atomism and Campanella attacked it – yet another conversation in which Galileo expressed himself more freely than he could ever do on paper.[13] But he was there on a mission. He had arrived carrying the completed manuscript of the *Dialogue*. He left having had an audience with the pope, and with a promise from Father Monster that he would need to make only a few small changes in order to get permission to publish. Everyone who could do so escaped from Rome during the hottest months of the summer, but Galileo probably felt that his business in Rome was finished. He would have left his manuscript behind, but Prince Cesi's health was poor.[14] He hoped to return a slightly revised manuscript (for which he anyway still had to write a preface and compile an index) to Rome in the autumn. Cesi would take charge of obtaining formal approval to publish and would then see the book through the press.[15] If necessary Galileo would return to Rome himself. The book would perhaps appear with a dedication either to the pope himself or to the cardinal nephew, and if not it could be dedicated to the prince.

There were three people Galileo had to convince in order to obtain permission to publish. First there was the cardinal nephew, Francesco Barberini. Barberini had had a long, private conversation with Castelli in February 1630, in which Castelli had outlined Galileo's theory of the tides as caused by the (purely hypothetical, of course) movement of the earth. Barberini had had one objection: if the earth moved, did not this mean that it was a star? And was that not theologically unacceptable? Castelli had reassured him: Galileo would certainly prove that the earth was not a star.[16]

This discussion is of foremost importance, as it is our best guide to Francesco Barberini's thinking. We need first to understand what was being said. In classical astronomy there were two types of star: fixed and moving. A moving earth would be a moving star – in other words, a planet. When Castelli reassured Francesco Barberini he was equivocating in relation to these two senses of the word 'star': Galileo could certainly prove that the earth was not a fixed star, but it had been precisely the argument of his letter to Ingoli that there was no apparent difference between the earth and a planet. But, Castelli reasoned, it was easy to see that no two planets were alike. Galileo could surely prove that the earth was different from Mars or Venus. Taken in this sense, there was no such thing as a typical moving star.

These were not matters of indifference to Barberini, and he must have given them considerable thought. The year before, his brother Taddeo had commissioned a vast fresco covering an entire ceiling in the Palazzo Barberini. It showed the earth hanging in space, a large globe on which the continents were clearly visible. At the centre is the noon sun, so bright that it dazzles the eye and the disc itself cannot be made out. Between the earth and the sun is Divine Wisdom, carrying, as Wisdom usually does, a mirror, a tribute to the Delphic oracle's 'Know thyself'.[17]

Since this fresco, which is the best guide to the astronomical thinking of the Barberini family, has been generally misunderstood, we need to pause over its interpretation. The painting is often said to represent a Copernican universe, but in fact it is perfectly ambiguous in its relation to the three cosmological systems – the Ptolemaic, the Tychonic and the Copernican – because we cannot tell from it where the centre of the universe is, or whether it is the earth or the sun which is stationary. This ambiguity is surely deliberate. The fact, for example, that we can see clearly the line where day turns into night in no way implies that the earth is rotating, as a Ptolemaic representation of the earth would have shown day and night in exactly the same way. But one thing is clear: the sun that illuminates the earth is not the same sun as is painted overhead shining down on everyone who enters the room, for that sun would not illuminate the side of the earth which faces the observer. It is clear, then, that for all its naturalistic exactness in the portrayal of the globe, the painting is not intended to represent the real configuration of the heavens – it is an allegory, not a cosmology.

The claim that the painting is implicitly Copernican depends on a further claim: that the allegory was constructed not by Andrea Sacchi, who painted it, but by Campanella. Urban VIII, we know, was terrified of eclipses, which, for reasons relating to the location of the sun in his birth chart, he believed might be portents of his death. We also know that he employed Campanella to perform astrological magic to ward off evil associated with an eclipse (an activity which caused a good deal of gossip).[18] The emphasis on the sun in this painting may indeed reflect Urban's own preoccupation with heliocentric imagery, and his belief that the sun was important for an understanding of the astrological conjuncture both of his birth and of his election as pope. But the astrological information provided by the painting is very limited; there is nothing to suggest that it has anything to do with eclipses, and there is no direct evidence that Campanella played a part in its design.

Campanella had met Sarpi and Galileo in Padua in 1592: at that meeting they told him of their atomism. Shortly afterwards Campanella himself was arrested by the Inquisition (he was merely passing through Venetian territory, so the Venetian state made no attempt to protect him) and the charge of atomism was one of many made against him. For five years he was held in a monastery, and soon after his release he led a revolt, in 1599, against Spanish rule in Naples. As a result he was tortured and held in a Spanish prison for twenty-seven years. In 1626 the pope arranged for him to be released from Spanish custody and transferred to one of the Inquisition's prisons in Rome. In 1629 he was given his freedom.

Campanella presented himself to Galileo as a convinced Copernican, but we know that he did not present himself to the Barberini family in the same terms.[19] Early in 1616, from prison, he had written a *Defence* for Galileo, intended to influence the outcome of the debate over whether to ban Copernicanism. The book had been published in Frankfurt, a Protestant city, in 1622. It was, Campanella later said to Galileo, written in cipher – that is, it concealed Campanella's support for Copernicanism behind a studied neutrality.[20] Ostensibly Campanella simply argued that there was nothing illegitimate about holding Copernican views; he never said that they were correct. Some believe that he was a more radical Copernican than Galileo himself – that (like Bruno) he thought that other planets might be inhabited.[21] Others believe that he never committed himself to Copernicanism, and he certainly assured Urban VIII that he was not a Copernican.[22] In Rome, as the pope's adviser, he encouraged the pope to present himself as the sun, and driven into exile in France in 1634 (the Spaniards accused him of continuing to plot against them, and the pope and the French ambassador helped him to escape) he carried with him the idea of a 'sun king', an idea soon to be adopted by Louis XIV. From Paris he sent word back to Galileo that all he needed for perfect happiness was the pleasure of his company.[23]

Taddeo Barberini, who had commissioned Sacchi's fresco, is unlikely to have been any more prepared than his brother Francesco to accept the view that the earth was simply a planet: theology required that the universe had been created in order to provide a home for mankind, and in order to play out the drama of salvation. If there were numerous inhabited planets, the Bible would cease to be the unique record of God's purpose, and Christ would cease to be the unique intermediary between the Father and the created universe. The point of the allegory of Divine Wisdom was not to show that the earth was merely a planet: to do that other planets would have had to be depicted, together with the earth. Rather it was to show that divine wisdom rules over the whole globe – a short step from claiming that the authority of the pope, as the embodiment of divine wisdom, ought to be acknowledged throughout the world.

In 1624 Francesco Barberini had been designing a state carriage for his uncle, recently elected pope. He planned a painting of the sun at the centre of the carriage's ceiling, with the twelve signs of the zodiac around it.[24] Galileo's friends were amused at the idea of the pope travelling around with the Copernican universe painted on the ceiling of his carriage – for did this image not imply, as Copernicus had claimed, that the sun was the centre of the universe? In 1633, in preparation for the condemnation of Galileo, Francesco Barberini removed armillary spheres that had stood in his library and that illustrated the Copernican system.[25] He was surely interested in Copernicanism, and but for his conversation with Castelli we might reasonably suspect that he was a Copernican too.

Fortunately for anyone seeking consistency, the fresco in Palazzo Barberini and the proposed decoration on the pope's carriage are both open to another interpretation. In the system of Tycho Brahe the sun functions as the centre of the heavenly bodies, but the earth is stationary and is quite different from the planets. If we look again at the fresco in the Barberini palace we see that the earth does not shine like a star, as Galileo claimed it would if one could see it from distant space. It does not even shine like the moon, as he said it would if seen from the moon. This suggests the Barberini preferred Tycho Brahe to Copernicus – and this may have been a crucial factor in Galileo's decision to write his *Dialogue* as an attack on Ptolemy rather than on Tycho. We certainly know that Riccardi favoured the Tychonic system.[26] Nevertheless, Galileo had dinner with Francesco Barberini and he evidently satisfied him that his book presented no serious problems.

The second person Galileo had to convince was Raffaello Visconti. Riccardi had already repeatedly assured Galileo that there would be no problems, but in practice there were issues that needed to be discussed, and he appointed his colleague Visconti to review the text, telling him both that he was to give it favourable treatment and that the grand duke of Florence had declared an interest in seeing it approved. Galileo and Visconti met and talked things through, and Visconti promised to do his best to persuade the pope that Galileo's theory of the tides was not objectionable.[27]

Finally there was the pope himself. He too was sympathetic. Campanella had been to see the pope in March and had told him that he had been on the point of converting some German Protestants to Catholicism when they had discovered that they would have to give up being Copernicans, and had baulked. The pope had replied, 'This was never our intention, and if the decision had been ours that decree would never have been issued.'[28] In 1629 Agostino Oreggi had published a book in which he reported a conversation between Urban, before he was elected pope, and a learned friend – someone we can safely identify as Galileo. The friend had outlined a Copernican model of the universe. Urban, in response, had asked whether this was the only model that would work. Before you answer, he had said, bear in mind that you need to show that any other model involves a logical contradiction, for God can do anything that is not contrary to the rules of logic.[29] Over and over again Urban and his intermediaries must have made it clear to Galileo that this was his position. Galileo must not claim that Copernicanism was the only possible theory that could explain the phenomena; he must acknowledge that God could devise ways of doing things that were beyond our comprehension and imagination. Galileo could therefore argue that the Copernican system was plausible, and that it made sense of the evidence, but he must not claim that it was true. This had been precisely Galileo's approach in the letter to Ingoli: there, for all the boldness of his pro-Copernican argument, he had firmly announced Urban's principle – that there are things we can never learn from experience and reasoning. A lengthy excerpt from the letter had been read to Urban by Ciampoli, and Urban had been delighted by it.[30]

The central idea of the book Galileo carried with him to Rome is carefully presented in the frontispiece, which Galileo had engraved by Stefano della Bella. There Aristotle, shown as a feeble old man, Ptolemy, wearing a turban because he comes from Egypt, and Copernicus, wearing the clothes of a Polish priest, stand on the shore of the Florentine port of Livorno debating questions of physics and astronomy. But Copernicus looks nothing like the image of Copernicus that appears in other sources, where he is always portayed as young and clean-shaven. Indeed the Latin translation of Bernegger soon corrected this 'error', providing a more accurate representation of Copernicus in a revised frontispiece to the second edition of the *Dialogue*. It seems unlikely that Galileo and della Bella could not have tracked down a representation of Copernicus; instead, they have chosen to portray him as Galileo. Over the heads of the three philosophers hangs what could easily be the curtain which rises at the beginning of a theatrical performance – a device that was in fact used by della Bella for the frontispieces of plays. The reader knows at once that Aristotle, Ptolemy and Copernicus can meet only in a fictive space, and the Florentine reader, familiar with Galileo's appearance, would have grasped that in this book Galileo is to play the part of Copernicus. Indeed Galileo's mouthpiece Salviati explains in the *Dialogue* that he has decided to play the part of a Copernican, even to disguise himself as Copernicus (to wear Copernicus's mask), but that this does not mean that he is really a Copernican. He reserves the right to take off his costume, and be a quite different person when off the stage.[31] But we never see the curtain fall – we never see Salviati off the stage.

Moreover, while the book is ostensibly a discussion between three people (Salviati, a convinced Copernican; Sagredo, an impartial intellectual; and Simplicio, a died-in-the-wool Aristotelian), none of whom can be identified with Galileo (thus enabling him to disclaim responsibility for Salviati's Copernicanism), a fourth voice is present in the marginal summaries which guide the reader. This fourth voice, the anonymous commentator, is indisputably a Copernican and can only be Galileo.[32] Not surprisingly, the marginalia were cited against Galileo by the theologians who read the *Dialogue* for the Holy Office.[33] Were these marginal comments present in the text Galileo brought to Rome, or were they added later? We cannot be sure. The manuscript of the *Dialogue* was sent from Florence to Rome in 1632, as evidence to be used in Galileo's trial. It never arrived, but was lost in the post.[34]

We can be sure, however, that in essence the text Galileo carried to Rome in 1630 was identical with the book that was finally published, so Galileo should have known that he had not genuinely respected the pope's one precondition. The whole point of his argument that the tides were caused by the movement of the earth (an argument of which there had been no mention in the letter to Ingoli) was to prove Copernicanism to be the only possible account of the structure of the heavens. A proof is a proof, whoever presents it; it makes no difference whether the words are spoken by Galileo himself, or by Salviati, a character in a dialogue, performing the part of a Copernican. Nevertheless, by the time he left Rome, Visconti and Riccardi had both read the book, and both were determined not to put obstacles in his way.[35] As for the pope, he had promised Galileo a pension.[36] Galileo had got away with it so far, and surely his luck would hold.

Galileo's luck did not hold. On 1 August Prince Cesi died, his death depriving Galileo of the patron whom he had hoped would supervise the publication of his book.[37] On 24 August Castelli sent the following words of advice: 'For numerous good reasons, which I do not want to write down here and now, beyond the loss of Prince Cesi, who we trust is in heaven, I think it would be wise that Your Most Illustrious Lordship had his book printed there in Florence, and printed as quickly as possible. I have asked Father Visconti if this would present a problem, and he replied that this would not be a problem at all, and that he very much wants to see your work in print.'[38]

What exactly had happened between Galileo's departure from Rome on 29 June and Castelli's letter of 24 August? Fortunately we can make a pretty good guess. We have seen that Galileo's plan to distribute the letter to Ingoli in 1624 had been halted by war in the Valtelline. Now again there was a war, between France, Spain and their proxies, on Italy's northern frontier, this time over the succession to the dukedom of Mantua. On 16 July Mantua fell to the forces of Spain and France was thrown onto the defensive.[39] Ten days earlier Gustavus Adolphus, the Protestant king of Sweden, had invaded Germany. Six months later he was formally to become an ally of France.

Urban's position was an unenviable one. He had spent crucial years as the pope's representative in France, where he had formed connections that continued to shape his thinking. Moreover France was the only power that could counterbalance Spain, and if Spain gained undisputed control of the Italian peninsula the papacy would lose all freedom of action. Urban was thus instinctively, consistently pro-French. But the Thirty Years War had begun in 1618, and in June of 1630 Sweden, encouraged by France, entered the war against the Habsburgs. If the pope was an ally of France, then he was also an ally of the Protestants and an enemy of Catholicism. This was not a comfortable position for a pope to be in. Nor was it good news for Galileo. Florence, like the pope, was keen to avoid any irreversible choice between the two sides, but if the pope was instinctively and consistently leaning towards France, then the grand dukes of Tuscany were, when the chips were down, allies of the Spanish; indeed, in 1632 two princes of the Medici family, Francesco and Mattias, were in Germany fighting the Protestants.[40] In 1624 Urban had been proud of his Florentine origins and had surrounded himself with fellow Florentines. Now he began to worry that their loyalties were divided, rumours circulating in 1630 that the he had fallen out with Ciampoli, the secretary who handled all his diplomatic correspondence.[41] Galileo and Urban were now potentially on opposite sides. Ciampoli, who knew every detail of the pope's diplomacy, had advised Galileo to come to Rome as a private person, without the support of the Florentine government.[42] Being close to the Florentine government was now not an asset but a liability.[43]

In 1630 there were bitter divisions in Rome between the supporters of France and Spain, the two sides coming to blows in the streets. The Spanish were putting pressure on the pope by every means at their disposal. Early in 1630 they managed – at least for a while – to persuade Urban and Francesco Barberini to stop meeting with Campanella, who was a known enemy of Spain. When Galileo arrived in Rome there were rumours that he and Campanella, the greatest astrologers of their day (to the lay mind there was no difference between astronomy and astrology), had predicted that the pope would die that summer. It was reported that Spanish cardinals were setting out from Spain on the long journey to Rome, in order to arrive in good time for the conclave that would elect the next pope. The Spanish were behaving as if Urban were already dead. As for the French, their cardinals also felt obliged to start towards Rome, fearing that a new pope would otherwise be elected in their absence. Urban, who took astrology seriously, was profoundly alarmed. He could not sleep at night. He had all the birds in the papal gardens killed because the dawn chorus woke him from his fitful slumbers.

Where had the astrological predictions originated? Francesco Barberini assured Galileo's friends that he knew Galileo had nothing to do with it: Campanella was, after all, the pope's astrologer and no friend of Spain, so the story was obviously false.[44] On 13 July the pope was satisfied he had traced the predictions to their source: Orazio Morandi, abbot of Santa Prassede (and a committed member of the anti-Aristotelian party) was arrested. The torture of a key witness on 26 August produced decisive evidence. And then on 7 November Morandi was found dead in his prison cell – some said he had been poisoned.[45]

In his letter of 24 August Castelli turned immediately from his advice that Galileo should publish quickly to saying that no one knew what was happening in the Morandi case. Some said the danger was great, others that everything would be all right. The only thing that was certain was (Castelli was aware that his letter might fall into the wrong hands) that 'our masters are as kind as could be'. Galileo had already been asking anxiously for news of Morandi's fate; the matter was so delicate that his correspondent would not put Morandi's name on the page.[46] For Galileo and Morandi were too closely connected.[47] They had known each other since 1611. Amongst Galileo's papers is a note from Morandi instructing him to attend Mass at Santa Prassede, as Visconti and the former chief inquisitor of Florence would be present. Morandi was obviously using his extensive influence to help Galileo's book slip past the censors – his influence deriving from a valuable library of forbidden books (both philosophical and erotic) and the fact that he ran a monastery where one could sample the libertine pleasures of debauchery, both homosexual and heterosexual. He was also preoccupied with astrology and magic, his papers include a horoscope drawn up for Galileo's nativity.[48] Since we know that Galileo kept the information required to calculate his horoscope a secret, we must assume that Morandi had considerable influence over him.

However close the association was between Galileo and Morandi, it was too close. It was Campanella who had denounced Morandi to the authorities, needing to act to protect himself. But Visconti, who had been appointed to approve Galileo's book for publication, was also part of Morandi's circle. When Morandi had produced horoscopes predicting the pope's death, Visconti had (or so he later claimed) dissented, putting in writing a prediction that the pope would have a long life.[49] But he had not denounced Morandi, and so he too was guilty by association. In November Castelli reported that he was in trouble for some astrological writing or other, and in March of 1631 the news was that

Visconti had been sent in disgrace to Viterbo. Riccardi had lost his colleague and Galileo his tame censor. But that was not the worst of it. Galileo faced a charge that no one voiced: that he had known about Morandi's prophecies and had done nothing to counter them. No wonder Castelli thought Galileo should publish away from Rome, and quickly.

Galileo obviously agreed; indeed he had begun looking for alternative places to publish on 6 August, before news of Cesi's death had reached him but surely after he had received news of Morandi's arrest.[50] He was reluctant to publish in Florence, for the quality of work done by the local printers was abysmal. He considered Venice – he even made enquiries – but there was a risk that Venice and Florence would end up on opposite sides of a war.[51] He also made enquiries in Genoa.[52] At the same time bubonic plague was being carried through Italy by soldiers retreating from the siege of Mantua. Quarantines were set in place, disrupting communications.[53] It soon became apparent that if Galileo was going to publish quickly he had no choice but to publish in Florence. On 11 September, without any fuss, he was granted an *imprimatur* by the Florentine ecclesiastical authorities – there was surely an understanding that the book would have to be cleared by Riccardi before it could be printed. Galileo must have been acutely aware that if he did not publish fast he might never publish at all: not only had his relationship with Rome deteriorated, but in October the plague took the life of his gardener, and in the months that followed the death toll in Florence rose inexorably. Galileo was now in a hurry – he was already at work on his next book.[54]

Alessandra Buonamici

In 1630, when Galileo was in Rome seeking permission to publish the *Dialogue*, he heard of the unexpected appearance in Florence of Alessandra Buonamici, the sister of his daughter-in-law Sestilia Bocchineri. In 1623, when in her late twenties, a widow employed as lady-in-waiting to the empress at the Viennese court, Alessandra had married for the third time. Her new husband, the Florentine Giovanfrancesco Buonamici, had soon departed on a mission to Spain; now he was back in Florence and Alessandra had travelled from Vienna to Florence, through war and plague, in only eighteen days. Galileo, who never went anywhere if he could help it, must have been astonished to hear of the derring-do of this unusual woman.

Within a few days of Galileo's return to Florence he must have met Alessandra. On 28 July she wrote to him, describing herself as 'affezionata' and letting him know that her husband would be away for the next few days.[1] Apparently the letter took eleven days (and eleven nights, says Galileo) to travel ten miles, and by the time it reached Galileo, Alessandra's husband had surely returned. Galileo was horrified at the thought that Alessandra might believe she had been spurned; and horrified too to have lost his chance to consummate their relationship. He was happy to claim that he had hurried back from Rome to meet her, and quick to insist that his greatest desire was to get to know her better.[2]

Of course these two letters are written in a polite code, and at first glance appear to be a purely innocent exchange of compliments. But neither of the recipients mistook them for such. A decade later we find Galileo, now blind, dictating a letter to her to say that he had purchased thirty yards of the material out of which petticoats are made because the cloth dealer had assured him the bolt of cloth had come from her – I bought it, he says, to have something of yours, but not because I was in any danger of forgetting our conversations of so many years ago, when you returned from Germany. You can't imagine, he tells her, how rare it is to meet a woman who talks as much good sense as you did. (They had spent, we later learn, less than two hours in each other's company, but evidently it was love at first sight.)[3] He has now, as he had then, an intense desire to serve her – and, naturally, her husband.[4]

Alessandra replied promptly to this letter, but her reply never reached Galileo, and a year later he was writing again, having learnt that she was dismayed not to have heard back from him. She replied at once, accusing her relatives of intercepting her letters, and expressing her desire to spend a day in his company, if some way could be found of doing it without provoking a scandal. Her husband, she happens to mention, is at the moment bedridden. Her separation from Galileo, she assures him, has been one of the worst experiences of her life. Can she send her carriage for him, and will he come and stay for a couple of days?[5]

Galileo, of course, was by this time, 1641, under house arrest, and replied that he could not come to her but she could come to him. Why not come (with her husband, of course) for four days? There was no need, he assured her, to worry about what people might say.[6] Alessandra, we must assume, thought this a step too far, and it seems they never met again. But the last letter Galileo dictated, when death was approaching, was a note to his beloved (the word is not, I think, too strong), thanking her for offering to sell him some cloth, and for keeping his interests in mind. By this time they both knew they would never meet again.

A river floods

In 1630 Galileo was already at work on his next book. But he also had to deal with a minor distraction. In December, in his capacity as the grand duke's mathematician, he was invited to express an opinion on a debate that had divided Florence's ruling establishment. The problem was the river Bisenzio, which meanders across the plain that stretches northwest from Florence. The Bisenzio had flooded during the previous autumn, and since many of Florence's wealthiest families owned land along its banks, and since the local landowners would pay the cost of the works required to prevent future flooding, this was a matter of considerable concern. Concern turned to anger when Florence's leading hydraulics expert, Alessandro Bartolotti, recommended cutting a new, straight bed for the river over the last five miles or so of its course. The new river would have a steeper gradient, and so the water would flow faster, and would be carried away before a flood could occur. The project would be expensive, and Bartolotti proposed that only those with land west of the river should pay for it. To these landlords it seemed clear: those living east of the river were trying to divert the problem onto their lands, endangering their crops, and to add insult to injury, placing the cost of the works not on those who would benefit from them but on those who would suffer if they proved unsuccessful. The opponents of Bartolotti's scheme argued that there was nothing wrong with a meandering river: all that was needed was that the river should be dredged here and there, the flood levees strengthened, and brushwood that had grown up between its banks cut away. These minor interventions would improve the flow of water, and were all that was needed to prevent the flood from recurring.

Hydraulics was already a matter of significant interest in Galileo's circle, Castelli having published a book on the subject, *On the Measure of Running Waters*, in 1628. Castelli's major contribution was to insist that it was wrong to think of rivers as if they were static. Not only width and depth but also rate of flow needed measuring, as increases to the speed at which water flowed would improve drainage every bit as effectively as expanding the cross-section of the river. Castelli, who was in Rome, was called upon by the opponents of Bartolotti's scheme, and it will have been at their instigation that Galileo was invited to write an official report on the contrasting proposals.[1]

Galileo's report is an eye-opener, telling us a great deal about his strengths and weaknesses as a scientist. For Galileo the issue was absolutely straightforward: a river flowing downhill is just a special case of a body sliding down an incline. Galileo had performed experiments in the first years of the century to assess how fast a ball would roll down a slope if there was no friction to slow it down. He had managed to get theory and experience to line up neatly: what mattered in determining the final speed was the total height through which the ball descended, not the steepness of the incline. A ball that slid down a slope of any length from a height of two metres would, in the absence of friction, be travelling at the end at exactly the same speed as a ball dropped from a height of two metres.

Referring explicitly to his work on falling bodies (there were evidently manuscript copies of this in circulation), Galileo maintained that this conclusion could be directly transferred to the Bisenzio. Straightening the river would increase the gradient, but it would not increase the speed at which the water flowed. According to Galileo's analysis the only thing that stops a river reaching a speed of hundreds of miles an hour as it descends from the mountains to the sea is that new, slower tributaries regularly join the faster water descending from upstream. Straightening the Bisenzio would, if anything, he argued, reduce the effectiveness of the river, since the river not only carried water away but acted as a sort of header tank, holding water that was neither in the fields nor yet in the river's destination, the Arno. A shorter river was equivalent to a smaller tank. He thus came down firmly on the side of Bartolotti's opponents: all that was needed was to dredge out silt and cut back brush, the two major obstacles to the water's flow.

Galileo's argument depended on the claim that a river running through its channel is equivalent to a perfectly round ball sliding down a frictionless slope. Water, he argued, slid by water without any resistance or friction. The channel a river ran through might not be smooth, but once the level of water in the river had risen to cover the rocks, then the water ran perfectly smoothly over the rocks, because, in effect, water was running over water. The Bisenzio meandered along its course, but a river bend was not like a sharp angle: it was a curve that the flow of water effectively met at a series of tangents. These infinitesimal deflections would have no effect on the speed of the water's flow (just as a pendulum is not slowed by its string). But Galileo had made a simple error: he was denying the possibility of turbulence.

Galileo's theoretical analysis was brilliant but flawed. The mistake he made was far from inevitable. A year later, two cousins, Andrea and Niccolò Arrighetti, Florentine nobles who had studied physics with Galileo (the only evidence we have that Galileo was teaching in these years), got into a bitter argument on precisely the question of whether the Bisenzio should be re-routed. Not knowing that Galileo had already taken a stand on this issue, they appealed to him to adjudicate between them. Their disagreement is a clear indication that you could easily reach very different conclusions when applying Galileo's law of fall to a river.[2] So too Galileo's friend Baliani had thought of applying his law of fall to rivers – and then realised that he lacked the conceptual tools required to analyse the behaviour of water.[3]

Most importantly, Galileo was so confident of his analysis that he made no attempt to see if it corresponded to reality. He did not go and inspect the Bisenzio, or ask for its route to be surveyed. He did not measure the flow of water through the streams that must have run through the gardens of his villa, or test whether water runs equally quickly through straight and curved pipes.[4] Empirical enquiry was unnecessary because he already knew the answers, or because he regarded this whole question as a distraction – he had books to publish, and little time to spare.

The commission to decide on Bartolotti's project was chaired by a close friend of Galileo's, Andrea Cigoli, the first secretary of state to the grand duke. Like Galileo, it came down firmly in favour of Bartolotti's opponents. But although it received Galileo's technical analysis, it made absolutely no reference to it. It seems to have had a perfectly healthy suspicion of Galileo's abstract theorising. The appeal to Galileo for his advice tells us that he was a respected figure at the Medici court; the failure to mention his contribution suggests that he was seen as erratic and unreliable in his judgement. Galileo was right – rivers are best left to find their own way downhill – but he was right for the wrong reasons. His report on the proposed straightening of the Bisenzio is a miniature case study in his intellectual arrogance, and in his deep-seated preference for deductive over empirical reasoning.

30

Publication

Galileo certainly had more important things to worry about. The wife of the Florentine ambassador in Rome – who was a good friend of Galileo's, and who could not be thought to represent the Florentine government – entered into negotiations with Father Monster on Galileo's behalf. At first Riccardi demanded that the final text of the book be sent to him, but then he agreed to sign off on the book on two conditions: he must see the beginning and the end, and the book must be reviewed in its entirety by a censor in Florence. He was happy for Galileo to suggest who would be suitable. Perhaps Caterina Niccolini charmed Riccardi; she was doubtless able to take advantage of the fact that he was her cousin, for he conceded too much (evidently in the belief that the pope was in favour of publication), Caterina certainly earning the gift of a telescope which Galileo had promised her.[1]

If the book was to be published in Florence, then full responsibility should have been transferred to the Florentine censors, and Galileo should not have been allowed to influence the choice of censor. The censor Galileo chose was a Dominican, Giacinto Stefani, and he chose well: according to Galileo, Stefani wept as he read the manuscript, so touched was he by Galileo's obedient submission to the requirements of the Church.[2] Riccardi, on the other hand, began to get cold feet. He was coming to think that Stefani was the wrong man for the job, and kept stalling when asked to give final approval to the beginning and end of the book, evidently regretting the concessions he had made to Caterina Niccolini.[3] In March 1631 Galileo asked the grand duke to intervene, and to put the ambassador himself to work on his behalf. His life, he complained, was slipping away, and he was, he frankly confessed, in a state of acute anxiety, and had been forced to take to his bed.[4] He could neither eat nor sleep.[5] He felt as if he was lost on an ocean without any prospect of making landfall. Galileo, of course, had never seen an ocean, but at this point he had a perfectly realistic fear that his life's work would never see the light of day.[6] To make matters worse he had become convinced that his intellectual powers were failing.[7]

And so the grand duke instructed the ambassador himself to come to Galileo's aid, and Riccardi agreed to write a memo to the Florentine inquisitor explaining what the pope required.[8] Reporting back, Francesco Niccolini explained that the fundamental problem was that Galileo's views were not popular with those in charge (presumably the Barberini themselves).[9] He might rather have said that Galileo himself was not popular: the shift in the official attitude towards Galileo that Castelli had predicted the previous August was the underlying reason for his difficulties. Moreover this was a bad moment for the Medici to try to exert influence in Rome. Between 1625 and 1631, Urban had, in a series of steps, annexed the duchy of Urbino to the Papal States. The heir to the last duke, who died on 23 April 1631, was Vittoria della Rovere, who was betrothed to Ferdinando II. In annexing Urbino, Urban was thus demonstrating complete disregard for the wishes of the Florentine government. Galileo, for his part, assured the Florentine government (in a letter which is a breathtaking misrepresentation of the facts) that his book was far from being a defence of Copernicanism, that in it Copernicanism was treated as a dream, a chimera, as logically contradictory, as futile, and that there was no reason at all why his opinions should not be welcomed by the authorities, because the opinions that were attributed to him were not his, his own opinions in fact being absolutely identical with those of St Augustine and St Thomas.[10]

Niccolini continued to apply pressure on Riccardi, and eventually, on 19 July, after Riccardi had been dragged along by his hair, as Niccolini put it, and the full weight of the grand duke's authority had been brought to bear (Riccardi was himself, after all, a Tuscan), Niccolini extracted from him the revisions he required to the preface to the reader.[11] Meanwhile printing had already begun (the various prefaces were, as usual, to be added last). Riccardi was to be given no opportunity to have second thoughts.[12]

The *Dialogue*

Galileo's *Dialogue* contains one major new argument, on the subject of sunspots. When the book appeared in February 1632 Galileo's Jesuit opponent Scheiner was convinced that this section had been based on material lifted without acknowledgement from his own *Orsini's Rose* (1630, named in honour of Scheiner's patron, Paulo Giordano Orsini), although we know (as Scheiner could not) that Galileo had not seen Scheiner's book until his own was well on its way through the press.[1] Valiant efforts have been made to find an innocent explanation for the close match between various details in Scheiner's account and in Galileo's. It has been suggested, for example, that Galileo suddenly and mysteriously grasped, at some point between April 1629 (when he said there was nothing new to say about sunspots) and September 1630 (when his book was licensed for publication), the significance of a letter he had received from his old opponent Sizzi in 1613 – a letter which should have alerted him to an important line of enquiry but which evidently did not when he first read it.[2]

Unfortunately the evidence is unequivocal: Galileo's argument follows Scheiner so closely that he must have had a copy of Scheiner's book in front of him as he wrote this section – Scheiner saw this clearly as soon as he read the *Dialogue*.[3] Since Galileo did not receive a copy of Scheiner's book until September 1631, a year after the *Dialogue* had been licensed for the press and three months after printing had begun, at which point this whole section must have been added, a second line of defence has been proposed: Galileo, it has been said, did not want to acknowledge his debt to Scheiner because he did not want to draw attention to the fact that he had altered his text after it had been approved for publication. But this will not work either: in the first place Galileo made little effort to conceal the fact that another section had been added to the book at the very last moment ('I have just seen a treatise which has astonished me'); moreover Scheiner's book had been published in June 1630, so no Church official would have responded with suspicion on finding it discussed in a book carrying an imprimatur dated 11 September 1630 (although Scheiner's publisher might have been a little puzzled, as he sent Galileo a copy only on 9 September 1631).[4] Galileo's discussion of sunspots is thus deliberately misleading in making no reference to Scheiner. Worse, Galileo claims that he had discovered the crucial new material that he lifts from Scheiner many years earlier and that he is arguing from his own observations; he thus not only plagiarises Scheiner but claims priority in making Scheiner's discovery.[5] Of course you might say that the *Dialogue* is a fiction, and not to be read as a statement of fact, but if there is one convention any reader of the *Dialogue* must accept, it is that references to 'the Academician' are to be taken as factually true statements about Galileo. Galileo has deliberately set out to efface Scheiner from the historical record and to deny his debt to him.

It is impossible to find any excuse for this behaviour. Of course, Galileo and Scheiner had long been involved in a series of disputes, which had become bitter after the controversy on the comets, and *Orsini's Rose* opens with a long, intemperate attack on Galileo. In those disputes, right had been on Galileo's side; but now he had put himself completely in the wrong. Why? I can think of only one reason. Scheiner, like many of the Jesuits, had made friendly overtures towards Galileo. Around Christmas 1625 he approached Giovanni Faber, a member of the Lincean Academy, in Rome, and told him that he agreed with Galileo on the system of the world – in other words that he was a Copernican. At least one person who was in a good position to know, the Jesuit Athanasias Kircher, also claimed this was the case.[6] And yet Scheiner, bound by the Jesuit obligation of obedience, had written an enormous tome, eight hundred folio pages, which was from start to finish an attack on Copernicanism. Galileo would have been justified in concluding that there was no point in arguing with him – but that did not release him from the obligation of acknowledging that everything he knew about the angular paths of sunspots across the face of the sun came from Scheiner.

In his earlier work on sunspots Galileo had confidently asserted that, as they rotate around the sun, sunspots follow a path which is parallel to the ecliptic – the line traced by the sun in its apparent annual orbit through the heavens. Sizzi (and others after him) had noticed that this was true only at certain times of the year; at others the sunspots appear to follow a curved line. This, Sizzi reasoned, was evidence that the sun's axis was tilted at an angle to the ecliptic (as is the earth's axis, which is the cause of the seasons). Scheiner had not only given a detailed account of this phenomenon, but had tried to reconstruct the movements of the heavenly bodies that would produce the observed effects – taking a geocentric system for granted.

What Galileo asserts in the *Dialogue* is that any geocentric model (and he clearly has the details of Scheiner's model in front of him) becomes impossibly complicated, while within a Copernican system explaining what movements are involved is perfectly straightforward: the path of the sunspots around the sun would look different depending on whether we are looking at the sun with its axis tilted towards us, away from us, to the left or to the right, and in the course of a year one would see it from every possible angle. Sometimes the sunspots would appear to trace a straight line, sometimes a curved one, sometimes curving upwards and sometimes curving downwards. But suppose that Ptolemy and Tycho are right and the earth is stationary at the centre of the universe, with the sun orbiting around it. What sort of movement of the sun would generate the required path of the sunspots? The answer is that the sun would have to be wobbling like a slowly spinning top, with the angle of its axis to the ecliptic constantly varying. Moreover this wobble would have to be timed so that it was completed exactly once in each year. This movement seemed to Galileo impossibly complex and completely implausible.

For all naked-eye observations the Ptolemaic, Tychonic and Copernican systems were observationally equivalent. The phases of Venus had provided decisive telescopic evidence against the Ptolemaic system, but not against the Tychonic. Galileo now believed that the path followed by sunspots on the disc of the sun was the first observation that was incompatible with the Tychonic system – Tycho had never envisaged a sun that wobbled on its axis during its annual progress through the heavens. Strictly speaking Galileo was not demonstrating the truth of Copernicanism; he was merely showing that the Tychonic system was problematic and implausible. Had he switched to this new argument for Copernicanism, rather than insisting on his old argument from the tides, no one could have complained that he had substituted demonstration for the hypothetical arguments required of him by the pope.[7] But of course even if this strategy occurred to him it was far too late: the argument about the tides lay at the very heart of the book that had just been licensed for publication, and he could not possibly back away from it now.

Perhaps the last passage to be written was the dedication to the grand duke. Here Galileo argues, as we have seen, that cosmology is the greatest of all subjects, and Ptolemy and Copernicus therefore the most important of all philosophers. But he pauses to make a nod in the direction of Plato: 'He who looks the higher is the more highly distinguished, and turning over the great book of nature (which is the proper object of philosophy) is the way to elevate one's gaze. And though whatever we read in that book is the creation of the omnipotent Craftsman, and is accordingly excellently proportioned, nevertheless that part is most suitable and most worthy which makes His work and His craftsmanship most evident to our view.' Any educated contemporary would have recognised the Craftsman as the demiurge of Plato's *Timaeus*.

Galileo has two motives here. In the first place he wants to suggest to his readers that his philosophy, like Plato's, is easier to reconcile with Christianity than is Aristotle's. Platonism is a sound choice for anyone claiming that their philosophy is compatible with the true faith.[8] But something else is going on. Every important manuscript of Plato known to the Renaissance was the property of the grand duke of Tuscany. In invoking Plato, Galileo was thus complimenting his patron and engaging in an elementary form of cultural patriotism.[9] But this prefatory invocation of Plato can give the false impression that Galileo's text is in fact Platonic in inspiration. Later references to Plato are much more cautious, and even implicitly critical. Had Galileo been publishing in Rome rather than Florence there might well have been no reference to the omnipotent Craftsman. And the rest of the book bears the mark of a text intended for Roman publication, for Galileo still refers to himself as 'the Lincean Academician'. In a book published under the patronage of Prince Cesi, as Galileo had originally intended, this would have been an obligatory reference to the Lincean Academy; now, with Cesi dead and the Linceans effectively disbanded, the reference to the Linceans, here and on the title page, amounted to a sentimental gesture; in addition, to rebut the complaints of those who claimed that he did nothing for his salary from the University of Pisa, he described himself for the first and last time on a title page as 'matematico sopraordinario dello studio di Pisa'.

Certainly the dedication is distinctly misleading in its claim that a study of cosmology demonstrates that the universe is 'excellently proportioned'. Perhaps the most radical arguments of the *Dialogue* are directed at undermining such providentialist assumptions. We have no way of knowing, Galileo insists, whether the universe exists for our sake, or for other quite different purposes, or whether much of it serves no purpose at all.[10] We have no way of judging whether it is to be thought of as large or small – assuming it is not, as it may be, infinite.[11] More remarkable still, Galileo's arguments proceed by considering the theoretical possibility of non-terrestrial life. Sagredo argues that there might be life on the moon, although we can be sure that if there is it can be nothing like life on earth:

> I am certain that a person born and raised in a huge forest among wild beasts and birds, and knowing nothing of the watery element, would never be able to frame in his imagination another world existing in nature differing from his, filled with animals which could travel without legs or fast-beating wings, and not upon its surface alone like beasts upon the earth, but everywhere within its depths; and not only moving, but stopping motionless wherever they pleased, a thing which birds in the air cannot do. And that men lived there too, and built palaces and cities, and traveled with such ease that without tiring themselves at all they could proceed to far countries with their families and households and whole cities. Now as I say, I am sure that such a man could not, even with the liveliest imagination, ever picture to himself fishes, the ocean, ships, fleets, and armadas.[12]

It is true that this is Sagredo speaking, but Salviati, after pointing out all the differences between the earth and the moon, cautiously agrees with him. Our understanding, Galileo insists, is limited by our immediate experience – even our imagination can work only upon the material provided by experience.[13] It follows that we cannot possibly judge the purpose served by the universe. Some, accepting the arguments of the astrologers, think the stars exist to influence life on earth, and then argue that the Copernican universe is too big for them to serve this purpose. But what do we know about the purpose of the celestial bodies? 'It is a great folly for us Terrestrials to want to be arbiters of their sizes and regulators of their local dispositions, we being quite ignorant of all their affairs and interests.'[14]

In place of a universe with humanity at its centre, a universe made to serve humanity's purposes, a universe designed to make possible humanity's salvation, Galileo offers a mysterious universe whose purposes are unknown, whose size is unfathomable and which may contain other beings quite different from ourselves. We are now simply 'noi Terreni', us terrestrials. The moon may be uninhabited, but you do not have to force the text to find in it Bruno's heresy: around other stars there may be other planets, other worlds which make the mistake of thinking that they are at the centre of everything, and that their experiences are the true measure of reality.

PART FOUR

PRISONER TO THE INQUISITION

My first businesse in London, was to seeke for Galileos dialogues; I thought it a very good bargaine, when at taking my leaue of your Lordship I vndertooke to buy it for you, but if yor Lo^p should bind me to performance it would be hard enough, for it is not possible to get it for mony; There were but few brought ouer at first, and they that buy such bookes, are not such men as to part wth them againe. I heare say it is called in, in Italy, as a booke that will do more hurt to their Religion then all the bookes haue done of Luther and Caluin, such opposition they thinke is betweene their Religion and naturall reason.

Hobbes to Newcastle (26 January [5 February] 1634)[1]

Unhappy the land where heroes are needed.

Bertolt Brecht, *Life of Galileo* (1938)

Maria Celeste and Arcetri

In the autumn of 1631, with his book finally cleared for publication, Galileo moved house. Since 1617 he had lived in a magnificent villa called Bellosguardo a couple of miles to the west of Florence. But now he was getting old, and he was going blind. He needed to think about a time when he would no longer be able to get about easily. What worried him most was that he would be cut off from his daughters who were in a convent in Arcetri, about an hour's walk away. He moved to be as near to them as possible. He gave up his villa and rented Il Gioiello, which was much more modest – the rent was a third of what he had paid before.[2] He left behind the trees he had planted. But he still had a good garden, and he would be able to continue to see his daughters.

Of the letters that Maria Celeste, his eldest daughter, wrote to Galileo between 1623 and 1633, 124 survive, the first being a letter of condolence on the death of his sister Virginia.[3] The letters stop in the summer of 1631, when Galileo moves in nearby, but resume while he is away in Rome and Siena during 1632–3. Amidst the nine volumes of Galileo's dreary correspondence, these letters shine, each one a tiny jewel. What makes Maria Celeste's letters quite astonishing – and so unlike her father's – is that she has an extraordinary capacity to describe the mundane and the particular in ways that make them seem perfectly real. This capacity is the more astonishing because she is often writing about things she has never seen. When Galileo was on trial in Rome she kept him constantly informed about developments at his villa. She wrote about the mule, who was contrary and a picky eater; about the lemons, and how those on the lower branches of the trees had been plundered by marauders; about the fat doves in the dovecot that had been attacked by a falcon. It comes as a shock to realise that Maria Celeste belonged to an enclosed order. She never left the nunnery. She never saw Galileo's villa, never met the mule, never heard the doves cooing in the dovecot. First she imagined it all, on the basis of the reports she received from servants and friends; then she wrote it down so precisely and carefully that nothing could be more real.

Maria Celeste had an authentic and profound piety. The two presents she wanted from Rome in the winter of 1632–3 were a small pair of paintings of the Virgin and the crucified Christ that she could always carry with her. But prayer was not the only thing on her mind. Although the Franciscans were under a vow of poverty, the nuns in the convent of San Matteo did not live by alms. Each nun brought with her into the convent a dowry which helped to fund it, and the nuns were also provided with small incomes by their relatives, but the convent was poor and the funds inadequate, so that the nuns were industriously engaged in commercial activities to bring in extra income. Maria Celeste worked as an apothecary in what she calls her 'bottega', or workshop, and she tells us that the nuns, who spent long hours in prayer, were often still hard at work at midnight.[4]

They also engaged in financial transactions with each other: Maria Celeste paid thirty-five scudi to one nun to take over her small cell, and then eighty scudi (equivalent to five years' rent on a modest house) to the abbess to move into more comfortable quarters.[5] They took it in turn to hold the nunnery's various offices, and when the office holders went into debt they had to borrow from the other nuns. So the nunnery was not so much a shelter from the world as a little world in itself. And Maria Celeste proved herself highly competent in dealing with the outside world as well, managing Galileo's affairs in the year he was away, and advising on the purchase of a house for her brother.

Of course the nuns were also active in the usual domestic tasks, and Maria Celeste sewed, mended and starched for Galileo (and, until his marriage, for her brother Vincenzo); the linen collars Galileo wears in his portraits will have been made, mended and starched by her. And she supplied him with a steady supply of sweetmeats such as baked pears and candied citrons. (A larger, sweeter lemon, the citron was a great delicacy because it did not grow locally but had to be imported from the south.) In return Galileo helped her out whenever she was in need of money. As we do not have his side of the correspondence, we can only guess at the support and encouragement he gave her.

Her correspondence, it must be admitted, is a Freudian feast. Her feelings for her father, she admits to him, are more than filial, and his for her, he says, are more than paternal.[6] She longs to outdo her siblings in his affection.[7] She is delighted when he complains that a post has arrived without a letter from her.[8] She wanted everything she made for her father – every candied quince and tartlet – to be exquisite.[9] Nothing could be too good for him. She was keen to die so that she could go to heaven to pray for him, and would cheerfully have given up her own life so that he could live.[10] She even took on the task, imposed on him as part of his punishment, of reciting once every week the penitential psalms – as if his prayers and hers were one and the same.[11]

Last but not least, she took an interest in Galileo's scientific work: she read *The Assayer* and borrowed a telescope so that she could see his discoveries with her own eyes. This mattered greatly to Galileo, who in Siena read his letters to her aloud to his friends, some of them apparently dealing with topics discussed in his *Dialogue*. (Admittedly he was also happy to amuse his friends at her expense: they all laughed when they learnt that because whole mozzarella cheeses were called 'buffalo eggs' she had actually imagined they were eggs laid by buffalos and had intended to turn them into a giant omelette.)[12] Indeed he even toyed with the idea of publishing some of his letters to her – or perhaps of constructing a book consisting of letters written as if to her.[13]

It is quite clear that Galileo and Maria Celeste had a close relationship. There is nothing to suggest that his relationship with her sister Arcangela, who lived in the same nunnery, was at all comparable. Arcangela was frequently depressed, and Maria Celeste had to intervene to make sure that she did not take her turn as cellarer, fearing she would drink her way through the cellar. There is no sign that Arcangela ever wrote to Galileo, or that they were at all close.

33

Trial

Publication of Galileo's book was completed on 21 February 1632, but the measures taken to prevent the spread of the plague meant that by the end of May only two copies had arrived in Rome, one of which was in Castelli's hands (although it had been presented to Francesco Barberini) and the other probably in Riccardi's.[1] It was not until July that six more copies arrived, three of which were given to Galileo's close friends and supporters. Within days the authorities were trying to seize every copy in Rome and the book had been referred to a special commission of theologians, who were to decide if there were grounds for bringing it before the Inquisition. The word in Rome was that opposition was coming from two quarters: from the Jesuits (one copy had indeed been given to a Jesuit) and from the pope himself.[2] The pope's complaint was that Galileo had failed to include at the end of his book the argument that he had insisted upon: that one cannot prove the truth of Copernicanism because God's power is such that he can achieve any natural effect by numerous different means, many of them beyond our comprehension. Galileo had made only a cursory reference to this argument, mocking the pope by giving it to Simplicio (who in the dialogue is always wrong).[3] There was also criticism of the fact that the preface (the text of which had been established through such prolonged negotiations between Galileo and Riccardi) had been printed in italics, thus distinguishing it from the rest of the book.[4]

The pope was also furious that the book had been published without his having received any prior warning.[5] For this he blamed Ciampoli, describing the whole issue as a 'ciampolata'.[6] Ciampoli had been the person from whom Riccardi had taken his instructions, and Ciampoli had assured Riccardi that the pope wanted to see the book published. Ciampoli was in the firing line because he had already fallen out of favour:[7] he was reported to have been less than complimentary about the pope's poetry (the pope had been encouraged by Campanella to think of himself as a major poet), and to have entered into secret dealings with the Spanish, who were bringing enormous pressure to bear on the pope in an attempt to break his ties with France. Since the French were supporting the Protestants in the Thirty Years War, and since Urban had failed to give his full support to the Hapsburgs, it was easy to claim that he was, in effect, giving his backing to heretics.[8] Ciampoli was already in a sort of limbo; he had not been stripped of his offices, but was not allowed into the pope's presence. His failure to alert the pope to the imminent publication of Galileo's book was to prove the last straw: he was soon sent into internal exile, being made governor of Montalto in the Marche, and was left to linger in one fly-blown dump after another until he died in 1643, at which point the Inquisition promptly seized all his papers.[9] There is good evidence that Ciampoli, like many of those closely associated with Galileo, was an admirer of Lucretius, an atomist and a materialist.[10]

The Florentine government immediately sprang to Galileo's defence: Galileo was a government employee, the book was dedicated to the grand duke, and of course the grand duke himself was already implicated, having pressed hard for the book's publication.[11] But it was clear from the beginning that the pope was not going to be satisfied with having the book prohibited. He wanted to see Galileo tried for heresy, and it was thought unwise to discuss the case with him because he was Galileo's most implacable opponent.[12] Indeed the pope became incandescent with anger whenever the matter was raised.[13]

In early September there were hopes that the matter would be resolved if the book was withdrawn and corrected, though the ambassador, having felt the full force of the pope's fury, was predicting disaster.[14] But matters took a turn for the worse when the records of the Inquisition relating to the condemnation of Copernicanism in 1616 were reviewed, and it was found that Galileo had been forbidden to enter into any discussion of Copernicanism in the future; also rediscovered was the old charge that Galileo had been (and perhaps still was) teaching heretical doctrines to his students in Florence; moreover the committee of theologians had ruled that the book amounted to a defence of Copernicanism.[15] On this there was general agreement – and as Francesco Barberini pointed out, Galileo expressed himself so brilliantly that he could hardly defend himself by claiming not to have had full command of his own language.[16] The special commission reported in mid-September, and the decision was taken to summon Galileo to Rome for trial.[17]

Galileo was left with no choice but to write to Francesco Barberini in October, pleading that, in view of his age and infirmities, he be spared the journey to Rome, and protesting his 'purissimo, zelantissimo e santissimo' affection for the Church and the Church authorities. No saint, he insisted, could have shown greater piety.[18] Similar pleas were made by the ambassador, who emphasised the risks to Galileo's health represented by the quarantine (reduced, as an exceptional favour, from the normal forty days to twenty) to which he would be subjected on the borders of the Papal States. But the pope was adamant: Galileo must come to Rome. A delay until warmer weather and the end of the plague precautions would not be permitted.[19] In Florence there was concern that Galileo (who had taken to his bed) was in such an acute state of anxiety that his health had deteriorated to a point where he was likely to die on the journey, or even before leaving. A month's delay was thus conceded, but no more.[20]

Galileo left Florence on 20 January 1633 and arrived in Rome twenty-five days later, having spent longer than he had expected in quarantine.[21] He travelled in a ducal litter, but there was uncertainty as to whether he had to contribute to the cost, and if so how much;[22] he was told he would be put up at the expense of the Florentine ambassador, but only for a month.[23] When he arrived he was instructed to avoid going out and meeting people (within a few days he was fretting over the lack of exercise) but was allowed his freedom; there were, he said, half joking, no signs of the threatened chains, prison cells and instruments of torture.[24] The threat had certainly been there: Niccolini had been unable to obtain an undertaking that Galileo would not be incarcerated, and the pope insisted that Galileo was being allowed his freedom only as a mark of respect for the grand duke.[25] But Galileo was far from being in good health, suffering from acute pains in his legs and being unable to sleep.[26] Indeed he was a less than ideal house guest, keeping everyone awake for two nights in a row with his cries and screams.[27] Concerned about his psychological condition, Niccolini often kept him in the dark about developments, reasoning that the more he knew about his situation the more alarmed he would become.[28] When on 8 April Galileo was finally informed that he was going to be required to move into the Inquisition's rooms (even though he was promised that he would be given comfortable lodgings and that no doors would be locked), his state of mental collapse was such that Niccolini feared for his life.[29]

Galileo reported to the Inquisition on 12 April and was held there until his release on 30 April. He was now completely isolated and alone, and at night he cried out in pain.[30] At first he held to his line that although he had discussed Copernicanism in the *Dialogue* he had not defended it – a claim that the Inquisition thought was manifestly untrue. Moreover he denied any knowledge of the injunction read to him by Seghezzi in 1616, but this cut no ice with the Inquisitors, who had the document in their hands.[31] So on 27 April Vincenzo Maculano, the Inquisition's commissary general, or prosecutor, held an informal meeting with Galileo to discuss the way forward.[32] We know that Maculano threatened Galileo with 'greater rigour' in the proceedings, and that he was concerned to reach a deal, presumably in order not to damage relations between Rome and Florence. And he had apparently devised a strategy that impressed the cardinals in charge, one that did not rely on simply engaging in debate with Galileo.[33] But what exactly was his strategy?

He surely will have explained to Galileo that if he continued to deny the charge he might be tortured; this was standard procedure in a case where it was important to establish the suspect's private views (although Galileo might have been able to claim an exemption on the grounds of age and ill health). He perhaps advised him that if he cooperated he could avoid both torture (a procedure that preceded a finding of guilt) and incarceration in one of the Inquisition's dungeons. He may have encouraged him to hope that even if the book was going to be banned he himself need suffer no punishment – certainly Galileo kept this hope alive.

Presumably Galileo on his part insisted that the book had been approved by the censors and that he had the support of the Florentine government, pointing out that the special treatment he was receiving was evidence enough that the Inquisition did not suspect him of any serious heresy. Back and forth the arguments went, Galileo still protesting his innocence and his inquisitor emphasising the consequences of sticking to this position, and at last Galileo gave way.

Why? The obvious arguments that Maculano could produce were all ones that would have occurred to Galileo already, which is precisely why the cardinals had initially been sceptical when Maculano proposed that he be given permission to meet informally with Galileo. There is only one explanation: Maculano must have threatened him with further charges. But what charges? The old charges that he had encouraged his pupils to think of God as material and to deny miracles had been rediscovered when the file on Galileo had been brought out of storage, but they were now stale, and it would have been impossible to proceed with them.

Here we have a puzzle, but it is one to which we can fortunately find a solution elsewhere in the Vatican documents relating to Galileo. In the papers of the Congregation of the Index there exists an anonymous, undated report surveying the evidence that Galileo was guilty in *The Assayer* of denying transubstantiation – a report first published in 2001.[34] From the handwriting, we can tell that the author of this report was Melchior Inchofer, a Jesuit, who had been appointed to the special commission to review Galileo's *Dialogue* in July 1632. He was also one of the theologians who reported to the Holy Office on its content: these reports, it seems, had been submitted by 22 April 1633. After Galileo's condemnation Inchofer was to publish a semi-official attack on Copernicanism, the *Summary Treatise* (1633), the title page of which shows the Barberini bees holding the earth firmly in place. This was ready for publication by the end of July 1633, and so work on it must have begun while Galileo's trial was in progress.

Inchofer had arrived in Rome from Messina in 1629: he came because a book he had written (on a letter supposedly written by the Virgin Mary to the people of Messina) had run into trouble with the Congregation of the Index, and the matter was only finally resolved in December 1630. We can be confident that he was not employed as a theological adviser before that date. In principle his critique of *The Assayer* might date to any time between early 1631 and the completion of the case against Galileo in June of 1633.[35] But it seems that between July 1632 and April 1633 Inchofer had been engaged in something of a research project involving Galileo. He had not only read the *Dialogue* with great care, but had also read a copy of the *Letter to the Grand Duchess Christina* (a clandestine text circulating in manuscript) and 'the little book published a long time ago' in which Copernicus is praised and defended – surely a reference to

The Starry Messenger. There is no evidence that he had scientific interests before 1629, and evidently he did not own, or have on long loan, the books he had been reading, for if he had had a copy of *The Starry Messenger* on his desk he would have given its title. He may well have been guided in his reading by fellow Jesuit Christopher Scheiner, to whose work he refers.[36] None of the other theologians felt any need to read around their subject – they stuck to the text in front of them. It would be natural, in the course of his reading, for Inchofer to turn his attention to *The Assayer*, particularly as his fellow Jesuits would have been well aware that it had been suspected of heresy. This, I believe, is the origin of the anonymous paper in the files of the Congregation of the Index, whose job it was to censor books.[37] It is also worth bearing in mind that some Jesuits sympathetic to Copernicanism (Grassi and Scheiner, for example) may have thought it better to proceed against Galileo's atomism than against his Copernicanism – against *The Assayer* rather than the *Dialogue*.[38]

The last sentence of Inchofer's paper on *The Assayer* (which in my view dates from between July 1632 and April 1633) calls for the matter to be handed over to the Holy Office for further investigation. Yet there is no trace of this new charge against Galileo in the Holy Office's file on him. How to explain this failure to act on a serious charge, made by a responsible authority, at a moment when any suspicion against Galileo was bound to be taken seriously? The answer, I suggest, is straightforward. Maculano brought Inchofer's denunciation to his informal meeting with Galileo on 27 April. He told Galileo that if he refused to cooperate, they would act upon Inchofer's denunciation. This would mean, at the very least, further investigations, a prolonged imprisonment, a lengthy trial. It might well result in conviction on a charge whose seriousness no one could doubt, transubstantiation being an absolutely fundamental doctrine distinguishing Catholics from Protestants. So Galileo and Maculano reached a deal: Galileo would cooperate, and Inchofer's denunciation of *The Assayer* would be left to moulder in the files of the Congregation of the Index.[39]

The deal they reached was highly favourable to Galileo, however. Proceeding with a charge against *The Assayer* would have been difficult: the book had been licensed and even praised by Riccardi, bore the papal arms on its frontispiece, and had been read aloud to the pope at mealtimes. In order to get Maculano to drop this charge Galileo did not have to concede much. He agreed only that anyone reading his *Dialogue* could easily form the impression that it had been written in defence of Copernicanism – he did not concede that this was his actual intention. In effect he agreed to plead guilty not to defending Copernicanism but to a lesser charge of appearing to do so.

Galileo was, he now admitted, guilty, of carelessness at least, and he was prepared to confess this to the court; he asked only for a little time to work out how best to frame his confession in order to minimise his guilt. Maculano had achieved what he was aiming at: he had made no binding commitment as to the punishment Galileo was to receive, but had set things up so that Galileo could be released without undermining the authority of the tribunal, which was the outcome desired by his superior, Francesco Barberini, the cardinal-nephew.[40] After this all that remained were formalities. Three days later Maculano drew up an official record of Galileo's confession, in which Galileo made no mention of their informal meeting, but said that he had reread his *Dialogue* and had been surprised at what he had found there. He had, he was forced to conclude, been led astray by pride – in other words, by intellectual ambition. As a result he had come to make the weaker argument seem the stronger.[41]

What I have offered is a conjectural interpretation which makes the best sense of the surviving evidence. We will never know for sure what Galileo and Maculano agreed at their informal meeting. But there is a tiny fragment of further evidence that supports my account. On 13 July Maria Celeste wrote to Galileo expressing puzzlement. She had heard earlier of the condemnation both of Galileo and of his book, and had expressed her profound dismay. Now she had received a letter from Galileo saying that the final outcome was a victory both for him and for his opponents. What could this mean, she wondered. In what sense could Galileo lay claim to a favourable outcome?[42] The only explanation is that Galileo took comfort in knowing that things could have been worse, that there were further charges that had been dropped. By holding out he had at least placed himself in a position to bargain.

Let us return to 30 April. That day, Galileo was released back into the care of the Florentine ambassador (who was now paying out of his own pocket the cost of housing him), and for almost two months he was in limbo.[43] On 10 May he made a further appearance before the Congregation, and handed in his formal defence, after which only a final judgement was required. Niccolini knew from his conversations with the pope that the final decree would condemn not only Galileo's book but also Galileo himself, but he withheld this information from Galileo for fear of upsetting him.[44] Galileo was clearly a very different character from another Florentine then being detained in the prisons of the Inquisition. Mariano Alidosi was being held incommunicado, but had been glimpsed in the distance cheerfully playing his Spanish guitar and singing, as if he were on holiday.[45] Galileo, who had always been treated with every consideration, had emerged from prison 'half dead'.[46] Niccolini drew the conclusion that being stupid was altogether preferable to being clever. Galileo was now effectively a prisoner in the ambassador's residence – he had to get permission to go out for exercise – but Rome's leading intellectuals were free to visit him.[47]

Finally on Tuesday 21 June Galileo was required to report to the Inquisition. There he was once more interrogated.[48] He denied ever having been committed to Copernicanism, although he said that before 1616 he had thought that either Copernicus or Ptolemy might be right. He insisted that his book should not be read as a defence of Copernicanism. He was threatened with torture and then

held overnight. The next day he appeared, dressed in the white robes of a peni-
tent, before the Congregation of the Inquisition at the monastery of Santa Maria
sopra Minerva. There he was declared guilty of having given grounds for vehement suspicion of having held Copernican doctrines and of thus being guilty of
heresy. Giving grounds for 'vehement suspicion' was a perfectly normal charge
in Renaissance law, used in cases where the evidence fell short of being conclusive; in this case, Galileo had confessed not to being a Copernican but to having
presented arguments in favour of Copernicanism with insufficient care. His
sentence was read to him and he was required to abjure Copernicanism. A copy
of his book, now banned, was burnt in front of him.[49] He was sentenced to the
prisons of the Holy Office at the pleasure of the pope, but on the Friday was
transferred to the Villa Medici, where he had stayed in 1616, and which he had
been visiting to take exercise. Niccolini travelled with him and found him
despondent – he had not foreseen any punishment beyond the banning of his
book. Two weeks later he was still profoundly shocked and dismayed.[50]

Galileo's dismay may seem surprising. Had he not known, from the moment
he had reached his deal with Maculano, that he would be condemned? The
answer is that his position had deteriorated in two respects between 27 April and
22 June. First, the Congregation had decided to brush aside his claim that he was
unaware of any injunction not to teach Copernicanism in any way whatever.
Second, and perhaps more surprising, the pope had decreed on 16 June that
Galileo was to abjure *de vehementi*, and in doing so had, apparently casually, clas-
sified Copernicanism as a heresy, which it had not previously been. Indeed the
pope had previously claimed credit for opposing efforts to make Copernicanism
a heresy in 1616. Now, however, a full-dress defence of the claim that
Copernicanism was heretical had already been commissioned and was shortly to
appear – the *Summary Treatise* of Melchior Inchofer.[51] Thus it was only the
letter of Galileo's bargain with Maculano that had been kept to, not the spirit.

According to a story first told in 1757, Galileo muttered, after reading his
recantation, 'E pur si muove' – and yet it (the earth) does move.[52] This story
conveniently appeared just at the time when the Catholic Church was beginning
to acknowledge that further opposition to Copernicanism was futile. But we can
be sure that Galileo never said anything of the sort. At no point in the trial had
there been any discussion of whether Copernicanism was true or not, for the
simple reason that Galileo conceded that it was false and claimed to have abandoned it in 1616. What may well be true, on the other hand, is a contemporary
report that Galileo insisted, before reading his abjuration, on declaring that he
was a Catholic and that he had dealt in good faith with the ecclesiastical censors
to obtain a licence to publish the *Dialogue*.[53] This report coincides exactly with
the position he had taken up during the trial, and it may be that his peculiar insistence that he was a Catholic, and intended to live and die in the Catholic faith –
an insistence which was apparently intended to shame his enemies – represents

his public response to the suppressed charge that he had denied transubstantiation in *The Assayer*, a charge that implied he was not a Catholic but a Protestant.

Inquisitors in university towns across Italy were soon ordered to summon together the local professors of philosophy and mathematics and to read to them Galileo's condemnation, so that there could be no doubt that Copernicanism was now a heresy and Galileo's book a forbidden book; nuncios abroad were instructed to make sure the news was disseminated.[54] The Florentine inquisitor and Father Monster were reprimanded for their part in approving Galileo's book.[55] Looking back in September, the Jesuit mathematician Orazio Grassi said that he had spoken up in Galileo's defence before he had been brought to Rome for trial, and had defended the arguments in his book, but that he felt that Galileo was the author of his own misfortune because he had fallen in love with his own arguments and had paid no attention to the views of others.[56] This seems an accurate account of what had happened. As for Grassi, he was deprived of all his positions within the Jesuit order and sent in exile to his home town of Savona, remaining in disgrace as long as Urban lived. Why? The only plausible explanation is that he was being punished for having spoken up, as he claimed, in Galileo's defence.[57]

On 30 June the pope agreed that Galileo might travel to Siena (within the territory controlled by Florence), to be held under house arrest by his friend the archbishop. Galileo arrived there on 9 July.[58] According to one report he was unable to sleep at night, but cried out and spoke to himself, so that there were fears for his sanity.[59] But he was also visited there by Campanella, who was facing ever greater hostility from the Spanish in Rome, and who would soon have to flee into exile in France. The two philosophers, both of whom had been held by the Inquisition, spent several days together in the archbishop's palace, sitting in rooms draped in the finest damask, eating the archbishop's exquisite food and drinking his noble wines.[60] Not surprisingly, Galileo put on some weight.[61]

A few months later the Inquisition in Rome received an anonymous denunciation against the archbishop. He was apparently telling people that Galileo was the greatest man of the age, that his condemnation was unjust, that the Inquisition should not have ruled on a question of science, and that posterity would vindicate the views Galileo had propounded in his *Dialogue*.[62] The accusation was left to lie on the file – a permanent memorial to the archbishop's good sense.

Galileo was kept waiting in Siena until long after the plague, which had been severe in Florence, had receded and it was safe to return. Eventually on 1 December the pope gave him permission to return to his villa at Arcetri outside Florence, where his movements were still to be restricted, and he was to avoid having more than one or two people at a time to visit him.[63] He arrived home on 17 December 1633. Almost immediately he was visited by the grand duke in what amounted to an official display of solidarity. He was to live for a further eight years, but he was scarcely ever to leave home again.

Eventually Galileo was given permission to travel to Florence to consult his doctors; on one occasion he was summoned to a meeting with the grand duke; on another he travelled, with permission, to meet the French ambassador to Rome.[64] Perhaps he occasionally slipped off – once when an official of the Inquisition came to call he found no one at home.[65] There appear to have been ways of getting round the restriction on meeting more than two people at a time. In 1636 a local clergyman invited him to attend a service to celebrate the feast of St John the Evangelist (27 December). The celebrations are normally confined to the church, he says, but if Galileo comes they will all go on celebrating at his residence – he is sure that everyone who attends will enjoy this more than anything else. It seems that Galileo is being invited to a party in his honour.[66] But until the end of his life he remained officially 'a prisoner to the Inquisition' (the wording is Milton's and is accurate: Galileo was not held by the Inquisition, but he was in law their prisoner, and could be returned to their prison at any time).

Galileo's social status was thus profoundly ambiguous. All his friends, including Castelli, who remained resident in Rome and answerable to the Barberini, stayed loyal to him, but some expressed their views frankly, saying that Galileo should never have been condemned, while others dared not put their view of the Inquisition in writing.[67] Galileo's friend and pupil Bonaventura Cavalieri, a priest in the papal city of Bologna, wrote to him on 17 December 1633, picking up a correspondence that he had dropped as soon as he had heard, a year earlier, that Galileo had been summoned to Rome. Cavalieri expected Galileo to understand (and Galileo does indeed appear to have understood) that it would have been impossible to communicate with him while he was under investigation or under guard.[68]

From the moment Galileo departed for Rome in 1633 until his death he and his friends always assumed that their correspondence might be intercepted.[69] Letters were generally sent to be delivered by hand, or to be included in a diplomatic pouch, or with an outward address to a safe intermediary (usually the grand duke's secretary and Galileo's relative, Geri Bocchineri, or, if the letter was coming from France, Galileo's relative in Lyon, Roberto Galilei). A cautious correspondent, such as Diodati, would ensure that Galileo's name appeared nowhere on the letter, and would sign it with a pseudonym.[70]

Fear of interception affected what Galileo was willing to write. In 1636 he was arranging for his complete works to be translated into Latin and published in Holland – except, of course, he wrote in a letter to Fulgenzio Micanzio in Venice, who served as his intermediary with Elsevier, the Dutch publisher, for his forbidden *Dialogue*. 'I read your letter aloud to Elsevier', Micanzio replied, 'and we had a good laugh. There would be little point in publishing the works without the *Dialogue*.'[71] Micanzio was secure under the protection of the Venetian government so he found the idea of inquisitors reading his letters rather amusing.[72] Galileo, whose wishes they had correctly interpreted, was not laughing, for he could never be sure that, if his continuing labours on behalf of Copernicanism were discovered, he would not be returned to the Inquisition's prisons. In 1638 he was formally told that if he discussed Copernicanism with anyone he would spend the rest of his life behind bars.[73] A failure to recognise the care with which he expressed himself on paper has led to the mistaken view that towards the end of his life he abandoned the theory of the tides that he had long thought to be the crucial proof of the movement of the earth.[74]

Galileo and his friends had also, of course, to be careful about what they said in person, and to whom. In March 1634 his friend Jean-Jacques Bouchard wrote to him to express his regret that Galileo could no longer even speak freely – although he felt sure that there were people like himself with whom Galileo could feel perfectly safe.[75] But how could one tell who could be trusted? In 1636 Famiano Michelini, who was known to be in Galileo's confidence, came to Rome, and Galileo's friends naturally pumped him for news. One of them, Raffaello Magiotti, wrote to Galileo, asking him to write to Michelini:

I do wish you would persuade him to be less cautious, not to say tight-lipped, with me when it comes to your affairs. I've waited two years for him to come here, and now when at last I get to talk to him I find him as silent as a fish. He won't say anything, except that he has never been so bold as to ask you, that he has seen little, that he hasn't had the opportunity, that he didn't write anything down and can't remember anything at all. And I stand there with my mouth open, while he blows me up with hot air. I say this because I would like him to realise that I am not naïve and can be trusted.[76]

Michelini's caution was perfectly sensible: Magiotti was altogether too close to the Jesuits, and Castelli had already become suspicious of him. Would anyone other than an agent provocateur have written such a letter? And would anyone other than an agent provocateur have tried to persuade Galileo that Mersenne was his enemy? No one else wrote letters in which they referred to Galileo by a code name in order to demonstrate their discretion. Certainly Galileo did not tell Michelini to be more forthcoming, and six months later Magiotti was still bitterly complaining that Galileo and Michelini were withholding texts from him, texts that were circulating widely.[77] Michelini could barely conceal his hostility, savagely mocking his pretensions as a mathematician. But they kept him in play: as long as they told him nothing he could do them no harm, and one day they might want to pass him some misinformation. Galileo had secrets to keep, and he kept them with considerable success.

As it turned out, their assessment of Magiotti was mistaken. He became a close friend of Torricelli and played a crucial role in Torricelli's vacuum experiments of 1644. When Mersenne arrived in Rome towards the end of that year,

Magiotti became (despite his earlier hostility) Mersenne's friend and collaborator; and Mersenne, on his return to France, spread the news of Torricelli's discoveries, which were soon replicated by Pierre Petit and Pascal. Magiotti, in the end, played a far more significant role than Michelini in the campaign against Aristotelian physics.[78]

34

The *Two New Sciences*

Within a few weeks of Galileo's arrival in Siena work began on a Latin translation of the *Dialogue,* which was eventually published in Strasbourg in 1635. The
project originated with Élie Diodati, a French Protestant who had long been a
friend of Galileo's, and he recruited Matthias Bernegger, a German Protestant
who had earlier translated Galileo's book on the sector into Latin. Bernegger
repeatedly claimed that he had been invited by Galileo himself to undertake the
translation, and it used to be argued that Galileo had contacted Diodati as soon
as he arrived in Siena to ask him to ensure publication – an interpretation of the
facts which implied that Galileo was immediately and blatantly in breach of the
recantation in which he had promised never to do anything that might imply
support for Copernicanism.

It seems that this story – although attractive – must be wrong, as Diodati
began to organise the translation before hearing news of Galileo's condemnation. But Bernegger was so convinced that he was acting at Galileo's request that
we must assume that Diodati had gently misled him – just as he had failed to
warn him in advance that the book he was to translate was not a short pamphlet
but a substantial volume. But whether or not Galileo asked Diodati to arrange
the translation, whether or not he specifically asked for Bernegger to carry out
the work, he certainly collaborated in the project, as he is the only possible
source of a correction which appears in the Latin edition, one that we know he
wanted introduced into any future edition. Galileo also thanked his Protestant
allies for their work. It thus seems fair to say that he was, more or less immediately, in breach of the recantation.[1]

We can be sure that as soon as he reached Siena Galileo got back to work on
the book he had been promising to write since at least 1609, the *Two New
Sciences* (Leiden, 1638). Although the text made no reference to Copernicanism,
the second of the new sciences, the science of local motion, was central to the
Dialogue's refutation of the standard arguments against Copernicanism. Galileo
first hoped to publish in Venice (where Micanzio soon discovered that any publication by Galileo, no matter what the subject, had been banned by Rome), then in
Vienna (within the world of Catholicism and of Medici diplomacy), then in Paris.[2]
Only when all else failed did he turn to the Dutch firm of Elsevier, who had recently

printed the Latin translation of the *Dialogue* and the first, bilingual (Italian and Latin) edition of the *Letter to the Grand Duchess Christina* in Strasbourg (the latter made possible because Galileo had arranged for the text of the letter to be smuggled out of Florence to Diodati; he was then sent the text of Diodati's preface, which he approved).[3] Micanzio acted as intermediary. He was alarmed to hear that, despite his failing eyesight, Galileo was copying the manuscript himself – he did not yet have an amanuensis he could trust – but the grand duke soon arranged for him to have the assistance of an able mathematician, Dino Peri.[4] Later he was to be assisted first by Marco Ambrogetti, then by Famiano Michelini, Clemente Settimi and Alessandro Ninci, and finally, particularly after Settimi was sent away, by Settimi's former pupil Viviani and Castelli's former pupil Torricelli.

The book was dedicated to François de Noailles, the French ambassador to Rome and a former student of Galileo's from the days of Paduan liberty. In the letter of dedication Galileo claimed that he had given the manuscript to Noailles at a meeting that had taken place when Noailles was returning from Rome to Paris, and that he had now learnt that Noailles had arranged for the Elseviers to publish the book – he describes this as 'unexpected and astonishing news'. Noailles had not in fact been given a copy – Galileo was going blind and had no amanuensis to make extra copies – though Viviani, in order to support Galileo's story, was later prepared to fake a letter from Diodati saying that he had been.[5] And Noailles had nothing to do with the publication, although he did give Galileo permission to use his name. But Galileo must have taken an exquisite pleasure in using his meeting with Noailles as a cover story to explain how his manuscript had ended up in the hands of a Protestant publisher, for Noailles had obtained authorisation for the meeting from the pope himself.[6] Galileo was, there can be no doubt, thumbing his nose at Urban.

We have seen that Galileo's first great innovation was to imagine a ball sliding for ever across a perfectly smooth sheet of ice; on the basis of this conceptual model he reached the conclusion that movement would continue indefinitely in the absence of any countervailing force. Since 1590 or so he had been working on his new science of movement, which he finally published in 1638. This new science was concerned with the acceleration of falling bodies, the isochronicity of the pendulum, the parabolic path of projectiles. All these discoveries, which constitute the foundation of modern physics, contain both an experimental and a theoretical or deductive component; in the published work theory is made, as far as possible, to predominate over experiment. They also involve the construction of models – both theoretical models and working models – which are idealised in exactly the same way as his first model of unimpeded movement. The idea of such an idealised model came from the study of mechanics, but Galileo had taken the idea from the study of artificial objects (pulleys, levers, balances, clocks) and applied it to natural objects. This was his first great innovation – to treat nature as if she were an artificial object.

How do you study movement? You construct working models in which friction is minimised, in which metal balls, for example, slide down chutes lined with polished parchment. And you construct theoretical models in which air resistance is excluded. You imagine a perfectly smooth surface, or the behaviour of a falling body in a vacuum. When Galileo says that the book of nature is written in the language of mathematics, he means that real nature embodies principles derived from this idealised nature.

Galileo was insistent that this process of idealisation did not imply falsification. Strikingly, he compared it to the calculations of a merchant engaged in double-entry bookkeeping:

> What happens in the concrete . . . happens in the same way in the abstract. It would be novel indeed if computations and ratios made in abstract numbers should not thereafter correspond to concrete gold and silver coins and merchandise . . . Just as the computer who wants his calculations to deal with sugar, silk, and wool must discount the boxes, bales, and other packings, so the mathematical scientist (*filosofo geometra*), when he wants to recognize in the concrete the effects which he has proved in the abstract, must deduct the material hindrances, and if he is able to do so, I assure you that things are in no less agreement than arithmetical computations. The errors, then, lie not in the abstractness or concreteness, not in geometry or physics, but in a calculator who does not know how to make a true accounting.[7]

Galileo's study of motion is a study of idealised motion in a theoretical universe. By constructing this theoretical universe he perfects the real world, eliminating friction and resistance, and discovering mathematical purity. Another way of describing his great innovation, then, is to say that he treats the sublunary world as if it were the superlunary world. The heavens had always been understood as a world of mathematical purity, but Galileo finds this purity precisely where all previous scientists had insisted it could not be found – on earth.[8]

By a remarkable reversal, Galileo then turns round (and there is evidence that he had started to do this even before the discovery of the telescope) and finds variation, imperfection, roughness rather than smoothness in the heavens: the lunar mountains, sunspots and novas prove there is change and imperfection where all previous scientists had seen only permanence and perfection. Galileo is very aware of this double movement: the rhetorical high point of *The Starry Messenger* is the claim that the earth is a heavenly body and that the heavens contain earthly bodies.

It is time to return to the starting point of any biography of Galileo. Galileo began on a path which was meant to lead to the practice of medicine. Instead he chose maths. Renaissance medicine is concerned with conflicting humours and bodily emissions – it is about the impermanent, the imperfect, the uncontained. It was the most empirical of all sciences. Galileo turned his back on this world because he preferred the abstraction, the perfection and the theoretical nature of mathematics. Here he was making a fundamental choice, a choice that remained with him throughout his life, and that he was still expressing in the *Two New Sciences*.

He could have stepped away from this choice. He could have become an experimental scientist, and studied, for example, vacuums. As soon as Galileo had a working telescope, he also had a working microscope, but he did nothing with the microscope until 1624, and never published on the invisible world – a world where what appears smooth turns out to be rough, where what appears fine turns out to be coarse, where what appears simple turns out to be complex.[9] This was not the world of Galileo's imagination. Writing to his colleague Giovanni Battista Baliani, who had founded a science of falling bodies on the basis of experimental evidence rather than theoretical deduction, Galileo did not hesitate to stress the difference between their approaches. The value of his own work, Galileo insisted, depended on its theoretical coherence, and so it would be a matter of indifference to him if it turned out to be of no use in predicting the behaviour of actual bodies in the real world.[10]

It is very easy to imagine that, as a great scientist and mathematician, Galileo was in some way bound to be the founder of a new physics. But there was nothing inevitable about it. He became the founder of a new physics because a particular way of thinking gave him satisfaction; and he became the founder of a new astronomy because what was in a sense the opposite way of thinking also delighted him. What was rough he made smooth, and what was smooth he made rough. No life is without conflict, but Galileo's mother thrived on dissension and delighted in it. This may have been why Galileo delighted in imagining a world all around him that was smooth, polished, friction-free, and another world, far away, where there was change and conflict, but change and conflict that would not affect him.

This might seem far-fetched, but consider his brother. In 1620 Michelangelo Galilei published a little book of lute music. The book was written in Italian, published in Germany and required a French lute – naturally few copies were sold. In any case the music was avant-garde (at least when it was first written, if not when it was eventually published). Michelangelo warned the reader, 'Do not think, because there are in this work many harsh sounds or dissonances, that they are printing mistakes, for they are what I intend; and be assured that I have looked over the whole book meticulously many times and I am certain that it is absolutely correct.'[11] Michelangelo too sought to bring harmony and dissonance into a new relationship, just as his brother sought to change the relationship between an imperfect reality and the perfect world of mathematics.

Galileo also delighted in exploring problems of scale, in inhabiting Gulliver's world (a world which would have been unimaginable without the telescope). Telescopes and microscopes are machines for altering the scale of the world.

The inclined plane was a tool for prolonging fall. And the first of Galileo's new sciences is entirely about problems of scale. The most elementary problems are those of extension. How far can you extend a beam before it breaks under its own weight? How long will a rope be before it breaks under its own weight? If you make the beam or the rope thicker and heavier, will it be longer or shorter at its breaking point? What profile of beam both in a cross-section and in a horizontal section will have the greatest strength for the least weight? In asking these questions (and we have seen that he had begun to ask them in 1592) Galileo was reversing the line of argument he relied on when thinking about machines. His machines were perfect; his architectural structures are so imperfect that they are on the point of collapse.

Galileo was inventing a new, abstract science of materials. Real beams, in Galileo's world, are made of wood or marble – they have cracks and fissures, imperfections and flaws. But Galileo's beams are imaginary beams, perfectly uniform through and through.[12] The core of Galileo's thinking is that weight increases with volume and resistance increases with surface area. An object which is twice as big in every direction generates four times as much friction falling through the air, but is eight times heavier. As a result it has a higher terminal velocity.

> For who does not see that a horse falling from a height of three or four braccia will break its bones, while a dog falling from the same height, or a cat from eight or ten, or even more, will suffer no harm? Thus a cricket might fall without damage from a tower, or an ant from the moon. Small children remain unhurt in falls that would break the legs, or the heads, of their elders. And just as smaller animals are proportionately stronger or more robust than larger ones, so smaller plants will sustain themselves better. I think you both know that if an oak were two hundred feet high, it could not support branches spread out similarly to those of an oak of average size. Only by a miracle could nature form a horse the size of twenty horses, or a giant ten times the height of a man – unless she greatly altered the proportions of the members, especially those of the skeleton, thickening the bones far beyond their ordinary symmetry.[13]

Galileo provides an illustration. He takes a small bone, and imagines increasing it in length three times. But how thick will the bone now have to be in order to support a creature which now weighs twenty-seven times as much as the original animal? See plate no. 25 for Galileo's answer.

We can summarise very simply, if anachronistically, the complex and subtle arguments of the first of the two new sciences: Gulliver's worlds are impossible. There can be no Lilliputians or Brobdingnagians. This was a remarkable scientific breakthrough, crucial for the design of machines and buildings, and for understanding biological form. It is a very different sort of breakthrough from the discovery of the law of fall. Galileo invented two new physical sciences and a new cosmology. These three transformative discoveries had little in common with each other.

Galileo was a great scientist because he was able to innovate in several, apparently unconnected, fields. What connected those fields was not some feature of the real world, but a feature of Galileo's own psychology – a preoccupation with resistance, which led him to be fascinated by fracture on the one hand, and by uninhibited movement on the other. Where did Galileo's preoccupation with resistance come from? From the same place as Michelangelo's preoccupation with the limits of harmony – from the conflicts of his childhood.

Vincenzo, son of Galileo

Vincenzo Galilei was born on 21 August 1606. He was four when Galileo moved to Florence, and naturally he stayed behind with his mother, Marina Gamba. When she died in 1612, Galileo made temporary arrangements for his care and then brought him to Florence.[1] In June 1619 he arranged for him to be officially made legitimate (although without entitlement to all the rights of citizenship), which opened the way for him to obtain an education at the University of Pisa: he graduated with a doctorate in law in 1628.[2] It was, unfortunately, a popular saying that a Pisan doctorate was only good for a laugh.

Sometime in 1623 Vincenzo did something that his father regarded as inexcusable – quite what we do not know, and indeed Castelli, who was taking care of him while he was studying in Pisa, came in the end to believe that Vincenzo's protestations of innocence might be genuine.[3] In the years that followed, Galileo seems to have effectively broken off all relations with his son, relying on Castelli to act as intermediary.[4] In 1627 Galileo obtained a pension for him from the pope, but Vincenzo refused to be tonsured, apparently out of hostility to the Church authorities, Galileo having to arrange for the pension to be transferred to Vincenzo Michelangelo Galilei, who was even more unsuitable, and who fled Rome before he could start receiving the money.[5] In 1628 Vincenzo was in some sort of trouble again. He was brought back to Florence from Pisa in a hurry, before anything worse could happen, Castelli regarding him as no better than his cousin Vincenzo.[6]

In January of 1629, newly qualified, Vincenzo made a surprisingly good marriage – to Sestilia Bocchineri of Prato, who came with a dowry of seven hundred scudi, and whose father agreed to pay Vincenzo a hundred scudi a year.[7] But Vincenzo still expected to be supported by his father, who had meanwhile acquired a small house in Florence for him to live in. (The house was even smaller than it seemed, because one room was directly over the cesspit, which rendered it uninhabitable.) There was a bitter exchange of letters in December 1630, Vincenzo complaining angrily that Galileo had failed to find him a job, was wasting money on other people's offspring, and was offering him no support in his hour of need. In the spring of 1631 Maria Celeste was distressed that there was no communication between Galileo and his son, despite the fact that

Galileo was looking after his grandson, Galileo (born in December 1629), Vincenzo and Sestilia having fled to Prato to avoid the plague.[8]

In fact Galileo – who had made his own way in the world without his father's support, and who had studiously avoided involvement in Florentine politics and patronage networks – was doing his best to provide for his son. He had taken out full rights of citizenship for himself in 1628 (a step which made him liable to new taxes), thereby becoming eligible for election to office. He then had his son recognised as also entitled to the rights of citizenship, supported his candidacy for office, and became briefly active in Florentine politics in 1629 and 1630.[9] Finally, after two years, this policy paid off, and in November 1631 Vincenzo was placed in charge of the finances of the little town of Poppi, thirty miles from Florence, a job in which he was soon complaining that he was overworked and underpaid. Within a few months he was in trouble for not being sufficiently assiduous, and in the autumn of 1633, when Galileo was under house arrest in Siena, he learnt that his son was about to be fired for failing to perform his duties adequately.[10]

After a good deal of lobbying, Galileo and his friends arranged that Vincenzo merely be transferred to another small town, San Giovanni Valdarno, in the spring of 1634 (San Giovanni was at least on the main road between Florence and Rome), and either Vincenzo's behaviour improved or Galileo's influence on his behalf became more effective, for in June of 1635 he was transferred back to Florence to manage the finances of the guild of merchants, a position he held until his death in 1649.[11] In the meantime, in the summer and autumn of 1633, under sustained pressure from Vincenzo's brother-in-law Geri Bocchineri and Maria Celeste, Galileo had agreed to purchase the house next to the one he had bought for Vincenzo in 1629, thus providing him with respectable accommodation near the centre of Florence.[12] He also agreed to clear some of his debts.[13]

The central charge against Vincenzo (since we do not know what he did to infuriate his father in 1623 and in 1628) must be that during the first four months of 1633, when Galileo was facing charges in Rome, he did not write a single letter to his father offering him moral support.[14] His first letter to him was sent on 2 May 1633, two days after Galileo had been released by the Inquisition – although it seems unlikely that news of this could have reached Vincenzo so quickly, for Poppi was not on the main postal route.[15] What he evidently had heard was that Galileo was being given special treatment by the inquisitors, and had not disappeared into the Inquisition's dungeons, from which he rightly concluded that there was little risk in identifying himself as his son. It is likely that Vincenzo had written only because Maria Celeste had urged him to do so. In the event he was lucky: he had re-established contact with his father just in time to take advantage of the house next to his own in Florence coming on the market,[16] and to mobilise Galileo's friends to protect his job.

It is hard to imagine that Vincenzo's behaviour had endeared him to Galileo, and his father would surely have continued to regard him with suspicion. But everything changed on 1 April 1634, when two dreadful things happened. That day Galileo learnt that the papacy had grown impatient with constant requests emanating from the Florentine government that he be released from house arrest. The message from Rome was blunt: if there were any more requests for an improvement in Galileo's conditions he would be taken into the Inquisition's prison. Galileo now learnt that Arcetri was to be his prison until his death, and that there was nothing more the Florentine government could do on his behalf.[17] Being a resourceful man he did not entirely give up. The Florentine government allowed him to open a channel through France to seek clemency, and as we have seen, Galileo even had papal permission for a meeting with the French ambassador, François de Noailles, when he was returning from Rome to Paris.[18] But his efforts met with no success.

On the same day as Galileo learned that Arcetri would be his prison for life, he also learnt that Maria Celeste was dying – she died of dysentery the next day.[19] She was his only reason for being in Arcetri, and it was her companionship that he had counted on for his old age. He was now a prisoner indeed.

On 28 April, less than a month after Maria Celeste's death, Galileo wrote to Geri Bocchineri, who had replaced Castelli as his intermediary with Vincenzo. He is worried about his health. His hernia has returned much enlarged. His pulse is irregular and he has palpitations. He is deeply depressed, cannot eat and wants to join his beloved Maria Celeste in death. He cannot get a wink of sleep. And now Vincenzo is planning to leave on business. Geri must talk to him, and persuade him to understand that his father needs him nearby.[20]

Galileo, who was fast going blind, now turned to Vincenzo for support, and there seems to have been no further conflict between them. They worked peacefully together, for example, as they tried to build a pendulum clock according to Galileo's design. Sestilia, Vincenzo's wife, even became friendly with Virginia, Galileo's unacknowledged granddaughter. But if father and son were reconciled they were not close. A Florentine inquisitor, assessing the suitability of Vincenzo's housing his father should Galileo be allowed to move temporarily to Florence to receive medical treatment, reported that one could be confident that Vincenzo would take good care of him: after all, he was worth a thousand scudi a year to him for every year he lived. In Galileo's mind his son's house (which legally belonged to Galileo but which he had never occupied) now became his home, from which he was excluded.[21] When Galileo was granted permission to visit Florence, the expectation was perhaps that he would effectively move in with his son, but whatever visits he made seem to have been short.

We have a description of Galileo, soon after he had gone completely blind, written by the Florentine inquisitor in February 1638, when the question of whether he should be allowed to go to Florence for treatment was being discussed. The inquisitor had visited him with a doctor to assess not only his medical condition but his occupations and attitudes:

I found him totally without sight and completely blind . . . he has a very severe hernia, constant pains in his guts, and a wakefulness such that, according to his own word and the reports of those who live with him, in twenty-four hours he never sleeps for a whole hour; and in other respects he is so reduced that he looks more like a corpse than a living person . . . His studies have been brought to a halt by his blindness, though he sometimes has something read to him, and he has few visitors, because, being in such poor health, all he normally has to talk about to anyone who does visit him is the dreadful pain he is in and the other conditions he suffers from.[22]

Of course it was in Galileo's interests on this occasion to make himself appear sick and pitiful, but we may well suspect that this account was fundamentally accurate, for we have plenty of complaints from Galileo himself of depression, sleeplessness and physical discomfort. Moreover, unable to sleep or eat, he was terrified that he might lose his grip on reality: he was afraid of slipping into a world of delusions.[23]

In one respect, however, Galileo was overstating his incapacity and hood-winking the Inquisition, for he was certainly still actively engaged in intellectual work. A week after the inquisitor's visit he dictated a long letter on the libration of the moon, a subject which had been much on his mind during his anxious and sleepless nights, and which he had been working on in the summer of 1637, as he struggled, for the last time, to use his failing eyesight (he finally lost the sight of his better eye, the right one, in July) to make astronomical observations.[24]

The moon is locked in its relation to the earth, so that we always see the same half of the moon's surface, or so it had always been assumed. As Galileo realised, if we are to the east or west of a line from the centre of the earth to the centre of the moon (and this happens every day as the earth turns), we can, as it were, see around the side of the moon, first on one side and then on the other – this is an example of parallax. In addition the moon does not trace the same route through the heavens each night, but rises and falls slightly during the course of a year, so that we can sometimes peek a little way over the top or under the bottom of the moon. This is what Galileo called the libration of the moon – the moon appears to sway in the sky, like scales ('libra' in Latin) swaying on a balance – and he had discussed this at some length in the *Dialogue*.[25]

But now Galileo had realised that there was something else going on. He could account for a daily and an annual libration, but there appeared to be a further swaying movement, which took place on a monthly cycle. We would explain what Galileo was seeing in the light of Kepler's discovery that planets orbit not in circles but in ellipses: the moon rotates steadily on its axis, but because its orbit is not circular, the hemisphere it shows the earth varies very slightly in the course of a month. Galileo may have thought it showed that the earth is not at the centre of the moon's orbit. But it is impossible to know exactly what he thought because he could not risk putting Copernican arguments on paper. It is clear, however, that he thought he was seeing something important, and that he wanted to record the fact that he had been the first to see it.

As he was working on this problem in 1637, and as he described it in letters of late 1637 and early 1638, after he had had to abandon his observations, what was Galileo looking for? Some tiny change in what was visible from which great consequences might be drawn, as he had drawn great consequences from the path of sunspots on the surface of the sun? He gave a clue in a letter to Micanzio, where he pointed out that the moon and the tides were alike in having daily, monthly and annual variations.[26] Since he had linked the variation of the tides to the Copernican theory, Micanzio thought he might have found, or at least have been looking for, new evidence in support of Copernicanism.[27] He was surely right. Galileo thought there were daily, monthly and annual variations in the tides caused by changes in the speed at which the surface of the earth was moving as it rotated around its axis and around the sun. He may have thought that a similar effect was causing the moon to wobble – that the monthly libration was caused by the same sort of force as caused the tides on earth, but working on a body rotating much more slowly (it takes the moon a lunar month to complete one rotation on its axis).

Galileo wrote up his new description of the moon's libration with the intention of circulating and perhaps even publishing it; consequently he could not so much as mention its possible relevance to the theories of Copernicus or of Kepler. He dared not even mention his previous discussion of the subject in the *Dialogue*, so he left unclear the relationship between his new and old observations. He was confident, however, that later astronomers would find what he had been seeking, and he wanted to leave a clear record of his having been there before them. Galileo had now adapted to being blind sufficiently to go on to do more new work: he produced an analysis of percussion in time for it to be added to the *Two New Sciences* as it went through the press.[28]

36

Galileo's (un)belief

The queer thing was that though she [Mme de Warens] did not believe in Hell she believed in purgatory... Another queer thing. Her system clearly destroyed the whole doctrine of original sin and redemption, and shook the complete basis of common Christianity, so that Catholicism, at any rate, could not subsist with it. Yet Mamma was a good Catholic, or claimed to be, and it is clear that her claim was made in very good faith. In her opinion the Scriptures were too literally and too severely interpreted ... In brief, she was true to the faith she had embraced, and sincerely accepted its whole creed; but when it came to the discussion of each separate article, it turned out that her belief was quite different from the Church's, though she always submitted to its authority.

Rousseau, *Confessions* (*c.*1766), trans. J. M. Cohen

This chapter presents a radically new account of Galileo's religious views, developing a line of argument previously touched on in chapters 16 and 22. Let me make clear, to begin with, that I do not think it is possible to answer the question whether Galileo was a believing Catholic in any straightforward way. Throughout his life he claimed to be a good Catholic, but then the consequences for him if he had failed to do so would have been extremely serious. The case of Mme de Warens is instructive: she was paid a pension because she had converted from Protestantism to Catholicism; she could not stop being a good Catholic without losing her income; and so she remained a good Catholic.

That much is clear; what is much less clear is whether she can be said to have done so sincerely and in good faith, and how you view this question will depend on whether you think you can believe something and dispute it at the same time. Certainly every community of believers contains within it a much wider diversity of actual beliefs than its leaders think it ought to, or than you might expect if you had regard only to official pronouncements. Believers construct their own idiosyncratic religions while subscribing outwardly to the doctrines required for membership of their particular religious community. Who would expect to find someone who denied the whole doctrine of original sin and redemption, and claiming to be a good Catholic? And yet Mme de Warens was surely not alone in her decision to hold two incompatible positions at the same time.

If belonging to a community is what counts, then Galileo and Mme de Warens were both good Catholics, for they were certainly members of Catholic communities. If agreeing with the fundamental teachings of the Church is what counts, then neither Galileo nor Mme de Warens was a Catholic at all. In these matters, however, one should (as Rousseau implies) allow people considerable freedom to bend and stretch and twist the language of belief if they wish to do so. If Mme de Warens wanted to think of herself as Catholic, then by all means let us respect that wish. What I want to argue here is that Galileo's relationship to Catholicism was rather like that of Mme de Warens; it may be easier to agree on that than on how to describe such (un)belief.

In developing this argument, I draw on several distinct types of evidence. I argue from the things Galileo said; from the things he did not say; and from the conversations of his disciples as they are reported to us. These three types of evidence establish, I think, a very strong presumption that Galileo was not a Christian; nevertheless they are not conclusive.

On the other hand Rousseau's account of the beliefs of Mme de Warens *is* conclusive: he knew her well, and wanted to give the most favourable possible description of her, while she would have had many opportunities, during their long and close relationship, to correct any misunderstanding he may have had. Rousseau's account may leave us in some uncertainty as to how best to describe Mme de Warens's (un)belief, but it makes the basic character of that (un)belief perfectly clear. His account is actually much more reliable than any statement we might have from Mme de Warens herself, because it summarises her views as he encountered them over a long period of time in many different circumstances. In this way the account of a privileged and sympathetic witness may be more reliable than a first-person statement, which is necessarily interested, partial and subject to revision.

We will see that there is one text in the Galileo archive which is similarly decisive: I maintain that it would be as futile to argue against it as it would be to argue against Rousseau's account of the religion of Mme de Warens. But it is of the nature of these questions that agreement is difficult to obtain, even under the most favourable of circumstances: if there were a scholarly literature on Mme de Warens I am sure there would be those who would be prepared to insist on the genuine character of her Catholicism. In the case of Galileo, where generations of scholars, particularly liberal Catholic scholars, have wanted to portray him as an innocent victim, whose genuine faith ought to have been a protection against any condemnation for heresy, there is now an enormous cultural investment in the idea of him as a good Catholic. Viviani was remarkably successful in establishing an account of Galileo's commitment to Catholicism which has survived largely unchallenged for more than three centuries. My task now is to offer, for the first time, a proper alternative: my argument is that, as Naudé is said to have remarked of Cremonini, 'Nihil habebat pietatis, et tamen pius haberi volebat.'[1]

Let us start with a book that Galileo loved: Giovanni Nardi's *Of Underground Fire* (1641). Nardi was a fellow Florentine and physician; he would soon produce a fine edition of Lucretius, in which he argued in favour of a germ theory of disease.[2] There need be no doubt that he was an atomist.[3] The purpose of *Of Underground Fire* was to argue that the centre of the earth is a great lake of molten lava, volcanoes allowing us to grasp the nature of this subterranean world. Galileo may have been particularly delighted by this book because it at last provided an alternative to Dante's vision of the underworld, a vision which Galileo had carefully studied in a pair of lectures which he delivered to the Florentine Academy as a young man.[4] Here, at last, was a scientific alternative. He urged all his friends to read it, and must have been dismayed that they could not stop themselves from laughing at it. In Nardi's world, they declared, as in the land of Cockayne, one would only have to dip macaroni in the lake to cook it, and birds would fall ready-roasted from the sky. What seemed to Galileo an admirable example of the new science seemed to them plainly nonsensical.[5] They had, after all, the evidence of their own senses to assure them that the earth was not on fire. Galileo's enthusiasm for Nardi tells us a good deal about his willingness to question his own day-to-day experience; but it also tells us that he was delighted to abandon the idea of the earth as being made in order to provide a habitation for man. The jokes about the land of Cockayne were precisely the opposite of the point: the puzzle about Nardi's earth is not that we have to kill and cook our own food, but that we ourselves are not killed and cooked. Thus theology is replaced by geology, and providence by a set of random factors that happen to make the earth habitable.

This is, at best, indirect evidence regarding Galileo's (un)belief. There is better evidence to be had. Between April 1639 and October 1641, one of the people closest to Galileo was Clemente Settimi, who was given permission to stay overnight at Arcetri.[6] Galileo's solitude had weighed heavily upon him in the years following the death of Maria Celeste; now he was no longer alone.[7] Settimi was a priest, but he was also a mathematician: he had offered instruction in geometry and had taught the young Vincenzo Viviani, who was to move in with Galileo in October 1639. In October 1641 Settimi was summoned to Rome to be investigated by the Inquisition, accused of having read Galileo's *Dialogue* and of believing that the universe had no creator. He was eventually cleared (the Tuscan ambassador had made clear that he was under the grand duke's protection), but he was not permitted to return to Florence, and was eventually transferred to Sicily.

The charge against Settimi came from Mario Sozzi, a member of the same religious order, the Piarist fathers. He made his accusations first to the Inquisition in Florence, and then, when no action was taken, to the Inquisition in Rome. He accused not only Settimi but other members of the order. Originally his accusation was that they had sought to protect Canon Pandolfo

Ricasoli, a relative of the grand duke's, who with an associate, Faustina Mainardi, had been condemned by the Florentine Inquisition to life imprisonment in 1641 for the heresy of teaching young orphan girls that sex without penetration was no sin, but the Piarists were cleared of these charges. So Sozzi proceeded to attack them for being pupils and admirers of Galileo. He claimed that the grand duke had provided scholarships so that they could educate gifted youngsters in Galileo's scientific beliefs, and that manuscript copies of Galileo's works would be found in the possession of their pupils.[8]

This charge was, in its essentials, correct. We know that the founder of the Piarist order, José de Calasanz, had arranged for a group of able young fathers to go to Florence so that they could acquire the principles of Galileo's new science and instruct others, and it is clear that this enterprise had the support of the grand duke. Settimi knew the grand duke well enough to confide in him his hatred for the Inquisition, and the original leader of the Piarist group in Florence, Famiano Michelini, had been transferred to Pisa so that he could teach algebra and Galileo's physics to Prince Leopoldo and his brother Cardinal Gian Carlo de' Medici.[9] Settimi's pupil, Vincenzo Viviani, had indeed been placed on a salary by the grand duke and recommended by him to Galileo.

According to Sozzi, Michelini was an atomist and a Copernican. Michelini, he said, believed that Galileo was the most important person in the world – a view not that far from Michelini's expressed opinion that Galileo was the greatest philosopher of the age.[10] It is not clear that Michelini was an able mathematician. He had presented himself to Galileo in 1629 with a letter of introduction from Baliani saying that he was better than average, but in Baliani's idiolect that may have been high praise.[11] Certainly he was utterly devoted to Galileo ('innamoratissimo', as Castelli put it).[12] According to Sozzi, Michelini believed that Galileo had been tried and his book condemned solely because he had placed Urban VIII's favourite argument in the mouth of Simplicio – and here we can be confident that he was reflecting Galileo's own views. Ambrogi Ambrogio, another of the Piarist fathers, was also an atomist, and denied that colour really existed (a view, Sozzi stressed, incompatible with belief in transubstantiation); Galileo, as it happens, described Father Ambrogio as a very close friend.[13] Sozzi had reached the conclusion that his fellow religious were closer to being atheists than Christians, although it is clear that they told him that the new science provided a way to know God. What would he have made of it, I wonder, if he had learnt that Michelini addressed his prayers to the Supreme Mathematician?[14]

Sozzi's accusations were based, he claimed, on numerous conversations: if any of these were with Famiano Michelini then he had been uncharacteristically loquacious, for we have seen that in Rome he had been as silent as a fish. He may have felt safer in Florence, where he had the grand duke's protection. The accusations were taken with great seriousness in Rome, and Sozzi was eventually made general of the order, against the wishes of its members and protectors, and despite the powerful opposition of the grand duke of Florence (who had banned him from setting foot on Florentine territory). Evidently he had successfully convinced the Roman authorities that the Copernican heresy had come to be widely disseminated within the Piarist order, but what he had discovered amongst Galileo's disciples was not only Copernicanism and atomism, but also a more profound scepticism about the truths of the Christian religion.

The standard response to accusations such as these is, of course, that they must be discounted; that they are born of private hostilities and personal grudges. Sozzi's charges, however, correspond too closely to the facts as we know them from other sources to be easily dismissed: indeed Giorgio Spini has concluded that on this occasion, at least, Sozzi 'spoke the truth'.[15]

Nor is Sozzi our only witness. Vincenzo Viviani, as we have seen, was first educated by Piarist fathers and then by Galileo. In 1646, a French scientist, Balthasar de Monconys, paid a visit to Florence. There he met with the little circle of Galileo's friends and disciples, including, of course, Viviani, and he acquired copies of almost all of Galileo's works; since he makes no mention of paying for these books, it seems clear that he was being inducted into the fraternity of *galileisti*. This meant not only providing him with the texts of Galileo, but informing him about the thinking that lay behind those texts. On 6 November de Monconys went for a stroll with Viviani, who told him that he believed that the sun is a star, that all things happen by necessity, that there is no devil ('la nullité du mal'), that the universal soul (the *anima mundi*) participates in all things, and that the universe may continue for ever ('la conservation de toutes choses').[16]

De Monconys's report of his conversation with Viviani is grammatically ambiguous: he may be telling us what Viviani thought, or what Viviani claimed Galileo thought. The marginal annotation states that these are the opinions of Viviani, but the text was published posthumously, and the marginal annotation is almost certainly attributable not to de Monconys but to his first editor. Since Viviani was so proud of being Galileo's faithful disciple, it is very likely that Galileo and Viviani thought alike on these matters; and de Monconys's ambiguity may be telling, since he may simply have assumed that Viviani was talking for Galileo as well as for himself. This certainly would be my suggestion – that de Monconys was being inducted into Galileo's esoteric teaching. It is important to note that de Monconys had no interest in falsifying his conversation: this was an entry in a private diary, written, with no expectation of publication, by someone who was an admirer of Galileo.

Let us deal first with the question of Viviani's religious beliefs. There is other evidence to suggest that these were far from conventional. On 9 June 1691 Magalotti wrote to Viviani about a book Viviani proposed to publish in which he would offer a geometrical proof of the truths of the Christian religion (the project reminds us of Michelini's prayers to the Supreme Mathematician).[17] Magalotti was deeply pious, partaking of the Blessed Sacrament once a week; there is ample evidence of his piety in a long letter he wrote to Robert Boyle urging him to convert to Catholicism, and confirmation too in the fact that he briefly joined the Oratory of St Philip Neri in 1691. (Indeed he was a member at the time of this correspondence.)[18] But Magalotti urged caution on Viviani: the proposed title of Viviani's book was *Testimonies* (in support of Christianity), and Magalotti was concerned that it might be taken to imply that Christianity needed the support of geometry, and that if geometry failed to speak up for it the case for religious belief would be lost. In other words, he doubted that Viviani was prepared to recognise that Christianity was a historical and revealed religion.

Viviani replied acknowledging the problem, and revising the title to *Essay on Moral Geometry*.[19] His intention, he told Magalotti, was to dedicate it to a Jesuit friend, Giuseppe Ferroni, who was a great admirer of Galileo. Viviani now also revealed that he was working on a longer work in Latin entitled *Moral Geometry*, and this he hoped to dedicate to Magalotti himself. Both works were already set in type – and yet neither was ever published. We can only assume that Ferroni and Magalotti dissuaded Viviani from going ahead with his peculiar project.

Of course it is almost impossible to guess at the content of Viviani's two books on moral geometry. But we may suspect that Magalotti's gentle warning was well aimed, and that Viviani did think that geometry (or science) was the final arbiter of truth. Viviani himself insisted that these works did not touch on any of the 'mysteries' of religion, in which case it is hard to understand how he could have thought of them as testifying for the truth of Christianity. We get a hint here, as with the Piarist fathers, of an esoteric religion, founded in philosophy and science, one that lives uncomfortably alongside an outward commitment to orthodox Catholicism.

Viviani, Settimi, Michelini, Ambrogio – these were amongst Galileo's closest disciples in the last years of his life. Much of what they said in their conversations corresponds perfectly with what we know of Galileo's own thinking – atomism, the sun as a star, the pope's role in Galileo's condemnation. Of course it does not follow that Galileo believed the universe to have no creator and perhaps no end, or that the *anima mundi* pervades everything, or that there is no such thing as evil. But it seems odd to imagine that all their other beliefs came from Galileo and these did not.

What I am suggesting is that in public, like Mme de Warens, Galileo was a good Catholic, while in private, in his conversations with his disciples, he questioned each and every one of the fundamental truths of the Catholic religion, just as Mme de Warens did in her conversations with Rousseau. What do I have to support this view? The testimony of Caccini and Lorini in 1616; of Sozzi in 1641; of de Monconys in 1646; a few of Galileo's marginalia; Galileo's strange silence on all questions of salvation and redemption. Something – enough, perhaps, for a conjectural history.

But there is more. What is strangest in Viviani's conversation with de Monconys is his references to the world soul, the *anima mundi*. Surely Galileo had no time for such an idea, which we normally associate with the semi-magical beliefs of neo-Platonism? But he did. There is one letter of Galileo's the content of which is unlike anything else he ever wrote. It was sent to Piero Dini on 23 March 1615, when Dini was serving as Galileo's Roman contact in the growing crisis over Copernicanism. In this peculiar letter, which needs to be read in its entirety, Galileo turns from a defence of Copernicanism and of his letter to Castelli to a reinterpretation of the biblical account of the creation.[20] His claim is that Copernicanism is more easily reconciled with the Bible than is the 'popular philosophy', and he is particularly interested in the Bible's account of the spirit that moved on the face of the waters (Genesis 1.2), and in the fact that light was created on the first day and the sun only on the fourth day:

> So I should say that it seems to me that there is in nature a very spirited, tenuous, and fast substance which spreads throughout the universe, pene-trates everything without difficulty, and warms up, gives life, and renders fertile all living creatures. It also seems to me that the senses themselves show us the body of the sun to be by far the principal receptacle for this spirit . . . One may reasonably believe that this is something very different from light since it penetrates and spreads through all corporeal substances, even if very dense . . . It seems to me that from Holy Writ we can acquire evident certainty that the solar body is, as I have said, a receptacle and, so to speak, a reservoir of this spirit and light, which it receives from elsewhere.

In other words, Galileo believes in the *anima mundi*, and wants to argue that this belief is compatible with the Bible; the main source for his discussion, apart from Plato's *Timaeus*, is Ficino's *On the Sun* (1494).[21]

Running through this letter there is a strange piety, for which there is also no parallel in Galileo's other writings: 'one must not doubt that Divine Love may sometimes deign to inspire humble minds with a ray of His infinite wisdom, especially when they are full of sincere and holy zeal'; the Bible verse 'the testi-mony of the Lord is faithful, giving wisdom to little ones' 'has given me hope that God's infinite love may direct toward my pure mind a very small ray of his grace, through which I may grasp some of the hidden meanings of his words'. It is this peculiar, Platonic and far from orthodox piety that we find reflected in Viviani's conversation with de Monconys and in his proposal for a geometrical faith.

Although I do not think that Galileo's epistemology was Platonic, many puzzles are resolved if we recognise that his cosmogony derived from Plato's *Timaeus*. This gave him an immaterial Divine Love, capable of inspiring human minds; both an immaterial creator, the Platonic demiurge, and lesser material gods, the creators of mankind; and an *anima mundi* ordering and vivifying the whole.[22] Understood in this way, Galileo's intellectual enterprise was the oppo-site of Robert Boyle's. Where Boyle argued in his *Free Enquiry* (1686) that the conventional idea of nature was superfluous, and that all one needed was the idea of an omnipotent deity who had created a mechanical universe, Galileo's central but unspoken claim was that if one had a proper idea of nature then one could dispense with the Christian idea of an omnipotent, providential God who had created the universe and would judge the souls of men, and replace it, on the one hand, with a Platonic idea of God as the Supreme Mathematician, indifferent to the affairs of men, and on the other, with nature as the *anima mundi*. The letter to Dini is the only occasion in his correspondence in which Galileo gives expression to his esoteric religious teaching, and of course it comes with an urgent request: 'I beg you not to let it come into the hands of any person who would use the hard and sharp tooth of a beast ... and so would completely mangle and tear it to pieces.' For the moment – and this moment was to last for the rest of Galileo's lifetime – these views were to be kept within the narrow circle of Galileo's disciples.

Thus we can pile conjecture upon conjecture. But it is time to come to a decisive document – the only decisive document we have. On 7 June 1639 Benedetto Castelli, Galileo's old friend, former pupil and long-time intellectual companion, wrote to him from Rome.[23] They had known each other for at least thirty-five years. They were so close that in 1620 Cavalieri had assumed that anything written to one of them would be known by the other.[24] Each had reason to trust the other completely. And in questions concerning the religion of Galileo we can trust Castelli, just as we can trust Rousseau in questions concerning the religion of Mme de Warens.

Castelli has heard news of Galileo that has made him weep with joy, for he has heard that Galileo has given his soul to Christ. Castelli immediately refers to the parable of the labourers in the vineyard (Matthew 20.1–16): even those who were hired in the last hour of the day received payment for the whole day's work. Then, having discussed the prophecies of a Sister Elizabeth (he evidently thinks that now Galileo is a believer he will have an appetite for such things), he turns to the crucifixion, and in particular to the two thieves crucified on either side of Christ (Luke 23.39–43). One confessed Christ as his saviour and was saved; the other did not and was damned. Soon he hopes to come to Florence, and they will be able to talk about these things, which are the only ones that count for the salvation of our souls.

Castelli's invocation of the parable of the labourers in the vineyard and of the two thieves crucified alongside Christ is clear and unambiguous. He believes Galileo is coming to Christianity at the last moment, but not too late to save his soul. There is no conceivable interpretation of this letter which is compatible with the generally held view that Galileo was, throughout his career, a believing Catholic. It will not do, for example, to suggest that Galileo had previously been a believer, but had been lax in the practice of his religion. The labourers did not merely put off starting work until the last moment: they were not *hired* until the eleventh hour. The thief crucified alongside Christ had no knowledge that Christ was, or even claimed to be, the Messiah until death was upon him. These texts, as generations of theologians have recognised, are about last-minute conversions, not (like the parable of the prodigal son, for example) about the amendment of a misspent life.

Castelli allows himself to discuss Galileo's unbelief only because he has been given to understand that he is now, at long last, a believer. There are no further letters like this one. There are certainly a few phrases in Galileo's letters at this time – asking for prayers to be said for him, for example – which may hint at a temporary piety.[25] But it is perfectly possible that Castelli and his informants had been taken in by a ruse – that Galileo hoped to improve his conditions by making a display of piety. Castelli's letter cannot tell us what really happened to Galileo in May 1639; but what is clear is what Castelli had always understood about his close friend: that he was no believer.[26] And if anyone was in a position to know if Galileo was or was not a believer it was Castelli.

This letter of 1639 casts a light backwards on the first letter we have from Castelli to Galileo. Written in 1607, this letter is well known because it is the first evidence of Galileo's having formulated an important doctrine – that movement is to be defined in relative rather than absolute terms, as the change of location of one object in relation to another; it also refers to Galileo's conviction that an object once set moving will continue moving indefinitely until some force intervenes to bring it to a halt.[27] But what interests Castelli is how these two theories bear on arguments for and against the existence of God. On the one hand, he maintains that there is a logical flaw in Aristotelian attempts to prove that the universe has always existed; there is nothing illogical in the Christian idea of a created universe. But he also accepts that Galileo's view of movement is compatible with the idea of an eternal universe, and that if the universe is eternal one can dismiss all arguments from the first cause or the first mover, get rid of God and become an atheist ('bestemmia scelerata' – wicked blasphemy – he hastens to add).

Why on earth would Castelli, who was always a pious Christian, even if he was also for much of his life a Copernican and an atomist, be interested in exploring with Galileo arguments that might lead one to abandon belief in God? The question is profoundly puzzling if this is seen as a correspondence between two pious scientists. But now that it is clear that Castelli knew Galileo to be an unbeliever the letter takes on a new significance. Castelli is exploring the relationship between Galileo's ideas on movement and the commonly acknowledged arguments for the existence of God because he believes that Galileo will already have addressed this question. What he is telling Galileo is that it is safe to discuss such questions with him, that he is prepared to accept Galileo's definition of movement even though it eliminates any need to believe in a first mover, and thereby undermines a standard argument for the existence of a providential God. Castelli is trying, delicately, to establish a basis for mutual understanding and respect between a believing scientist and an unbelieving scientist. That mutual understanding kept his relationship with Galileo intact through thirty-five years of intense stresses and strains, and even through the disappointment of meeting with Galileo for the last time in 1641 (Urban insisted that Castelli be absolutely forbidden from entering into any discussion of Copernicanism, on pain of damnation) and discovering that, if he had indeed undergone a religious crisis, the effects had been only temporary.

Let me make clear that I am not suggesting that Galileo was, properly speaking, an atheist or even an agnostic. All the Greek philosophers known to him, even the Epicureans, had had some sort of belief in the gods.[28] But Galileo went further in the direction of religious faith than Epicurus or Lucretius: the book of nature must, after all, have an author. To see the world as mathematical is (in Galileo's terms) to see it as rational, and to see it as rational is to see it as the product of an intelligent mind. But Galileo was certainly convinced that most arguments intended to show that the universe exhibits a providential order are fallacious. In the margin of one of his books he mocked the idea that God had created dry land for the benefit of animals, seeing that it was a *post hoc ergo propter hoc* fallacy: if there were no dry land, then there would be no animals.[29] Since we cannot tell when providential ordering is absent, we cannot tell when it is present. Here he parted company decisively with the Platonic tradition, which insisted that the universe had been designed to fit humanity's capacity for understanding.[30] Above all, there is no evidence prior to 1639 that Galileo believed that Christ died to save our souls from damnation. To employ a terminology that is inherently anachronistic, Galileo was neither an atheist nor a theist. If we must find a label for him, we had better call him a deist.

It may be thought that the picture of Galileo I have drawn – of Galileo the (un)believer – is unconvincing because it requires us to believe in two Galileos, the public Catholic and the private sceptic. Such internal divisions, however, were entirely commonplace in Galileo's world, as they had been throughout the Renaissance. According to Vasari, the artist Perugino, who made his living producing religious paintings, could never convince himself of the immortality of the soul.[31] Galileo's friend Paolo Sarpi was Venice's state theologian in public, a Catholic when dealing with Catholics, but at the same time a Protestant when dealing with Protestants, while in his notebooks he made clear that he had no time for revealed religion. 'I am compelled to wear a mask', he wrote, 'perhaps there is nobody who can survive in Italy without one.'[32] Galileo's respectable opponent Antonio Rocco, a narrow-minded Aristotelian philosopher, accepted Aristotle's views on the mortality of the soul. Rocco, a priest, was also the author of a remarkable pornographic text, still read today, *Alcibiades as a Schoolboy*.[33] Equally instructive is the case of the Jesuit theologian Melchior Inchofer, who as we have seen wrote the major theological opinion condemning Galileo at his trial and went on to publish a sustained critique of Copernicanism. At Galileo's trial and afterwards, it was Inchofer who defined what orthodoxy was. Yet he was also the author of a bitter pseudonymous satire on his fellow Jesuits, *The Monarchy of the Solipsists*, which was published in Venice in 1645 and repeatedly reprinted and translated in the seventeenth and eighteenth centuries.[34] In the world of Perugino, Sarpi, Rocco and Inchofer, there was nothing unusual about concealing unbelief beneath a veneer of orthodoxy and dissent beneath a facade of conformism. There were also, of course, philosophers like Cremonini and Girolamo Borro (who was teaching in Pisa when Galileo was a student) who barely bothered to conceal their hostility to orthodox religion, but they were the exceptions, and their rarity should not be taken as indicating the true extent of dissent and (un)belief.[35]

Does it matter whether Galileo was or was not a good Catholic? In one sense it matters not at all: he presented himself to the world as a good Catholic and it was as a good Catholic (except as regards Copernicanism) that he was condemned by the Inquisition. If Galileo did have an esoteric teaching on the subject of religion, however, then that is of some importance for understanding the place of the new science in seventeenth-century Tuscan culture.

Yet the real importance of reassessing Galileo's (un)belief lies elsewhere. In reply to nineteenth-century accounts of the inevitable clash between science and faith, much recent historical writing has been committed to exploring the Christian roots of the scientific revolution.[36] Boyle and Newton have come to occupy a prominent place in the new histories, their Christianity, it has been argued, being inseparable from their science. Rethinking Galileo's (un)belief is an important step towards re-examining current orthodoxies regarding the intellectual and cultural origins of the scientific revolution.

The cosmography of the self

Galileo was the first modern scientist, arguably the first true scientist. He constructed a new generation of instruments to be used for scientific purposes: he invented the thermometer, the micrometer and the first accurate timepiece, the pendulum clock, and he transformed the telescope and the microscope from toys into tools for the discovery of invisible worlds.[1] Galileo's ambition was to be a new Archimedes, but almost by accident he came across a new way of learning about nature, the experimental method, and as a result invented what we now call science. Before he could invent science he had to devise and defend an idea that seems to us so elementary that we take it entirely for granted: the idea of the fact. These inventions brought him into conflict with the most powerful intellectual institution of his age, the Catholic Church.

When Galileo began writing, the word for 'world', in Latin and the various modern languages, meant 'the universe', yet by the time he died every educated person was aware of the possibility that this world was merely a planet, one of many; some even understood that the sun might simply be a star. 'World' and 'universe', once the same thing, had begun to be understood as two quite different things.[2] 'Cosmography', a word which had meant 'geography' or the study of the world, came to mean the mapping of the new, Galilean universe, a study quite separate from geography. The competition between the new Galilean universe and the old Aristotelian one gave birth to a new word: cosmology. 'Cosmos' and 'cosmology' are words that appear in English in the 1650s, in response to the publication in 1635 of the Latin translation of Galileo's *Dialogue* under the novel title *Systema cosmicum*. The title itself was an afterthought, for when the frontispiece was commissioned the intention was to call the book the 'Systema mundi', or System of the World, in line with Galileo's own title 'Sistemi del mondo'.[3]

Behind this revolution in astronomy (and in our understanding of our place in the cosmos) lay a revolution in physics: the construction of an idea of infinite space, with neither up nor down, left nor right, centre nor periphery. In Galileo's new physics movement became a relative condition: the earth moves, but not for us. Size too became relative: Galileo was impatient with the argument that the new universe was too big and contained too much empty space. The idea of the universe being 'big' was meaningless in the absence of any standard of comparison.[4] But, if pressed, he thought it most probable that the universe was not only big but infinite.[5]

Indeed much of Galileo's work as a scientist is concerned with issues of scale and infinity. His law of the acceleration of falling bodies is true for any unit of measurement. One of his 'two new sciences' is concerned with the way changes in scale alter the strength of physical objects. In the same work, he argues that there are infinitesimal vacuums within all physical objects, from which they derive much of their strength. If Galileo was the first philosopher to exploit the telescope's capacity to provide new information, it was perhaps because he had already thought about the distinction between cases where a change of scale makes no difference (falling bodies) and those where it makes all the difference (the span of a bridge). His friend Fulgenzio Micanzio thought the central theme of the *Dialogue* was 'immensity and infinity', and the great French philosopher Gassendi said that when he read Galileo's work he felt his mind was 'unchained and free, wandering through the immensity of space, once the barriers represented by the popular theory of the cosmos have been broken down'. Galileo himself said of the *Two New Sciences* that it would allow the best minds to 'diffondersi in immenso' – to lose themselves in the unbounded.[6]

Galileo was not only able to tolerate the idea of the infinitely large and the infinitesimal; he also held that falling bodies pass through an infinite number of degrees of speed in a finite time, thus reinventing Zeno's paradox (in which no arrow can ever hit a target, and no hare can ever overtake a tortoise), and turning every falling apple into a logical conundrum. Similarly he could conceive of a circle as simply a polygon with an infinite number of sides.

These revolutions required a fertile and undaunted imagination (Galileo emphasised that his new sciences were concerned with an imaginary world, one that could be compared to a dream or a fantasy), and understanding Galileo means understanding his inner world, and not just his account of the world around him.[7] This is far from easy, for this inner world was always heavily fortified: it is no accident that Galileo wrote brilliantly on fortification, or that one of the best books about him, Dava Sobel's *Galileo's Daughter*, looks at him not through his own eyes but through those of his daughter. If Galileo thought about what one of his readers, Sir Thomas Browne, called 'the cosmography of myself', he did not entrust his thoughts to paper. Just as he had to construct his account of the universe out of a few scraps of information, just as he could imagine far more than he could prove, so the exploration of his inner life, of his patterns of behaviour and habitual responses, requires making the best, and sometimes the boldest, use of what information there is.

What were the fundamental principles of Galileo's inner life? They were principles of internal conflict, and in this he was no different from anyone else. But Galileo was a far less simple or straightforward character than most people, and often the key to understanding him is to recognise that he was determined to stand for contradictory principles. One fundamental conflict was over the nature of knowledge. The primary thrust of Galileo's new science was to argue that knowledge must be grounded in sensory experience – that seeing is believing. But almost equally powerful was the conviction that knowledge is about abstractions, and that it can go beyond and even against the evidence of our senses. Thus Galileo praised Copernicus for holding firm to his conviction that the planets (including the earth) go round the sun, even when the evidence of his senses directly contradicted his theories.[8]

These two views of knowledge were rooted in two opposing worlds: on the one hand the world of his father, and of his early training as a medical student, both of which fostered Galileo's empiricism; and on the other the world of Euclid and of Archimedes, a world he discovered in a moment of adolescent rebellion and which fostered his love of abstraction and deductive reasoning. One might reasonably think that Galileo ought to have chosen between these two views of knowledge, and that his failure to do so meant that he was caught up in inconsistencies and contradictions. But his failure to choose, his attempt to have it both ways, gave him an extraordinary intellectual flexibility, and was one of the main reasons he was so successful as a scientist.

Galileo's internal conflict over the nature of knowledge is best understood as an ever deferred settling of accounts with his father. He took after his father and he took against his father in equal measure. His mother's influence on him was quite different. His mother was alarming and frightening; she knew no boundaries, and never recognised that her son had an independent existence. Galileo sought to escape from a world in which his mother loomed too large by discovering the insignificance of humankind: far from being at the centre of a universe built especially for them, human beings were insignificant creatures within the vast expanses of an inhuman cosmos. His friend Fulgenzio Micanzio told Galileo that his discussion of atoms in *The Assayer* had forced him to acknowledge the immensity within every tiny thing, as well as our own insignificance, in spite of our belief in our own importance.[9] Another contemporary complained that if Galileo was right the world is a football that has been kicked high in the air and we are ants crawling on its surface.[10] In other words, Galileo, like Jonathan Swift, thought of those around him as Lilliputians.

If Galileo's enlargement of the universe diminished other people, he simultaneously experienced it as an enlargement of himself. This is apparent from a remarkable letter to Élie Diodati, written at the very beginning of 1638, in which he described what it was like to go blind:

Alas, a month ago your dear old friend and servant Galileo became completely blind. Now imagine in what state of affliction I find myself, while I think that this sky, this world and this universe, which I with my marvellous observations and clear demonstrations enlarged a hundred, nay a thousand, times beyond what had been commonly seen by wise men of all the past centuries, now for me is so diminished and restricted that it is no greater than that occupied by my own person.[11]

If Galileo's blindness meant that the universe now extended no further than his own skin, with the telescope he must have felt that he had been able to expand himself to fill the universe.

Thus it was natural for Galileo to see himself as a Gulliver, a giant amongst men, an eagle amongst birds, almost a god. In the *Dialogue* he claimed that when we understand geometry our thoughts are identical to God's thoughts – a claim that aroused suspicions of heresy, although it was hard to define exactly where the heresy lay.[12] Caught up in an unending struggle with his tyrannical mother, Galileo wanted simultaneously to belittle human beings in general and to magnify himself. Hence his preoccupation with questions of scale, and hence his ability to contemplate the infinitely large and the infinitely small without experiencing a Pascalian vertigo.

Galileo may not have suffered from vertigo, but he did suffer from anxiety. His letters are almost devoid of vivid descriptive passages, but there is one striking exception.[13] One foggy night in May 1608, in pouring rain, Galileo set out by gondola at around eleven o'clock to deliver an extremely valuable package to the Florentine ambassador in Venice. He had been paid a great deal of money for the package, and was very anxious in case anything should go wrong – he was already late, and worried that he had delayed too long. But the canal on which the ambassador lived was long, and there was no way of identifying his house. In the near darkness he banged on door after door, to be met either with echoing silence or with torrents of abuse from wakened householders. Galileo's description of being lost in the dark on an interminable canal is an account of a nightmare come alive. Every time he banged on a door he thought he was about to justify the prize to which he had already laid claim; every time success lay outside his grasp. Galileo's whole life, like the letter in which he describes what had happened, was a re-enactment of this nightmare.

Indeed he writes the letter describing his experience, a month after the dreadful event, because he finds himself re-experiencing exactly the same anxiety.[14] He has received two messages from Florence. In one he has been told that the grand duke requires his presence that summer to continue teaching mathematics to prince Cosimo; in the other that, if he happens to be in Florence, he is welcome to visit the court and continue the prince's instruction. One message means that he is a person of importance – his presence is required because he is indispensable. The other means that he is a person of no significance – if he fails to appear no one will notice or care. Awkwardly, he writes to ask which is the right message: is he important or is he insignificant? Is the door to the court standing open to him, or is it barred and the building shuttered? Will his knocking meet only with silence or, worse still, expletives?

Over and over again Galileo grasped at the bird in the bush and let slip the bird in the hand. We are bound to think as we read his anxious enquiry as to whether he has been invited or commanded to attend on the grand duke (an enquiry that seems almost designed to invite a rebuff) that success frightened him even more than failure. Nothing provoked his mother (or rather his internal mother) like success; failure generated anxiety, guilt and despair, but the consequences of success were (in his imagination) even worse. Galileo's friends could not help but notice his self-destructive behaviour, his determination to ensure that success lay always just outside his grasp.

These two conflicts – the paternal conflict between experience and reason and the maternal conflict between power and impotence – shaped Galileo's internal life, and constitute the cosmography of his self. Without reference to these conflicts we cannot hope to understand the path he followed, the path that brought him into conflict with the pope and the Inquisition.

For Galileo's life, like all our lives, was built around a series of choices. Some of those were intellectual choices. He chose not to become a careful experimental scientist, although it was easy for his admirer Mersenne, following in his footsteps, to turn physics into an experimental science.[15] He chose not to study (except in theory) vacuums, although it was easy for his disciple Torricelli, following in his footsteps, to provide experimental proof of their existence (and so to inflict a mortal blow on Aristotelian physics); soon Boyle would make the vacuum pump the emblem of a new experimental science. Galileo was reluctant to engage in any purely empirical enquiry. He chose to do very little with the microscope: where the telescope could be used to demonstrate that the universe was vaster than previously suspected, the microscope does not alter the relative size of human beings when compared to the world around them – to magnify a flea is also to magnify the human hair it clings to.[16] These were intellectual paths not chosen. And he chose to abandon Venice, to fall out with the Jesuits and to earn the displeasure of Urban VIII, on each occasion because he felt belittled by the treatment he had received. For the last forty years historians of science have aspired to write a social history of intellectual change, but no social history can explain Galileo's crucial choices. We need an inward history.

Such an inward history has above all to grasp Galileo's willingness to be different. In trying to think about this we face a double obstacle. First, we know that on many of the topics where he disagreed with his contemporaries Galileo was right: we underestimate the psychological stress of holding to a position where everyone (or almost everyone) disagrees with you. Because the position in question (Copernicanism, for example) happens to be one we hold, and hold without any sense of anxiety, we underestimate the difficulty of holding it in the face of universal opposition. Second, we live in a world of rapid change, and in our world radical differences of intellectual opinion are commonplace. Disagreement is often rewarded with success. This was not Galileo's world. Hobbes said that William Harvey was the first intellectual to propose a major new theory (the circulation of the blood) and see it triumph in his lifetime. Galileo did not have this experience: Copernicanism became generally established only in the 1660s. We underestimate the difficulty of innovation in a world as intellectually conservative as Galileo's.

I do not mean to suggest that Galileo was isolated. He had close friends who admired him: Sagredo, Salviati, Castelli. He had six portraits of friends on the wall of his sitting room when he died, and he had been prepared to pay a good deal for them.[17] But after Salviati died in 1614 Galileo never again had the prospect of living in daily contact with a close friend who was at all near to him in age. In the last decade of his life he was surrounded by a small group of young admirers who thought he was the most important man in the world. We catch a glimpse of them gathering over drinks in the evening to read his long-awaited *Dialogue* aloud, too impatient to wait for it to appear in print.[18]

But what struck most people was his extraordinary willingness to stand alone. This was not something they often mentioned when writing to him, any more than they would have told him that he was bad-tempered or ugly. But we catch sight of it when a clergyman lambasts him to his face for his arrogance in daring to believe in a vacuum,[19] and there is an indirect admission of it when Galileo confesses his guilt to the Inquisition, saying that he has been led astray by pride: this is evidently how he thinks others explain his aberrant behaviour. We best see how he is regarded in fleeting remarks: when a supporter writes to him regretting that Galileo has given up writing because he has been crushed under the enormous weight of opposition; or when an opponent writes a placatory letter saying that he and Galileo differ in that Galileo is prepared to stand alone while he prefers whenever possible to agree with other people (which entails, of course, disagreeing with Galileo).[20]

Outwardly, Galileo's path is easy to trace. For the first half of his life he was an obscure mathematician who published little, and nothing of importance. Yet during this time he made most of his great discoveries in physics, discoveries that he was to publish only at the end of his life. The second, and shortest, period in Galileo's life began in 1609, when he turned a telescope towards the sky. He discovered new worlds and transformed astronomy; it was this revolution which gave rise to our modern conviction that what scientists do is make discoveries. Relying on his new discoveries, Galileo set out to convince the world that the earth was flying through space: that the earth is in fact merely one of the planets and the sun merely one of the stars. (In private he had already been a Copernican for many years.) These arguments were condemned in 1616, bringing the second period of his life to an end.

Yet throughout the rest of his life Galileo struggled to make the case for Copernicanism, drawing on the discoveries he had made before the invention of the telescope. His *Dialogue Concerning the Two Chief World Systems*, the first great work of modern science, was published under licence in 1632 but condemned by the Inquisition in 1633: Galileo spent the rest of his life under house arrest, and there was a general ban on the publication of any of his works in any Catholic country. In 1636 he secretly arranged for the publication in Lutheran Strasbourg of the *Letter to the Grand Duchess Christina*, an argument for intellectual freedom, and in 1638 he arranged for the publication in Calvinist Holland of his *Two New Sciences*, the two sciences being the science of projectiles and the science of the strength of materials.

Galileo puzzled over the hostility his work engendered, and suggested a psychological explanation. The difficulty, he thought, was not simply his Copernicanism, which implied that we cannot trust our own senses, in that the world around us seems to be standing still but is in fact travelling at immense speed through space. It lay, he suggested, in his sustained efforts to deny the validity of the traditional distinction between the sublunary realm, the world of change, of generation and of death, and the superlunary realm, the world of perfection, of permanence and so of immortality. As far as Galileo was concerned the earth and the moon are heavenly bodies like other heavenly bodies, and other heavenly bodies are like the earth and the moon. He found imperfection in the moon and change in the sun, and he declared that the earth, seen from afar, would shine like a star.

It did not follow, he stressed, that because heavenly bodies were imperfect the universe was therefore fragile or doomed to destruction. Change did not necessarily mean death, but his readers, he suspected, feared that it did.[21] If the Fall had begun the reign of death on earth, then Galileo had extended death's reign throughout the heavens. His version of Copernicanism represented a second, intellectual Fall, a second expulsion from Eden. One complaint was that by making the earth a heavenly body he had placed hell in the heavens; another was that in claiming there were other, uninhabited worlds he had made nonsense of the notion that this world existed because it was made for living creatures.[22] Galileo was trampling down the barriers that kept apart God and the devil, the sacred and the secular, the mortal and the immortal.

In the early modern period there was a fundamental intellectual commitment that was deeper than the commitment to Aristotle or to a literalist reading of the Bible, a commitment that underpinned both these explicit, openly avowed commitments. That commitment was to dualist modes of thought – Carnival and Lent, time and eternity, superlunary and sublunary, heaven and hell, the Mass and the sabbat, God and the devil. By denying the distinction between sublunary and superlunary, Galileo made it impossible to locate heaven and hell, angels and devils, good and evil, on a map of the cosmos. Indeed he offered a view of the cosmos in which humankind, and the things that matter to humankind – love and hatred, virtue and vice, mortality and immortality, salvation and damnation – were irrelevant. Far from embodying a scheme of values, far from embodying a *telos* or purpose, Galileo's universe appeared to be indifferent to moral and metaphysical issues, and even indifferent to our own existence.[23] It is not hard to sympathise with those who recoiled from this new vision. Galileo's greatest and at the same time most disturbing achievement was to recognise that the universe was not made for the sake of human beings, and that it teaches us nothing about right or wrong, and offers us neither salvation nor damnation.

John Milton, who had secretly met with Galileo when, old, famous and blind, he was confined to house arrest, understood better than anyone that this was the fundamental question raised in his work. Here again, as with Donne, we catch the faint echo of a conversation that is forever lost.[24] *Paradise Lost* describes the creation of a Ptolemaic universe, with the earth at the centre, but hedges its bets as to whether Galileo may not be right, conceding that he is indeed right about the scale of the universe, and the consequent loss of humankind's sense that it holds a privileged place within it. There Adam says (and Milton has Galileo's *Dialogue* in mind),

> When I behold this goodly Frame, this World,
> Of Heav'n and Earth consisting, and compute
> Their magnitudes, this Earth a spot, a grain,
> An Atom, with the Firmament compar'd
> And all the numbered Stars, that seem to roll
> Spaces incomprehensible (for such
> Their distance argues and their swift return
> Diurnal) merely to officiate light
> Round this opaceous Earth, this punctual spot
> One day and night; in all their vast survey
> Useless besides, reasoning I oft admire
> How Nature wise and frugal could commit
> Such disproportions . . .[25]

To accept Galileo's new science it was necessary to abandon the view that nature (at least as seen from our perspective) is 'wise and frugal'.

Coda: Galileo, history and the historians

Galileo died quietly on 8 January 1642. He was seventy-seven. As soon as the news reached Rome instructions were sent that there was to be no tombstone, or any memorial within a church. (Viviani ended up turning the front of his house into a private memorial to Galileo, and left money in his will for a fine tomb.) A modest plaque was permitted in 1673, and a proper tomb finally allowed in 1737. (When Galileo's tomb was opened to move his body to its new location, the body of an unknown woman was found buried – perhaps reburied in 1674 – with him. Readers of this book will not be entirely surprised to discover that there is one more woman in his life about whom nothing is known. It was at this point, as his body was being moved, that three fingers, a vertebra and a tooth were removed to be fashioned into secular relics.)[1]

Galileo's work marked the beginning of a scientific revolution, before which even the Catholic Church had eventually to give ground: the first legal reprinting of Galileo's *Dialogue* in the original Italian was in 1744, albeit accompanied by the condemnation of 1633 and Galileo's recantation. The Catholic Church did not formally permit the teaching of Copernicanism until 1820, although over the previous century restrictions on the discussion of his work had slowly broken down.[2] And it was only in 1834, the year before Copernicus and Galileo were finally removed from the Index of forbidden books, that the English word 'scientist' first appeared in print.

Because the word 'scientist' appears so late, and the word 'science' acquires its modern sense in English (referring to the natural sciences) only in the eighteenth century, historians of science generally regard reference to seventeenth-century 'science' and 'scientists' as anachronistic. But in this respect Italian was far in advance of English. Galileo published a book called *Due nuove scienze* (1638), two new sciences, following the example of Tartaglia's *La nova scientia* (1537).[3] An opponent writes to him of 'dottori e scienziati', and supporters write of the 'universal delli scienziati' and the 'republica scienziata'. In the Romance languages 'science' referred then, as it does now, to knowledge in general, and the 'dottori e scienziati' referred to by Galileo's opponents may simply be learned men in general, but Galileo's supporters were specifically discussing experts in the new natural philosophy.[4] We should therefore feel free to follow this contemporary Italian usage and describe Galileo as a scientist.

Since 1835 one question above all others has dominated the literature on Galileo: why was he wrongly condemned? The key sources have all been available since 1877, and the question has received three basic answers. The first reiterates the view expressed by the Catholic Church itself in 1616 and in 1633: modern science, Galilean science, is fundamentally at odds with Christianity. This view was particularly prevalent in the nineteenth century, embodied in works such as John Draper's *History of the Conflict between Religion and Science* (1874) and Andrew Dickson White's *History of the Warfare of Science with Theology in Christendom* (1896). But Draper and White were fighting against an imaginary enemy. Since there were no modern defenders of Ptolemaic astronomy or Aristotelian physics, no Christian still held that Galileo's science was incompatible with Christianity. The real battle was, and still is, over the relationship between Christianity and Darwinian biology.

The view that the clash between Galileo and the Church was inevitable has thus been replaced in the course of the last century by two more subtle attempts to explain what went wrong. One approach stresses that Galileo was never intended to be prevented from using Copernican theories as a mathematical hypothesis; he was simply forbidden to say that Copernicanism was true. Given the state of knowledge at the time, this, it is argued, was a perfectly reasonable limitation. The classic exponent of this view is the great historian of science (and practising Catholic) Pierre Duhem: his *To Save the Appearances* was published in 1906. A similar approach was taken by Arthur Koestler, in *The Sleepwalkers* (1959). Koestler argued that the truly important discoveries in astronomy were made by Johannes Kepler; Galileo overstated his own achievements in the *Dialogue* and thus provoked the Church into condemning him. This view, which presents Galileo as an overreacher, seems to me essentially correct.

The alternative view is that the Church made a fundamental error of judgement. In his *Letter to the Grand Duchess Christina* (written in 1615) Galileo clearly demarcated different spheres for religion and science. There was, he insisted, no need for religious authorities to interfere with the enquiries of scientists. Many Catholic historians and philosophers would now argue that Galileo was right, and that the Church has finally come to adopt his position. This is the view expressed, for example, in Annibale Fantoli's study, and in a recent collection of essays edited by Ernan McMullin, *The Church and Galileo* (2005).

There are two problems with the liberal Catholic view of Galileo. The first is that the issues it addresses are much more relevant to the condemnation of Copernicanism in 1616 than to the trial of Galileo in 1633. It may be argued that, given the attitude the Catholic Church had adopted towards tradition and authority in its struggle against Protestant innovation, Galileo was asking the impossible when, in the letter to Christina, he claimed that scientists must always have the right to disagree with established authority, and indeed there could be no scope for compromise between Bellarmine's understanding of the Church's position and Galileo's claims on behalf of the new science. Impossible or not, Galileo was right to ask for intellectual freedom, and Bellarmine wrong to deny it. But if the liberal Catholic position is helpful for understanding the events of 1616 it is much less helpful for understanding those of 1633. Urban VIII, unlike Bellarmine, was certainly willing to envisage the triumph of an anti-Aristotelian science – one of his first actions on being elected pope had been to bring Benedetto Castelli to teach at the pontifical university. His accession meant that Galileo had a genuine opportunity to win a space for the new science within Catholic culture; his failure to make the most of that opportunity is a result of his unwillingness to compromise and his inability to work within limits imposed by others.

The second problem with the liberal Catholic view is that it accepts without question the claims made on behalf of modern science. Yet over the last half-century the nature of science has been a subject of bitter debate, and Galileo has been at the centre of those debates. One of the founders of modern history of science, Alexandre Koyré, made his reputation by presenting Galileo as a theo-retical mathematician and a philosopher rather than the first modern scientist. A series of science studies scholars (Kuhn, Feyerabend, Biagioli) have sought to minimise the importance of the new facts discovered with the telescope. In opposition to these two dominant traditions in the history of science, a powerful body of scholarship (Drake, Settle, Palmieri) has sought to present Galileo as the first great experimental scientist. The liberal Catholic tradition has yet to provide a satisfactory account of just what sort of scientist Galileo was, and yet this must be central to any discussion of the clash between Galileo and the Church.

At the heart of all the debates about Galileo is a fundamental question about the status of his commitment to Copernicanism. On one side we have Duhem, the founder of modern history of science; Koestler, once the most widely read of all historians of the scientific revolution; and Kuhn and Feyerabend, the founders of the science studies school. For them Galileo's Copernicanism was an unproven hypothesis. On the other side are Drake and the best recent Catholic scholarship. For them Galileo's determination to defend Copernicanism was a straightforward expression of his commitment to the scientific method. Both sides make the following assumption: that the question of Galileo's Copernicanism really begins when he becomes a public figure in 1610. The result is that the literature on Galileo, like the surviving evidence, is tilted towards the later decades of his life.

Yet Galileo himself claimed in 1597 that he had been a Copernican for a number of years. To the question, Was his Copernicanism an unproven hypothesis or a scientific truth? the answer has to be that his initial commitment to Copernicanism went far beyond the available evidence. (Here I part company with Drake, who sought to defend Galileo against the charge of being an over-zealous Copernican.) One of my contentions in this book is that we need an explanation for Galileo's premature Copernicanism, and that such as explanation cannot be purely scientific: it must also be psychological.

We can debate endlessly at what point it became 'rational' to adopt Copernicanism: the earliest date would be 1540, when the first summary of Copernicus's views, the *First Report*, was published; the latest would be 1838, when the movement of the earth was first reliably demonstrated by the measurement of stellar parallax – that is, by the demonstration that some stars alter their relative position when seen from opposite sides of the earth's annual orbit of the sun. The Foucault pendulum, which allows one directly to see the earth moving, came later, in 1851, but stellar aberration, which is a visual distortion caused by the earth's movement, had been discovered in 1729, and this was arguably the first conclusive proof that the earth is moving. In addition, an eastward deviation of bodies falling from a high tower, the opposite of the westward deviation predicted by the Aristotelians (the result of the fact that a body at the top of a tower is moving faster as it rotates around the centre of the earth than the foot of the tower is moving), had been measured in 1792.[5]

Galileo was of the view that there were more than enough good arguments in Copernicus's *On the Revolutions* to convince an intelligent reader. Indeed he said that you could not understand Copernicus without agreeing with him. In 1612 he was convinced that the arguments in favour of Copernicanism were over-whelming.[6] He would surely have claimed that the publication of his own *Dialogue* in 1632 marked the crucial moment at which any sensible person should have accepted the truth of Copernicanism. But in making such a claim he would have been profoundly wrong.

Galileo had no decisive astronomical evidence that showed that the Copernican system was superior to the system of Tycho Brahe: the one argument he had, from the line of movement of sunspots across the face of the sun, was and still is highly contentious. He thought he could sidestep this fundamental difficulty. If the evidence provided by astronomy could never be decisive, then he would introduce a quite different type of evidence into the argument. He believed he could use a fundamental phenomenon on earth to prove that the earth was in movement, and consequently that the Ptolemaic and Tychonic systems must be wrong.[7] This strategy was perfectly sound in principle: the Foucault pendulum really does show that the earth is moving. But unfortunately Galileo did not devise a Foucault pendulum, relying rather on a mistaken theory of the tides. (It is interesting to reflect that there was nothing to prevent him from devising the Foucault pendulum. The physical phenomenon isolated for observation by Foucault was noticed by Viviani, who simply failed to grasp its significance. Galileo, had it been brought to his attention, might have been more perspicacious.)[8]

Galileo's theory of the tides was disastrous, both for him and for Catholic Europe, because, armed with it, he saw no need to respect the limits on debate imposed by the Church. In 1624 he was given permission by his friend the newly elected Pope Urban VIII to reopen the debate on Copernicanism, provided he respected certain conditions. By respecting the letter but not the spirit of these conditions, Galileo brought catastrophe on himself, but more importantly he reinforced rather than diminished the developing conflict between the new science and the Catholic Church.

Galileo's mistaken theory of the tides was not an isolated intellectual slip; it was a natural consequence of his understanding of the scientific enterprise. In the debate between those who believe that he was an experimentalist and those who believe that he dealt in mathematical abstractions, it is evident that there is right on both sides. Galileo practised both an experimental science and an abstract, mathematical science. In combining the two, however awkwardly, he is representative of the leading figures of the scientific revolution, such as Simon Stevin, Evangelista Torricelli, Blaise Pascal, Christiaan Huygens and Isaac Newton – and indeed, we may now add René Descartes to this list.[9] But he always thought deductive science was superior to experimental science. The consequence of this is an interesting paradox: Galileo was the founder of modern science, but he was always a reluctant experimentalist, and consequently he never produced an unambiguous defence of the experimental method. Worse, because he believed in the possibility of a deductive science of nature, he was prepared to gamble everything on his theory of the tides, even when it was seriously incomplete.

But the theory of the tides does not merely expose a flaw in Galileo's understanding of science. His insistence that he had found a definitive argument also represents a flaw in his personality. Between 1610 and 1616 Galileo had struggled to persuade the Catholic Church to give its support to the new science. In 1616 that struggle ended in defeat. Then in 1624, entirely unexpectedly, Galileo was offered the official encouragement he had always sought. But because the terms were not of his own choosing, because the support was conditional, he betrayed the pope, who was both his ally and his friend. This was self-destructive behaviour – and if 'self-destructive' seems a hopelessly anachronistic term, it was precisely how Galileo was seen by his friends, who accused him of 'stubbornly and perfidiously assassinating himself'.[10]

All sides within these debates accept Galileo's claim to be a faithful Catholic, and his Christian belief was not disputed at his trial. In 1983 Pietro Redondi published the original Italian edition of *Galileo Heretic*, the first major study to free itself of the intellectual framework inherited from Viviani and Favaro. Redondi maintained that the unspoken charge against Galileo was not Copernicanism but atomism.[11] Atomism, as expressed by Galileo and by later seventeenth-century scientists and philosophers such as Descartes, Hobbes and Locke, was grounded in a rejection of the Aristotelian doctrine of forms or essences. What makes something what it is, is not some immaterial form which it expresses, but simply the way in which it is organised in space. Thus for Descartes an animal is merely a machine – there is no life, or form, or intelligence, or soul which inhabits it. Shape, movement and resistance are the three crucial characteristics of the components of a machine, and these are the fundamental features or primary qualities of matter; different substances are made up of tiny invisible building blocks, atoms, which slot together to form different shapes with different capacities for movement – solids and fluids, for example – and different degrees of resistance – hard and soft, rigid and flexible. Other characteristics, which Aristotelians had taken to be equally 'real' (colour, taste and sound), are purely subjective: colours, tastes and sounds are not in the objects themselves, but (like the sensation of being tickled) in human sensory perceptions. Sounds, for example, are merely vibrations, movements transmitted from one atom to another. There would be no such thing as sound in the absence of ears to hear. No seventeenth-century philosopher asked the question, If a tree falls in the forest and no one is there to hear it, does it make a sound? but the atomists would have been certain that the right answer was no.

Atomism was a threat to religion in general in that it did not presume that the world was the rational expression of a design or a purpose – it was anti-teleological, and thus potentially atheistical – although one could certainly formulate a Christian form of atomism. But there was a particular and fundamental tension between atomism and Catholicism. Catholics believe that in the Mass, the liturgical re-enactment of the Last Supper, the bread and wine continue to have the appearance of being bread and wine but in their essence are transformed into the body and blood of Christ. This is the doctrine of transubstantiation, which all Protestants dispute. It does seem almost impossible to defend the doctrine of transubstantiation while talking in terms of primary and secondary qualities, and some contemporaries therefore insisted that Galileo's atomism implied a rejection of the doctrine of transubstantiation.[12] The Redondi thesis (which was met with a storm of criticism) thus represented a fundamentally new way of arguing that there was an inescapable disagreement between Galileo and the Catholic Church.

Redondi's account of the trial was not conclusive. There were, I would argue, two concealed charges against Galileo, not one: the denial of transubstantiation and the denial of God's omnipotence. It was also profoundly puzzling, however, for it centred on the claim that Galileo was properly suspected of being guilty of a major heresy (and only escaped serious punishment because he benefited from the pope's protection) without ever discussing his attitude to Christianity. In later work, Redondi has made clear that he shares the general consensus that Galileo was a believing Christian, if not an orthodox one. This consensus, in my view, is simply mistaken.

Urban VIII regarded the argument of the *Dialogue* as not only disloyal but impious. Here, as elsewhere, his judgement was sound. Galileo always acknowledged the authority of the Church and always claimed to be a pious Catholic.[13] But a distinction needs to be drawn between his official position and his private convictions. In the twenty volumes of his works there is a very striking absence of evidence suggesting any private piety. Reading his letters, there is no sign – or almost no sign – of his saying his prayers, listening to sermons, or reading either the Scriptures or the fathers of the Church.[14] There is no indication that he believed in sin, contrition and redemption. He avoids all mention of Jesus. Galileo was no Christian: we can see well enough behind the public persona to be fairly sure of this, and we have the confirmatory testimony of Castelli. Galileo's Copernicanism, his scientific method and his unbelief were, indeed, mutually supporting: all three represented a rejection of the traditional view that the world was made for man, and that man was made in the image of God. Rather, Galileo argued, we need to recognise that the world is profoundly imperfect, that we can understand only fragments of it, and that humankind appears irrelevant to its purposes, supposing it has any. Galileo sought to live with the idea that we do not know what the universe is for, even though certain aspects of it suggest that it was designed for a purpose.[15]

My account of Galileo in this book has been novel in three significant respects: I have emphasised his early Copernicanism, his reluctant empiricism and his private irreligion. I have also stressed his extraordinary intellectual ambition, his enormous vanity and his capacity for self-destruction: Galileo was no secular saint, although he was capable of pretending that he was. These new insights bring us, I believe, much closer to seeing Galileo as he saw himself, but they also make it clear that Koestler (who was wrong about many things) was right to portray him as the architect of his own downfall.

The story of Galileo has not one ending but two: it ends both with his condemnation by the Church and with the intellectual triumph of Copernicanism. It would be easy to write an account of his life in which the emphasis from beginning to end fell on one or other of these – all the easier now that we know that the Copernicanism and the accusation of heresy go back to a far earlier point in his life than has usually been acknowledged.

Yet one of the obligations of the biographer is to respect the open-endedness of a life as it is lived. From the very beginning, Galileo aimed to excel as a mathematician and a scientist. But success came late. His first two major projects were abandoned unfinished. He carefully kept his Copernicanism to himself. Had he been asked in 1608, he would have said that he was going to make his name through a new account of the movement of falling bodies – a technical project which would have won him few readers. He was forty-five before it became clear that he was destined to be famous, and he never shook off the fear of failure that had grown within him over the years. Indeed that fear simply became more acute as his ambitions grew ever greater. In the end, as so often happens, he brought about the very catastrophe that he had most feared: he was mocked for defending Copernicanism in the public arena.

Hindsight can also easily mislead us when it comes to understanding Galileo's condemnation by the Church. It is easy to portray this as an inevitable clash between two sides committed from the outset to opposing principles. But the clash was far from inevitable, and the outcome far from certain. After the condemnation of Copernicanism in 1616, Galileo could simply have turned his attention elsewhere. He had, after all, the material for the *Two New Sciences* already prepared. Then, after the election of Pope Urban VIII, there was a real possibility that the Church might come to terms with Copernicanism. The clash, when it came, was not between an impersonal institution, the universal Church, on the one hand and a dedicated scientist on the other. Rather it was a falling out between friends, a betrayal, a just punishment. Galileo was indeed a heretic; but worse (for heresy was much more common than historians have realised), he was disloyal and ungrateful. In the world of Counter Reformation Italy heresy often went unpunished; disloyalty and ingratitude, on the other hand, were never tolerated.

For over a hundred years there has been a serious scholarly literature on Galileo. But he has been much misunderstood. The greatest and most influential historians and philosophers who have written about him – Duhem, Koestler, Feyerabend, Biagioli – have misjudged both his strengths and his weaknesses. Biagioli, for example, writes in praise of Galileo's skills as a courtier, to which, he says, he owed all his success.[16] Yet Galileo himself tells us that he had no taste for courtly life, and it should by now be apparent that it was precisely his incapacity as a courtier that was responsible for his trial and condemnation.

Fortunately by far the best-known account of Galileo's life is Brecht's brilliant play *Life of Galileo* (1938, with a second version in 1945–7 and a third in 1956), a play which, for all its fictions, deserves to be taken seriously as a historical interpretation.[17] Brecht's Galileo, like all Brecht's protagonists, is not a hero but an anti-hero. (In presenting him as such Brecht was taking a political stand, for Copernicus and Galileo were represented as heroic figures by Nazi propagandists such as Ludwig Bieberbach.) In the second version of the play Galileo's recantation prepares the way for a dreadful collaboration between science and power, a collaboration that had led, Brecht believed, to Hiroshima. In the first version, however, the recantation – Galileo's failure either to go into exile or to die for his principles – turns out to be for the best, because it makes possible the publication of the *Two New Sciences*. Brecht, who had been in exile from Nazi Germany since 1933, was profoundly uncomfortable with this conclusion, which appeared to condone those who had gone into an 'inner emigration' under Nazism. Revisiting the play in 1944, before he began to rework it, he wrote in his diary, 'Precisely because I was trying to follow history here and had no personal moral interest, a moral emerged, and I am not too happy about it.'[18]

The Enlightenment philosopher La Mettrie, driven into exile in Berlin, wrote, 'Happy the man who cares nothing for success; for whom nothing is more important than freedom of thought; and for whom the true fatherland is the land where one is free to speak one's mind.' (La Mettrie himself was far from happy: according to Voltaire he missed France so badly he cried like a baby.)[19] Galileo cared greatly for success; he cherished Florence as his true home; and he was prepared, if awkwardly and reluctantly, to compromise his freedom of expression. We would admire him more if he had been prepared to face death bravely, like Bruno, who told the judges who condemned him that they were more frightened than he was. We would approve of him more if he had gone into exile, as so many Italian Protestants had been prepared to do. But it is unreasonable to ask him to have been braver or bolder, for as Brecht was obliged to recognise, he was brave enough and bold enough to complete before he died the great project on which he had embarked half a century earlier.

For knowledge was not enough for Galileo. He also insisted on communicating that knowledge. Let this be his epitaph:

If a man could ascend to heaven and get a clear view of the natural order of the universe, and of the beauty of the heavenly bodies, that wonderful spectacle would give him little pleasure, though one can imagine nothing more delightful if he had just one person to whom he could describe what he had seen.

Archytas of Tarentum (428–347 BC), according to Cicero,
On Friendship (44 BC)

Notes

ABBREVIATIONS

A&C Antonio Favaro, *Amici e corrispondenti di Galileo*
C&G Ernan McMullin, ed., *The Church and Galileo*
Comets Stillman Drake and Charles Donald O'Malley, eds, *The Controversy on the Comets of 1618*
Dialogue Galileo Galilei, *Dialogue Concerning the Two Chief World Systems*
DV Sergio Pagano, ed., *I documenti vaticani del processo di Galileo Galilei (1611–1741)*
EDG Stéphane Garcia, *Élie Diodati et Galilée*
GA Maurice A. Finocchiaro, *The Galileo Affair*
GC Mario Biagioli, *Galileo Courtier*
GH Pietro Redondi, *Galileo Heretic*
GIC Mario Biagioli, *Galileo's Instruments of Credit*
LCF José Montesinos and Carlos Solís, eds, *Largo campo di filosofare*
MC Galileo Galilei, *Sidereus nuncius: Le messager céleste*, ed. Isabelle Pantin
MM Galileo Galilei, *'On Motion', and 'On Mechanics'*
OG Galileo Galilei, *Le opere di Galileo Galilei*
SHPS *Studies in History and Philosophy of Science*
SN¹ Galileo Galilei, *Sidereus nuncius* (Venice: Thomas Baglioni, 1610)
SNᵉ Galileo Galilei, *Sidereus nuncius*, ed. and trans. Albert van Helden
TNS Galileo Galilei, *Two New Sciences*

Wherever possible I cite from the Favaro edition of Galileo's *Opere* in the form *OG* [volume no.] [page no.], e.g. *OG* x 57. The first edition (1890–1909) is available online at *www.gallica.fr* (and large parts of it, in searchable form, are at *www.liberliber.it*), but I use the 1968 reprint which reproduces the appendices added in the edition of 1929–39; the first edition also contains supplements to vols xix and xx. If there is a convenient English translation I then give a reference to that (except when I am quoting directly from the English translation, in which case the reference to it comes first); in all other cases translations are my own.

A website accompanies this book: *www.watcheroftheskies.org*. There you will find links to other websites and to a range of texts available on the internet.

INTRODUCTION: CONJECTURAL HISTORY

1. Quoted from Alistair C. Crombie, ed., *Scientific Change* (London: Heinemann, 1963), 847–8. I owe this reference to Pietro Redondi: cf. *GH* ix.
2. *A&C* 1015–19.

3. The full story of Galileo's manuscripts is told in Antonio Favaro, 'Documenti inediti per la storia dei manoscritti galileiani', *Bullettino di bibliografia e di storia delle scienze matematiche e fisiche* 18 (1885), 1–112.

4. On Favaro, see Giuseppe Castagnetti and Michele Camerota, 'Antonio Favaro and the *Edizione Nazionale* of Galileo's Works', *Science in Context* 13 (2000), 357–62. A new supplement to Favaro is under preparation: *Galilæana* 4 (2007), v. For a survey of new documentary evidence regarding Galileo's trial, see Francesco Beretta, 'The Documents of Galileo's Trial: Recent Hypotheses and Historical Criticism', in *C&G* 191–212.

5. Because Galileo was obliged to rely on an amanuensis once he had gone blind it was possible for the recipients of his letters to be in some doubt as to whether the views expressed in them were really his own: *OG* xviii 202–3.

6. Paolo Galluzzi, 'I sepolcri di Galileo', in Roberto Lunardi and Oretta Sabbatini, eds, *Una casa per memoria* (Florence: Edizioni Polistampa, 2009), 203–55, at 208–13; Pietro Redondi, 'La nave di Bruno e la pallottola di Galileo: Uno studio di iconografia della fisica', in Adriano Prosperi, ed., *Piacere del testo* (Rome: Bulzoni, 2001), 285–363, at 343–8.

7. Redondi, 'La nave di Bruno', 345–7.

8. Paolo Galluzzi and Maurizio Torrini, eds, *Le opere dei discepoli di Galileo Galilei: Carteggio* (2 vols, Florence: Giunti-Barbèra, 1975–84), ii, 305–6.

9. See below, p. 230.

10. Favaro, 'Documenti inediti', 51.

11. *EDG* 231–9.

12. Antonio Favaro, 'Sulla pubblicazione della sentenza contro Galileo, e sopra alcuni tentativi del Viviani per far rivocare la condanna dei Dialoghi galileiani', in Favaro, *Miscellanea galileiana inedita* (Venice: Giuseppe Antonelli, 1887), 97–156, at 115–16.

13. *OG* xix 309–10.

14. 'I began with the desire to speak with the dead': Stephen Greenblatt, *Shakespearean Negotiations* (Berkeley: University of California Press, 1988), 1.

15. The element of fire: John Donne, *The Epithalamions, Anniversaries and Epicedes*, ed. W. Milgate (Oxford: Clarendon Press, 1978), 140; the phoenix: *OG* vi 237; *Comets* 189.

16. R. C. Bald, *John Donne: A Life* (Oxford: Clarendon Press, 1970), 148–52. When Donne died he owned portraits of Sarpi and of his close associate Fulgenzio Micanzio.

17. On George Rooke, see Antonio Favaro, *Galileo Galilei a Padova* (Padua: Antenore, 1968), 277–8; on John Wedderburn, see Giorgio Spini, 'Un difensore scozzese del *Sidereus nuncius*', in Spini, *Galileo, Campanella, e il 'Divinus poeta'* (Bologna: Il Mulino, 1996), 79–90; and Antonio Favaro, *Scampoli galileiani*, ed. Lucia Rossetti and Maria Laura Soppelsa (2 vols, Trieste: Lint, 1992), ii, 702; on Richard Willoughby, see *A&C* 1001–5; and on Thomas Seget, see *A&C* 937–74 and Favaro, *Scampoli galileiani*, ii, 665–6, 700–1.

18. Mary Laven, *Virgins of Venice* (London: Penguin Books, 2003), 149.

CHAPTER 1: HIS FATHER'S SON

1. Loosely translated from Lodovico delle Colombe, *Risposte piacevoli* (Florence: Giovanni Antonio Caneo, 1608), f. 111r.

2. On the date of Galileo's birth, see David Wootton, 'Accuracy and Galileo: A Case Study in Quantification and the Scientific Revolution', *Journal of the Historical Society* 10 (2010), 43–55 at 44–7.

3. *OG* xix 22, 26–9.

4. Maria Luisa Righini Bonelli and William Shea, *Galileo's Florentine Residences* (Florence: Istituto e Museo di Storia della Scienza, 1979), 8–9.

5. *OG* vi 160.

6. On Galileo's artisitic abilities, see Horst Bredekamp, 'Gazing Hands and Blind Spots: Galileo as Draftsman', *Science in Context* 13 (2000), 423–62.

7. On the quality of Galileo's lenses, see Vincenzo Greco, Giuseppe Molesini and Franco Quercioli, 'Optical Tests of Galileo's Lenses', *Nature* 358 (1992), 101.

8. *OG* xiv 73, 91–2, 106–7, 117–18, 120, 123, 140, 145–8, 150, 165–6.
9. A. Schiavo, 'Notizie riguardanti la Badia di Passignano estratte dai fondi dell'Archivio di Stato di Firenze', *Benedictina* 4 (1955), 31–92, at 44.
10. *OG* xix 422–3.
11. *OG* xix 527.
12. Vincenzo Galilei, *Dialogue on Ancient and Modern Music*, ed. Claude V. Palisca (New Haven, Conn.: Yale University Press, 2003).
13. *OG* xix 170.
14. Daniel Chua, 'Vincenzo Galilei, Modernity and the Division of Nature', in Suzanna Clark and Alexander Rehding, eds, *Music Theory and Natural Order from the Renaissance to the Early Twentieth Century* (Cambridge: Cambridge University Press, 2006), 17–29.
15. For Vasari's emphasis on progress in art, see Paul Feyerabend, 'Progress in Philosophy, the Sciences and the Arts', in Feyerabend, *Farewell to Reason* (London: Verso, 1987), 143–61.
16. Favaro was unacquainted with this work (the first volume of which appears to be very rare), hence his false reference at *OG* xvi 310.
17. *OG* xvi 305; compare xviii 356, and Campanella at xiv 367.
18. *OG* xvii 65, 80. William Harvey had visited Rome a few months earlier, hence the sudden interest in the circulation of the blood: see Charles Schmitt and Charles Webster, 'Harvey and M. A. Severino', *Bulletin of the History of Medicine* 45 (1971), 49–75, at 55. The word I have translated as 'revolutionise' is 'rivolgere' – turn upside down. A particularly clear idea of progress is present in the work of Galileo's friend Giovanni Ciampoli: see Federica Favino, 'Scetticismo ed empirismo: Ciampoli Linceo', in Andrea Battistini, Gilberto de Angelis and Giuseppe Olmi, eds, *All'origine della scienza moderna: Federico Cesi e l'Accademia dei Lincei* (Bologna: Il Mulino, 2007), 175–202.

CHAPTER 2: FLORENCE

1. In 1590, however, he was accorded only the lesser rank of *cittadino*: *OG* xix 47.
2. *OG* xix 32.
3. *OG* xiii 458.
4. *OG* xix 311.
5. *OG* xix 61. According to James Reston, *Galileo: A Life* (London: Cassell, 1994), 27, prior to his obtaining his first university post Galileo was in despair: 'Together with a young Florentine friend Ricasoli Baroni, he began to look to distant shores, to the Middle East and beyond. Perhaps there, among the heathen Turks, he could find his place as a scientist.' This seems to originate in a misreading. Giovambatista Ricasoli believed that he was fleeing Florentine justice (see below, pp. 30–2) and told Galileo that the long arm of the Medici would catch up with him (Ricasoli) unless he went as far as Cairo or England (*OG* xix 55). There is nothing to suggest that Galileo thought of going with him.
6. *OG* xii 237, 260.
7. *OG* xii 278.
8. *OG* xii 258.
9. *OG* v 311.
10. *OG* xii 294.
11. *OG* ix 32.
12. *OG* x 327.
13. *OG* xi 458, 482, v 190.
14. *OG* xvi 443, 445. See also his response to the prospect of an English translation: xvi 355. Even allowing for the caution with which Galileo expresses himself in his letters to Micanzio (see especially 385), the fear that his book will be forgotten seems authentic: we already find it in his letter to Peiresc of 12 May 1635. See Galileo Galilei, *A Long-Lost Letter from Galileo to Peiresc on a Magnetic Clock*, trans. Stillman Drake (Norwalk, Conn.: Burndy Library, 1967).
15. In a letter intended for a distinguished foreigner, on the other hand, Galileo expressed pride in having supporters throughout the northern hemisphere: *OG* xvi 474–5.

CHAPTER 3: GALILEO'S LAMP

1. Fulgenzio Micanzio, *OG* xvi 209.
2. *OG* xix 604, 637.
3. Albert O. Hirschman, *Exit, Voice, and Loyalty* (Boston, Mass.: Harvard University Press, 1970).
4. *OG* xix 603.
5. Stillman Drake, 'Renaissance Music and Experimental Science' [1970], in Drake, *Essays on Galileo and the History and Philosophy of Science* (3 vols, Toronto: University of Toronto Press, 1999), iii, 190–207; Claude V. Palisca, 'Vincenzo Galilei, scienziato sperimentale, mentore del figlio Galileo', *Nuncius* 15 (2000), 497–514.
6. Paolo Palmieri, *Reenacting Galileo's Experiments* (Lewiston, N.Y.: Edwin Mellen Press, 2008), 5–9, 221–69. Thomas B. Settle, 'Experimental Sense in Galileo's Early Works and Its Likely Sources', in *LCF* 831–50, at 834, refers to 'a standard steel lute string', but there was no such thing. All lute strings were made of gut. And Galileo never specifies use of a steel string – see, for example, *OG* xvii 91.
7. See, for example, Simon Schaffer, 'Glass Works: Newton's Prisms and the Uses of Experiment', in David Gooding, Trevor Pinch and Simon Schaffer, eds, *The Uses of Experiment* (Cambridge: Cambridge University Press, 1989), 67–104.

CHAPTER 4: EUREKA!

1. *OG* ii 215.
2. W. Roy Laird, 'Archimedes among the Humanists', *Isis* 82 (1991), 628–38, at 628; *OG* i 300; see also *OG* xvi 399, xvii 90. Galileo's friends recognised that he aspired to equal Archimedes: e.g. x 240–1. A valuable study of Archimedes and his significance is Lucio Russo, *The Forgotten Revolution* (Berlin: Springer, 2004).
3. Koyré is generally reported as arguing that Galileo was a Platonist, which he does – and this is misleading: see, for example, Thomas P. McTighe, 'Galileo's "Platonism": A Reconsideration', in Ernan McMullin, ed., *Galileo, Man of Science* (New York: Basic Books, 1967), 365–87. But Koyré also argues that Galileo is a follower of Archimedes – and this is true: see, for example, Koyré, *Galileo Studies* (Atlantic Highlands: Humanities Press; Hassocks: Harvester Press, 1978), 33–8. Koyré's account of Galileo is therefore closer to the truth than is generally recognised. See Corrado Dollo's chapters 'L'uso di Platone in Galilei' [1989] and 'L'egemonia dell'archimedismo in Galilei' [1989], in Dollo, *Galileo Galilei e la cultura della tradizione* (Soveria Mannelli: Rubbettino, 2003), 23–62 and 63–86. Favaro, writing in 1915–16, had no time for the notion that Galileo was a Platonist: Antonio Favaro, *Adversaria galilæiana*, ed. Lucia Rossetti and Maria Laura Soppelsa (Trieste: Lint, 1992), 10–11.
4. *OG* i 209–20; partially translated as: Galileo Galilei, 'The Sensitive Balance', in Laura Fermi and Gilberto Bernadini, *Galileo and the Scientific Revolution* (New York: Dover, 1961).
5. See Wootton, 'Accuracy and Galileo'.

CHAPTER 5: SEEING IS BELIEVING

1. One of these mathematicians was Giuseppe Moletti (b. 1531), whom Galileo was later to succeed in the chair of mathematics at Padua. Laird, 'Archimedes among the Humanists', 637, argues that Moletti had never read a word of Archimedes, and must have found Galileo's work profoundly puzzling.
2. On Guidobaldo, see Mary J. Henninger-Voss, 'Working Machines and Noble Mechanics: Guidobaldo Del Monte and the Translation of Knowledge', *Isis* 91 (2000), 233–59.
3. Valerio first published his work in 1604, but he and Galileo first met in 1590, so Galileo may have discovered that he was already working on the topic at that date, and decided then to move on to another subject.
4. *OG* i 187–90; *TNS* 261–3. On Clavius, see James M. Lattis, *Between Copernicus and Galileo* (Chicago: University of Chicago Press, 1994).

5. *OG* x 24, 34. See Paolo Palmieri, 'Mental Models in Galileo's Early Mathematization of Nature', *SHPS* 34 (2003), 229–64.

6. *OG* x 27–9.

7. *OG* x 37.

8. *OG* x 29, 35: 'due volte mi ha replicato che *petit principium*'. See also the concerns of Pietro Francesco Malaspina: xii 314–15.

9. A new answer to this puzzle has been propounded by Rolf Willach, *The Long Route to the Invention of the Telescope* (Philadelphia: American Philosophical Society, 2008), who stresses both the very slow improvement in the quality of lenses, and the invention of the diaphragm to stop down the objective lens. The first argument is surely true, but the second must be overstated. Galileo's telescopes had diaphragms (cf. *OG* x 278, 501) and it is possible that one can be seen in Porta's drawing of August 1609 (x 252); but it seems Santini's telescopes did not (x 485), and they did the job; nor did the first telescope made by Kepler (x 414).

10. Vasco Ronchi, 'The Influence of the Early Development of Optics on Science and Philosophy', in McMullin, ed., *Galileo, Man of Science*, 195–206, at 197. Ronchi's argument has been powerfully criticised in David C. Lindberg and Nicholas H. Steneck, 'The Sense of Vision and the Origins of Modern Science', in Alan G. Debus, ed., *Science, Medicine and Society in the Renaissance* (2 vols, London: Heinemann, 1972), i, 29–45, and receives no support from Stuart Clark, *Vanities of the Eye: Vision in Early Modern European Culture* (Oxford: Oxford University Press, 2007). But I persist in thinking it is basically correct: see Lucien Febvre, *The Problem of Unbelief in the Sixteenth Century* [1942] (Cambridge, Mass.: Harvard University Press, 1982), 423–37; Catherine Wilson, *The Invisible World: Early Modern Philosophy and the Invention of the Microscope* (Princeton N.J.: Princeton University Press, 1995), 215–18; Robert Jütte, *A History of the Senses* (Cambridge: Polity, 2005), 61–71; and Galileo's own lecture notes, *OG* i 157–8, translated in Galileo Galilei, *Galileo's Early Notebooks: The Physical Questions*, trans. William A. Wallace (Notre Dame: University of Notre Dame Press, 1977), 226–7.

11. *OG* xii 294. See John Wedderburn, iii 172.

12. On the intellectual framework that made such moments of mutual incomprehension inevitable, Ugo Baldini, 'The Development of Jesuit "Physics" in Italy, 1550–1700: A Structural Approach', in Constance Blackwell and Sachiko Kusukawa, eds, *Philosophy in the Sixteenth and Seventeenth Centuries: Conversations with Aristotle* (Farnham: Ashgate, 1999), 248–79, at 256–60.

13. In 1615, Cesi insisted that not all Linceans were Copernicans, but all were committed to the principle of 'libertà di filosofare *in naturalibus*': *OG* xii 150. Partial exceptions were Johannes van Heeck, who defended an Aristotelian cosmology: Saverio Ricci, 'I Lincei e le novità celesti prima del *Nuncius sidereus*', in Massimo Bucciantini and Maurizio Torrini, eds, *La diffusione del copernicanesimo in Italia (1543–1610)* (Florence: Olschki, 1997), 221–36; and Luca Valerio, who resigned in 1616 because of the Copernicanism of his fellow Linceans. The key statement of principle is Federico Cesi, 'Del natural desiderio di sapere', in M. L. Altieri Biagi and B. Basile, eds, *Scienziati del Seicento* (Milan: Rizzoli Editore, 1969), 53–92. On *libertas philosophandi*, see *EDG* 247, 348–53; Robert B. Sutton, 'The Phrase "Libertas philosophandi"', *Journal of the History of Ideas* 14 (1953), 310–16; and below, ch.23 n.42.

14. *OG* xix 204.

15. *OG* xvi 420.

16. *Dialogue* 107–8.

17. *OG* vii 133–4; Galileo Galilei, *Dialogo sopra i due massimi sistemi del mondo, tolemaico e copernicano*, ed. O. Besomi and M. Helbing (Padua: Antenore, 1998), ii, 347–8.

18. *OG* vii 355, 362–3, 367; *Dialogue* 328, 334–5, 339.

19. Paolo Palmieri, 'A New Look at Galileo's Search for Mathematical Proofs', *Archive for History of the Exact Sciences* 60 (2006), 285–317.

20. *OG* xix 638.

21. *OG* xi 168.

22. See the discussion on primary and secondary qualities: *OG* vi 350; *Comets* 311–12.

23. Plutarch, 'Life of Marcellus,' in *Lives* (London: Loeb, 1917), v, 487.
24. *OG* x 35–6.

CHAPTER 6: A FRIEND IN NEED

1. *OG* xix 44–108.
2. William A. Wallace, 'Galileo's Pisan Studies in Science and Philosophy', in Peter K. Machamer, ed., *The Cambridge Companion to Galileo* (Cambridge: Cambridge University Press, 1998), 27–52, at 29, says that Galileo lectured in Siena in 1585–6; taught in Vallombrosa in the summer of 1585; and gave his lectures on Dante in 1588. What we actually know is that he lectured in Siena before March 1588; taught in Vallombrosa in the autumn of 1588; and gave his lectures on Dante before August 1594. The rest is mere conjecture.
3. *OG* xix 48.
4. Bernard Weinberg, 'The Accademia degli Alterati and Literary Taste from 1570 to 1600', *Italica* 31 (1954), 207–14.
5. *OG* xix 54.
6. *OG* xix 56.
7. *OG* xix 62.
8. Galileo Galilei, *Contro il portar la toga* (Pisa: ETS, 2005), ll. 136–7; *OG* ix 217.

CHAPTER 7: JUVENILIA

1. Adriano Carugo and Alistair C. Crombie, 'The Jesuits and Galileo's Ideas of Science and Nature', *Annali dell'Istituto e Museo di storia della scienza di Firenze* 8 (1983), 3–68; William A. Wallace, *Galileo and His Sources: The Heritage of the Collegio Romano in Galileo's Science* (Princeton: Princeton University Press, 1984); Crombie and Carugo, review of Wallace, *Times Literary Supplement*, 22 November 1985, 1319–20. The Wallace thesis continues to be influential: see, for example, William E. Carroll, 'Galileo Galilei and the Myth of Heterodoxy', in John Brooke and Ian Maclean, eds, *Heterodoxy in Early Modern Science and Religion* (Oxford: Oxford University Press, 2005), 115–44, at 116–23.
2. *OG* xviii 234, 245, 248–9. For Galileo's (fully deserved) reputation as an anti-Peripatetic, see, for example, *OG* iv 317.
3. *OG* xi 423. Thus Salviati recommends Baliani to Galileo: 'Lui filosofa sopra la natura, si ride di Aristotile et di tutti i Peripatetici' (xi 610).
4. Galileo, *Galileo's Early Notebooks*; Galileo Galilei, *Galileo's Logical Treatises*, trans. William A. Wallace (Dordrecht: Kluwer Academic Publishers, 1992).
5. Baldini, *Legem impone subactis* (Rome: Bulzoni, 1992), 371.
6. The point is well made in Corrado Dollo, 'Galilei e la fisica del Collegio Romano' [1992], in Dollo, *Galileo Galilei e la cultura della tradizione*, 87–128. For a critique of claims that Galileo's scientific method is Aristotelian in origin, see Nicholas Jardine, 'Galileo's Road to Truth and the Demonstrative Regress', *SHPS* 7 (1976), 277–318.
7. Sven Dupré, 'Ausonio's Mirrors and Galileo's Lenses: The Telescope and Sixteenth-Century Practical Optical Knowledge', *Galilæana* 2 (2005), 145–80, at 148–9. Galileo refers to a manuscript on falling bodies and projectiles being in common circulation early in 1631: *OG* vi 631.
8. Dupré, 'Ausonio's Mirrors', 149.

CHAPTER 8: THE LEANING TOWER

1. *OG* xix 39; Galileo, *Contro il portar la toga*, 47. In principle there were three distinct gold coins, the florin, the ducat and the scudo. In practice the three terms seem to have been interchangeable. Thus in 1598 Galileo's salary is denominated in florins, while in 1610 it is in ducats; he receives payments in ducats but gives receipts for the same number of scudi.
2. Frederick Purnell Jr, 'Jacopo Mazzoni and Galileo', *Physis* 14 (1972), 273–94, at 273–4.

3. See Michele Camerota, 'Borro, Girolamo' and 'Buonamici, Francesco', in *New Dictionary of Scientific Biography* (New York: Scribner, 2007). The key study is Michele Camerota and Mario Helbing, 'Galileo and Pisan Aristotelianism: Galileo's *De motu antiquiora* and the *Quaestiones de motu elementorum* of the Pisan Professors', *Early Science and Medicine* 5 (2000), 319–65. On religion, see Spini, *Galileo, Campanella*, 25, and Michele Camerota, 'Galileo, Lucrezio, e l'atomismo', in M. Beretta and F. Citti, eds, *Lucrezio, la natura e la scienza* (Florence: Olschki, 2008), 141–75, at 143–5.

4. Antonio Favaro, 'La libreria di Galileo Galilei descritta ed illustrata', *Bullettino di bibliografia e di storia delle scienze matematiche e fisiche* 19 (1886), 219–93, at 259.

5. *MM* 88; *OG* i 273; Ernest A. Moody, 'Galileo and Avempace: The Dynamics of the Leaning Tower Experiment', *Journal of the History of Ideas* 12 (1951), 163–93, 375–422, at 418–19 – but note that Moody gives the wrong date for the first publication of Pereira (and of Buonamici). On Galileo's knowledge of Pereira, see Camerota and Helbing, 'Galileo and Pisan Aristotelianism', 347–8, 356–7.

6. There is an excellent introduction to Aristotelian physics in Maurice Clavelin, *The Natural Philosophy of Galileo* [1968] (Cambridge, Mass.: MIT Press, 1974).

7. Camerota, 'Borro, Girolamo'.

8. *OG* iv 242.

9. *OG* xix 606.

10. *OG* i 273, 334; *MM* 37–8, 107. Simon Stevin reported in 1605 that he had performed a similar experiment 'long ago': Lane Cooper, *Aristotle, Galileo and the Tower of Pisa* (Ithaca: Cornell University Press, 1935), 14; and Moletti reported similar experiments in an unpublished manuscript: W. Roy Laird, 'Giuseppe Moletti's *Dialogue on Mechanics* (1576)', *Renaissance Quarterly* 40 (1987), 209–23.

11. Lane Cooper, *Aristotle, Galileo, and the Tower of Pisa* (Ithaca, NY: Cornell University Press, 1935); Alexandre Koyré, 'Galilée et l'expérience de Pise', *Annales de l'Université de Paris* 12 (1937), 441–53; Carl G. Adler and Byron L. Coulter, 'Galileo and the Tower of Pisa Experiment', *American Journal of Physics* 46 (1978), 199–201; Byron L. Coulter and Carl G. Adler, 'Can a Body Pass a Body Falling through the Air?' *American Journal of Physics* 47 (1979), 841–6; Herman Erlichson, 'Galileo and High Tower Experiments', *Centaurus* 36 (1993), 33–45.

12. *OG* viii 108–9. He also predicts that, leaving to one side certain aspects of air resistance, an ebony (or wood) ball and a lead ball will reach the ground four inches apart. But he does not claim that this is confirmed by experiment, and he does not claim to be able to make calculations covering all the relevant factors: these limitations on his claim are missed, for example, by Adler and Coulter, 'Galileo and the Tower of Pisa Experiment', and Erlichson, 'Galileo and High Tower Experiments'.

13. *OG* xviii 305, 310. Similar experiments were being reported in Rome in 1617 – but these are possibly Galileo's own experiments.

14. Erlichson, 'Galileo and High Tower Experiments', 38; *OG* xviii 305–6. Erlichson does not seem to grasp that this undermines his whole argument, which is that Galileo never conducted any high tower experiments.

15. Peter Dear, 'Narratives, Anecdotes, and Experiments: Turning Experiences into Science in the Seventeenth Century', in Dear, ed., *Literary Structure of Scientific Argument* (Philadelphia: University of Pennsylvania Press, 1991), 135–63, at 146–52; also Alexandre Koyré, 'An Experiment in Measurement' [1953], in Koyré, *Metaphysics and Measurement* (London: Chapman Hall, 1968), 89–117.

16. Thomas B. Settle, 'Galileo and Early Experimentation', in Rutherford Aris, H. Ted Davis and Roger H. Stuewer, eds, *The Springs of Scientific Creativity* (Minneapolis: University of Minnesota Press, 1983), 3–20, at 12–14.

17. *OG* i 296–302; *MM* 63–9.

18. Moody, 'Galileo and Avempace', 415–18.

19. *OG* i 263; *MM* 27. Contrast Galileo's firm insistence on the primacy of experience over reason in his later work: *OG* v 139, vii 80, xviii 247.

20. *OG* i 273, 301, 302; *MM* 37–8, 68, 69. See also *OG* i 298–9; *MM* 65.

CHAPTER 9: INERTIA

1. *OG* i 299–301; *MM* 65–8, 170–1.
2. On the difference between impetus and impressed force theories, see Moody, 'Galileo and Avempace', especially 394–5, 402–10.
3. See, for example, Raymond Fredette, 'Galileo's *De motu antiquiora*: Notes for a Reappraisal', in *LCF* 165–82, at 180; also Stillman Drake, 'Introduction' in M. Stillman Drake and I. E. Drabkin, eds, *Mechanics in Sixteenth-Century Italy: Selections from Tartaglia, Benedetti, Guido Ubaldo and Galileo* (Madison: University of Wisconsin Press, 1969), 3–60 at 36–8 (a reference missed by Fredette).
4. Moody, 'Galileo and Avempace', 401–2, for fourteenth-century statements of this idea. Wallace went to some lengths to argue that Galileo might have encountered these medieval ideas through a reading of Domingo de Soto: William A. Wallace, 'Domingo de Soto and the Iberian Roots of Galileo's Science' [1997], in Wallace, *Domingo de Soto and the Early Galileo* (Aldershot: Ashgate, 2004). The case is far from proved, however; for a sceptical assessment of Galileo's dependence on medieval sources, see Edith Sylla, 'Galileo and the Oxford *Calculatores*', in William A. Wallace, ed., *Reinterpreting Galileo* (Washington, D.C.: Catholic University of America Press, 1986), 53–108.
5. For a somewhat different account of his intellectual difficulties, see Camerota and Helbing, 'Galileo and Pisan Aristotelianism', 355.
6. See Moody, 'Galileo and Avempace', 418–19.
7. See, for example, Koyré, *Galileo Studies*, 30, 31, 33, 34, 37, 40 n. 24, 69.
8. Koyré, *Galileo Studies*, 74.
9. Koyré was sure that this thought experiment was taken by Galileo from Benedetti (57 n. 118), yet in an autobiographical fragment Galileo presents it as entirely of his own making: *OG* vii 731–3; Paolo Palmieri, ' "Spuntar lo scoglio più duro": Did Galileo Ever Think the Most Beautiful Thought Experiment in the History of Science', *SHPS* 36 (2005), 223–40.
10. For an account of the networks for the transmission of information in Galileo's world, see Filippo de Vivo, *Information and Communication in Venice: Rethinking Early Modern Politics* (Oxford: Oxford University Press, 2007).
11. Paolo Sarpi, *Pensieri naturali, metafisici e matematici*, ed. Luisa Cozzi and Liberio Sosio (Milan: Riccardo Ricciardi, 1996), 265, 350.

CHAPTER 10: NUDISM

1. *OG* xix 109–10.
2. *OG* x 47.
3. *Pace OG* x 54. Paul F. Grendler, *The Universities of the Italian Renaissance* (Baltimore: Johns Hopkins University Press, 2002), 34, 76.
4. Anne Reynolds, 'Galileo Galilei's Poem "Against Wearing the Toga" ', *Italica* 59 (1982), 330–41.
5. For Galileo's continuing commitment to egalitarian principles, see *OG* xi 327, xii 361.
6. *OG* xi 605.

CHAPTER 11: COPERNICANISM

1. *OG* xi 361.
2. *OG* xviii 209.
3. Giovanni Bordiga, *Giovanni Battista Benedetti* [1926] (Venice: Istituto Veneto di Scienze, Lettere ed Arti, 1985), 667.
4. Georg Rheticus published in favour of Copernicanism *before* 1543; others, such as Gilbert, *after* 1593; others in the relevant period, such as Erasmus Reinhold, had used Copernican tables without committing themselves to Copernicanism; others still, such as Christoph Rothman and Michael Maestlin, had written or lectured in defence of Copernicanism but had not published. On Diego de Zuñiga, see Irving A. Kelter, 'The Refusal to Accommodate: Jesuit

Exegetes and the Copernican System', in *C&G* 38–53, at 38–42. For longer lists of early Copernicans, see Sosio 'Fra Paolo Sarpi e la cosmologia', in Sarpi, *Pensieri*, xciii–xciv, and Robert S. Westman, 'The Copernicans and the Churches', in David C. Lindberg and Ronald L. Numbers, eds, *God and Nature* (Berkeley: University of California Press, 1986), 76–113, at 85. Galileo discussed semi-Copernicanism in his Paduan lectures: *OG* ii 223–4.

5. 'In Copernico non ho cosa alcuna che mi apporti un minimo scrupolo': *OG* xii 35. Cesi began to consider Copernicanism seriously only in June 1612, but despite discussing the matter with Galileo moved rapidly to adopt Kepler's account: xi 332–3; 344–5; 366–7. This last seems to have met with no reply from Galileo, other than a promise that all would be made clear in the great work in defence of Copernicanism: 370, 403. But see *Dialogue* 455, which seems to be implicitly a reference, perhaps even a sympathetic reference, to Kepler's work.

6. Massimo Bucciantini, *Galileo e Keplero* (Turin: Einaudi, 2007), 93–116.

7. I owe this point to Isabelle Pantin. On Bruno's influence on Galileo, see in particular Redondi, 'La nave di Bruno', 318–20.

8. *OG* x 67–8.

9. *OG* x 68–9.

10. *OG* x 47–50.

11. Aldo Stella, 'Galileo, il circolo culturale di Gian Vincenzo Pinelli e la "patavina libertas" ', in Giovanni Santinello, ed., *Galileo e la cultura padovana* (Padua: CEDAM, 1992), 325–44; Bucciantini, *Galileo e Keplero*, 33–4. On the fate of Pinelli's library, see Anthony Hobson, 'A Sale by Candle in 1608', *The Library* 26 (1971), 215–33.

12. Translation from Annibale Fantoli, *Galileo: For Copernicanism and for the Church* (3rd rev. edn, Vatican City: Vatican Observatory Publications, 2003) 60 (revised).

13. The interpretation of this passage was a matter of dispute between Favaro and Emil Wohlwill: see Favaro, *Galileo Galilei a Padova*, 267–8.

14. In 1630 Galileo claimed he had been working on the *Dialogue* for thirty years (*OG* xiv 85) – not quite long enough for my account, but too long for the conventional one.

15. See, for example, Michael Sharratt, *Galileo: Decisive Innovator* (Cambridge: Cambridge University Press, 1994), 70: 'That Galileo had been a Copernican for *many* years is at least as doubtful as his claim that he had discovered many effects explicable only by the earth's motions.' See also William R. Shea, 'Galileo the Copernican', in *LCF* 41–60, which mentions neither the letter to Kepler nor the letter to Mazzoni, and effectively begins the story of Galileo's Copernicanism with the telescope. The orthodox view is that of Stillman Drake, 'Galileo's Steps to Full Copernicanism, and Back' [1987], in Drake, *Essays on Galileo*, i, 351–63, and Maurice Finocchiaro, 'Galileo's Copernicanism and the Acceptability of Guiding Assumptions', in Arthur Donovan, Larry Laudan and Rachel Laudan, eds, *Scrutinizing Science* (Baltimore: Johns Hopkins University Press, 1988), 49–67. Finocchiaro questions whether there is unequivocal Copernicanism in *The Starry Messenger*, but Galileo's opponents certainly thought there was: see, for example, Francesco Sizi (or Sizzi), *Dianoia* (Venice, 1611), in *OG* iii 205–50, at 227. Works such as Fantoli, *Galileo: For Copernicanism and for the Church*, seem to me not to grasp that the main objections to Copernicus were from physics rather than astronomy, and that Galileo was in a position to overcome these objections once he had an account of the path of projectiles. Views closer to my own are those of Maurice Clavelin, 'Galilée astronome philosophe', in *LCF* 19–40, at 20 n. 3; R. H. Naylor, 'Galileo, Copernicanism and the Origins of the New Science of Motion', *British Journal for the History of Science* 36 (2003), 151–81; and Winifred L. Wisan, 'Galileo's De systemate mundi and the New Mechanics', in Paolo Galluzzi, ed., *Novità celesti e crisi del sapere* (Florence: Giunti-Barbèra, 1984), 41–9. The line of argument pursued here is foreshadowed in Jürgen Renn, Peter Damerow and Simone Rieger, 'Hunting the White Elephant: When and How Did Galileo Discover the Law of Fall?' *Science in Context* 13 (2000), 299–419, at 352.

16. For information relevant to the dating, see *OG* xiv 386, 395, and Renn, Damerow and Rieger, 'Hunting the White Elephant', whose results are recapitulated in Peter Damerow, Gideon Freudenthal, Peter McLaughlin and Jürgen Renn, 'Proofs and Paradoxes', in Damerow,

Freudenthal, McLaughlin and Renn, *Exploring the Limits of Preclassical Mechanics* (2nd rev. edn, New York: Springer, 2004), 135–279, at 158–79.

17. *TNS* 142–3; *OG* viii 185–6.
18. *TNS* 143; *OG* viii 186.
19. Renn, Damerow and Rieger, 'Hunting the White Elepehant', 92–104, 115.
20. 'This conclusion . . . is the keystone both to Galileo's mechanics and to his astronomy': Winifred L. Wisan, 'Galileo and the Process of Scientific Creation', *Isis* 75 (1984), 269–86, at 270. It involved a break with the approach of Guidobaldo: see Drake in note to *MM* 166–7; W. Roy Laird, 'Renaissance Mechanics and the New Science of Motion', in *LCF* 255–67, at 260.
21. Alan Chalmers and Richard Nicholas, 'Galileo on the Dissipative Effect of a Rotating Earth', *SHPS* 14 (1983), 315–40.
22. *OG* xiv 386.
23. Renn, Damerow and Rieger, 'Hunting the White Elephant', 56.
24. *A&C* 136.
25. Bucciantini, *Galileo e Keplero*, 58.
26. *OG* x 372. Of course this might be a manuscript copy of, for example, Calcagnino; or even – the hypothesis, though bold, is irresistible – Galileo's lost work, *Considerazione astronomica* [*de motu terrae*], a copy of which he had given to Guidobaldo just a year before the latter's death, calling it a book, although it was presumably a manuscript rather than a printed volume (xx 598).
27. The issue reappears in *TNS* 119–20; *OG* viii 161–2.
28. 'I have almost abandoned the study of natural and mathematical matters', Sarpi wrote in 1609: Paolo Sarpi, *Lettere ai gallicani*, ed. Boris Ulianich (Wiesbaden: Franz Steiner Verlag, 1961), 179; translation from Eileen Reeves, *Galileo's Glassworks* (Cambridge, Mass.: Harvard University Press, 2008). Reeves's attempt to explain this statement away (131–2) does not convince me; nor am I convinced that Galileo had a mirror-based instrument for observing the heavens before the telescope (141). For an account of Sarpi's preoccupations in 1609–10, see Corrado Pin, 'Progetti e abbozzi sarpiani sul governo dello stato "in questi nostri tempi turbolenti" ', in Paolo Sarpi, *Della potestà de' prencipi* (Venice: Marsilio, 2006), 89–120.
29. Sarpi, *Pensieri*, 391–405; *OG* x 114. Sosio thinks that Sarpi was a Copernican, though he acknowledges that the evidence is far from conclusive: 'Fra Paolo Sarpi e la cosmologia', in Sarpi, *Pensieri*, xci–cxciv, especially cxxx–cxxxviii.
30. *OG* viii 116–8; *TNS* 76–7.
31. *Dialogue* 211–18, *OG* vii 237–44; Maurice Finocchiaro, 'Physical-Mathematical Reasoning: Galileo on the Extruding Power of Terrestrial Rotation', *Synthèse* 134 (2003), 217–44.
32. Cf. *OG* xii 35.
33. *OG* xix 616.
34. *OG* x 72.
35. Sarpi, *Pensieri*, 424–7.
36. On Galileo's denial of Copernicus's third movement, see Galileo Galilei, *Il saggiatore*, ed. Ottavio Besomi and Mario Helbing (Rome: Antenore, 2005), 256 (*OG* vi 326) and note, 594–5.
37. Paolo Palmieri, 'Re-Examining Galileo's Theory of Tides', *Archive for History of Exact Sciences* 53 (1998), 223–375.
38. Thomas Digges had already tested this idea: Francis R. Johnson and Sanford V. Larkey, 'Thomas Digges, the Copernican System, and the Idea of the Infinity of the Universe in 1576', *Huntington Library Bulletin* 5 (1934), 69–117, at 93, 99; and it had been taken up by Bruno in 1584: Daniel Massa, 'Giordano Bruno and the Top-Sail Experiment', *Annals of Science* 30 (1973), 201–11. The experiment was performed prior to 1628 by Jean Gallé, claiming to vindicate Ptolemy (Massa, 'Giordano Bruno', 208); in 1634 by Jean-Baptiste Morin, who obtained the results predicted by Digges but claimed they proved the falsity of Copernicanism (Massa, 'Giordano Bruno', 209), Bernard Frénicle de Bessy (Piero Ariotti, 'From the Top to the Foot of the Mast on a Moving Ship', *Annals of Science* 28 (1972),

191–203, at 202) and Mattia Naldi (Favaro, *Scampoli galileiani*, i, 317–22); and in 1639 by Baliani (*OG* xviii 103). Had they been able to read Galileo's reply to Ingoli they would have known that Galileo had already performed it: *OG* vi 545. It is clear that Galileo conducted other similar experiments at an early date: see the marginalia to Lodovico delle Colombe's 'Contro il moto della terra' (1610–11, cf. *OG* iii 12), *OG* iii 256–8, and Stelluti's letter of 8 January 1633 (see below, pp. 176–7). Stelluti also reports a number of experiments conducted by Annibale Brancadoro to test Galileo's claims regarding the behaviour of projectiles fired from a fast-moving carriage and boat. See now Pietro Redondi, 'La nave di Bruno'.

39. *SN^e* 53, *SN^1* D2r, *MC* 19, *OG* iii 72.

40. *SN^e* 27; *A&C* 659 – we surely catch an echo of conversations with Galileo when Gualterotti contrasts Aristotle with the book of nature; Henry Stevens, *Thomas Hariot the Mathematician, the Philosopher, and the Scholar* (London: privately printed, 1900), 120.

41. Sarpi, *Pensieri*, 37–8, 363; Eileen Reeves, *Painting the Heavens: Art and Science in the Age of Galileo* (Princeton: Princeton University Press, 1997), 32–4, 105–6. There is good evidence that Sarpi felt some hostility to Galileo around the time of the publication of *The Starry Messenger*, presumably because he felt that his own contribution to the invention of the telescope had been effaced, because he was not pleased with its dedication to Cosimo II or indeed the book's overt Copernicanism, and perhaps in addition because Galileo had not acknowledged his priority in explaining the phenomenon of earthshine. It is to be noted that Sarpi presents the explanation of earthshine as his own in his letter to Leschassier of 27 April 1610 (*Lettere ai gallicani*, 79). On these tensions, see Reeves, *Painting the Heavens*, 104–7, and Gaetano Cozzi, 'Galileo Galilei, Paolo Sarpi, e la società veneziana', in Cozzi, *Paolo Sarpi tra Venezia e l'Europa* (Turin: Einaudi, 1979), 135–234, at 182–7. But his remarkable claim not to have read the book, despite having had a copy in his hands (*Lettere ai gallicani*, 79, 80, 239), can be explained only by his having discussed its contents in such detail with Galileo (as he says) that he felt confident he knew everything of importance that it contained. It is these conversations which presumably led to a significant revision of the text of *The Starry Messenger* as it went through the press – the introduction of a discussion of why the moon does not look like a cog-wheel (*SN^1* C3v–D1r; *MC* 15–17; *OG* iii 69–71; *SN^e* 48–51), which Galileo presents as a response to criticism.

42. It is important that one cannot 'see' the mountains on the moon; one can only deduce that they are there. As Galileo puts it, 'How, then, do we know that the moon is mountainous? We know it not simply by the senses, but by copying and combining discourse with observations and sensory appearances' (*OG* xi 183, translation from Rivka Feldhay, *Galileo and the Church: Political Inquisition or Critical Dialogue?* (Cambridge: Cambridge University Press, 1995), 242). In 1637 Galileo sent his translator Bernegger a pair of lenses for a telescope. Bernegger complained that the objective lens had been stolen and an inferior lens substituted. His evidence: he could not *see* the mountains on the moon (*OG* xvii 55).

43. Galileo Galilei, *Galileo against the Philosophers in His Dialogue of Cecco di Ronchitti (1605) and Considerations of Alimberto Mauri (1606)*, trans. Stillman Drake (Los Angeles: Zeitlin and Ver Brugge, 1976). Mauri's *Considerazioni* is not in *OG* and there is no modern edition, but it is quoted verbatim in Lodovico delle Colombe, *Risposte piacevoli* (Florence: Giovanni Antonio Caneo, 1608), here ff. 73r–74r. On naked eye observation, see Peter Grego, *The Moon and How to Observe It* (London: Springer, 2005), 98–9.

44. *OG* ii 197–202.

45. On Mazzoni, see Purnell, 'Jacopo Mazzoni and Galileo'.

46. He was still dealing with this argument at the end of his life: *OG* xiv 417, xviii 314–15.

CHAPTER 12: MONEY

1. *OG* x 61.

2. *OG* xix 571.

3. *OG* xix 212–13.

4. *OG* x 81–2, 84–5, 193; cf. *OG* xix 201.

5. *OG* x 88, 89, 100–1, 103–4, xix 214.
6. *OG* x 142, 145, 202, xix 209–11, 214–17.
7. *OG* xix 111–15, x 159. For the exchange rate between Venetian and Florentine currencies, see, for example, Richard S. Westfall, 'Science and Patronage: Galileo and the Telescope', *Isis* 76 (1985), 11–30, at 12 n. 4; Galileo Galilei, *Operations of the Geometric and Military Compass*, trans. Stillman Drake (Washington, D.C.: Burndy Library, 1978), 50; and *OG* xii 454–5. In his account books Galileo takes one Venetian ducat to be worth 5 lire, and one Florentine ducat to be worth seven lire, but the dowry contract specifies a rate of 6.2 lire to the ducat.
8. *OG* xix 115–17.
9. *OG* xix 218–20.
10. *OG* xix 131–49, 166–7.
11. *OG* xix 139–40, 153–5, 159–61. In 1607, by contrast, after the Interdict, he had only one student boarding with him, and was offering to undercut the competition: x 178–9. Matteo Valleriani, 'A View on Galileo's *Ricordi autografi*: Galileo Practitioner in Padua', in *LCF* 281–91.
12. Claudio Bellinati, 'Integrazioni e correzioni alle pubblicazioni di Antonio Favaro, studioso di Galileo Galilei', in Santinello, ed., *Galileo e la cultura padovana*, 275–90, at 278, 281.
13. *OG* x 350.
14. *Naudæana et Patiniana* (Paris: Florentin et Pierre Delaulne, 1701), n 45.
15. *OG* xii 167, 191–2, 313, 365–6, 369, 392–3, 447, 480. It is amusing to find Cremonini far too busy with his lecturing to find time to pay his bills. Galileo finally cleared his debts in 1610 by arranging to be paid two years' salary in advance: x 374.
16. *OG* xi 172, 447.
17. *OG* x 91.
18. *OG* x 165, on Acquapendente.
19. *OG* x 165–6.
20. *OG* xix 233, 255, 261–2, 422–3.
21. *OG* xix 507.

CHAPTER 13: FIELDS OF FIRE

1. Sharratt, *Galileo*, 66.
2. *OG* xix 126–9.
3. Jürgen Renn and Matteo Valleriani, 'Galileo and the Challenge of the Arsenal', *Nuncius* 16 (2001), 481–503.
4. *OG* x 55–60.
5. As Bertoloni Meli puts it, Galileo's response 'betrayed a lack of familiarity with practical issues'. Contarini should not be thought of as a naïve interlocutor: he had earlier engaged in an important exchange with Guidobaldo del Monte on whether theoretical results regarding equilibrium in pulleys could in principle be confirmed in practice: Domenico Bertoloni Meli, *Thinking with Objects: The Transformation of Mechanics in the Seventeenth Century* (Baltimore: Johns Hopkins University Press, 2006), 32–4, 68.
6. See, for example, *OG* xi 392–3, xii 291–5, xiv 202–4, xvi 470–1; *Dialogue* 105; *EDG* 305–13.
7. *OG* ii 363–424; Galileo, *Operations of the Geometric and Military Compass*.
8. There was still considerable demand for Galileo's sector – or at least for the instructions on how to use it – in the 1630s: *OG* xvi 445.
9. *OG* x 146.
10. *OG* ii 79–146.
11. *OG* ii 147–91. The standard edition is now Galileo Galilei, *Le meccaniche*, ed. Romano Gatto (Florence: Olschki, 2002).
12. See, for example, Mario Otto Helbing, 'Galileo e le *Questioni meccaniche* attribuite ad Aristotele', in *LCF* 217–36.
13. *MM* 65; *OG* i 298.

14. Koyré, *Galileo Studies*, 37–8; Ernan McMullin, 'Galilean Idealization', *SHPS* 16 (1985), 247–73; Peter K. Machamer, 'Galileo's Machines, His Mathematics, and His Experiments', in Machamer, ed., *Cambridge Companion*, 53–79.

CHAPTER 14: THE EXPERIMENTAL METHOD

1. For 'experimentum', *OG* iii 75, viii 197; for 'esperimento', *OG* vi 194, 339 (twice), 364, 367; for 'sperimentare', *OG* xviii 77. It is clear that Galileo generally avoids the word 'esperimento' as a neologism; as far as I know he never uses the alternative word, 'cimento'. Drake's translations need to carry a health warning, as he frequently translates 'esperienza' as 'experiment'. On the phrase 'sensata esperienza', which has sometimes been misinterpreted to mean 'experiment', see Neal W. Gilbert, *Renaissance Concepts of Method* (New York: Columbia University Press, 1963), 230–1 and Gabriele Baroncini, 'Sulla Galileiana *esperienza sensata*', *Studi secenteschi* 25 (1984), 147–72. On experience, see Giorgio Stabile, 'Il concetto di esperienza in Galileo e nella scuola galileiana', in *Experientia: X colloquio internazionale* (Florence: Olschki, 2002), 217–41.
2. Charles B. Schmitt, 'Experience and Experiment: A Comparison of Zabarella's View with Galileo's in *De Motu*', *Studies in the Renaissance* 16 (1969), 80–138, at 115–23.
3. James MacLachlan, 'A Test of an "Imaginary" Experiment of Galileo's', *Isis* 64 (1973), 374–9. The classic replication of a Galilean experiment is Thomas B. Settle, 'An Experiment in the History of Science', *Science* 133 (1961), 19–23. Paolo Palmieri (personal communication, 8 December 2009) has now replicated the experiment with waves of water in a glass whose rim is being rubbed, which had been dismissed as fictional by, for example, D. P. Walker, 'Some Aspects of the Musical Theory of Vincenzo Galilei and Galileo Galilei', *Proceedings of the Royal Musical Association* 100 (1974), 33–47, at 46.
4. *OG* xiv 343. For a useful discussion, see R. H. Naylor, 'Galileo's Experimental Discourse', in David Gooding, Trevor Pinch and Simon Schaffer, eds, *The Uses of Experiment* (Cambridge: Cambridge University Press, 1989), 117–34.
5. For examples of his use of this word in Italian, see Galileo, *Il Saggiatore*, 309, 312; *TNS* 282; *OG* viii 322.
6. According to John Wedderburn, Galileo's pupils were discussing the relevance of Alhacen to an understanding of the telescope in 1609: *OG* xiii 175. Galileo refers to Alhacen in the marginalia of his copy of Sizi's *Dianoia* (1611): *OG* iii 239, 244. On Galileo's knowledge of optics, see Alistair C. Crombie, *Styles of Scientific Reasoning* (3 vols, London: Duckworth, 1994), 582–4; Dupré, 'Ausonio's Mirrors'. For a strong statement of Alhacen's influence, at least in the Middle Ages, see David C. Lindberg, 'Alhazen's Theory of Vision and Its Reception in the West', *Isis* 58 (1967), 321–41, though I find it difficult to see how this argument could be applied to Galileo. In a crucial exchange with Sagredo, it was Sagredo rather than Galileo who showed an advanced grasp of optics: see *OG* xi 350, 356, 371, 379, 398. It is striking that Galileo's alter ego, Mauri, holds an extromission theory of sight, while his bitter opponent, delle Colombe, holds an intromission theory.
7. He did, however, request a copy, and one was sent to him in 1610: *OG* x 441, 493.
8. *OG* x 91.
9. *Dialogue* 400.
10. *OG* x 101. Antonio Favaro, *Galileo Galilei e lo studio di Padova* [1883] (2 vols, Padua: Antenore, 1966), ii, 78–9. For Sarpi's enthusiasm for Gilbert, see Sarpi, *Lettere ai gallicani*, 256 n. 4.
11. Galileo and his close associates were deeply concerned that his new science of motion was founded on an unproven axiom. In 1638 Galileo eventually produced a proof (*OG* xvii 399–400, xviii 95, 126) that appeared in posthumous editions of the *Two New Sciences* (viii 214 n. 1).
12. *OG* x 97–100.
13. Christiaan Huygens, *Horlogium oscillatorium* [1673], published in English as *The Pendulum Clock* (Ames: Iowa State University Press, 1986).
14. Palmieri, *Reenacting Galileo's Experiments*, 5–9, 221–69.

15. *OG* x 115–16. Many years later Galileo stressed how difficult and lengthy was the process of getting from the recognition that the path of a projectile is a parabola to formulating the law of fall: *OG* xiv 386 (which is hard to reconcile with the argument of Renn, Damerow and Rieger, 'Hunting' the White Elephant).

16. *Dialogue* 400–7; *OG* x 222.

17. G. E. L. Owen, ' "Tithenai ta Phainomena" ' [1961], in Jonathan Barnes, Malcolm Schofield and Richard Sorabji, eds, *Articles on Aristotle 1: Science* (London: Duckworth, 1975), 14–34.

18. Cf. his summary of his two new sciences in 1609: *OG* x 228–30.

19. *OG* vii 75–6; *Dialogue*, 51.

20. *TNS* 89; *OG* viii 131.

21. He does once use the word 'experimentum' in Latin: *OG* viii 197.

22. Paula Findlen, 'Fare esperienza', in Findlen, *Possessing Nature: Museums, Collecting, and Scientific Culture in Early Modern Italy* (Berkeley: University of California Press, 1994), 194–240.

23. 'Il padre degli esperimenti e di ogni loro essattezza': *OG* xvi 232–3. For Castelli's reflections on the experimental method, see *OG* xvii 150–69.

24. W. E. Knowles Middleton, *The Experimenters: A Study of the Accademia del Cimento* (Baltimore: Johns Hopkins University Press, 1971).

25. *Dialogue*, 3–4; *OG* vii 27.

26. René Descartes, *Principia Philosophiae* (1644), III ii–iii; *Les principes de la philosophie* (Paris: Michel Bobin, 1668), 114–15.

CHAPTER 15: THE TELESCOPE

1. *OG* x 134–5; he later seems to have revised it (xx 597–8), but he still did not publish.

2. Stillman Drake thought that Cecco's dialogue was entirely the work of Galileo; others think it entirely the work of Spinelli; still others believe it to be a collaboration between Galileo and Spinelli. The most recent authorities favour sole authorship by Spinelli: Marisa Milani, 'Galileo Galilei e la letteratura pavana', in Santinello, ed., *Galileo e la cultura padovana*, 193–217; Ludovico Maschietto, 'Girolamo Spinelli e Benedetto Castelli', in ibid., 453–67; Marisa Milani, 'Il "Dialogo in perpuosito de la stella nuova" di Cecco di Ronchitti da Brugine', *Giornale storico della letteratura italiana* 170 (1993), 66–86.

3. That Galileo was thinking within a Copernican framework is made clear by two drawings relating to the nova, first published in the 1929–39 reprint of *OG* (ii 621–2).

4. Bucciantini, *Galileo e Keplero*, 34.

5. *OG* x 176; Galileo, *Galileo against the Philosophers*, 55–71.

6. Galileo, *Galileo against the Philosophers*, 76.

7. Galileo, *Galileo against the Philosophers*, 67; see below, p. 94. For myself, the main reason for hesitating in attributing Mauri's book to Galileo is that delle Colombe pointed out in his *Risposte piacevoli* that Mauri's Tuscan prose style is not very good (e.g. f. 81r v). On the other hand, Galileo had at this point been living 'abroad' for fourteen years, and delle Colombe would have taken pleasure in claiming that he could no longer write proper Tuscan.

8. The classic study is Albert van Helden, 'The Invention of the Telescope', *Transactions of the American Philosophical Society* 67:4 (1977), 1–67.

9. On Badoer, see Reeves, *Galileo's Glassworks*, 125–8.

10. *OG* x 250.

11. *OG* x 184–91, 194–5, 197–8, 199–202, 205–13.

12. *OG* x 250–1.

13. *OG* x 251, 255. Galileo's view seems to have been that he was not bound by this agreement because the contract had not yet begun: 373. Similarly he seems to have believed that he was under no obligation to resign his position in Padua because his existing contract terminated at the end of September. The view in Venice was rather different: 384.

14. *OG* x 307.

15. Favaro, *Galileo Galilei e lo studio di Padova*, ii, 303–4.

Chapter 16: Mother

1. *OG* x 268–9.
2. *OG* x 270.
3. *OG* x 279.
4. *OG* x 313, xii 108.
5. Antonino Poppi, *Cremonini e Galilei inquisiti a Padova nel 1604* (Padua: Antenore, 1992), 58–61.
6. *OG* x 202.
7. *OG* xii 139. See Edward Muir, *The Culture Wars of the Late Renaissance* (Boston: Harvard University Press, 2007), for the context of this accusation.
8. *OG* xii 494.
9. *OG* xiii 406.
10. *OG* x 43–4.
11. *OG* xix 43.

Chapter 17: *The Starry Messenger*

1. Stevens, *Thomas Hariot*, 116.
2. *SN¹* E1r; *MC* 28; *OG* x 277. Mario Biagioli argues persuasively that the letter of 7 January was probably never sent: *GIC* 91. I cite the first edition of the *Sidereus nuncius*, or *SN¹*, by signature. Links to copies of this available on the web will be found at www.watcheroftheskies.org. The pagination of the first edition is reproduced in the best modern edition, *MC*. For a detailed discussion of Galileo's revisions to the text during publication, see David Wootton, 'New Light on the Composition and Publication of the *Sidereus nuncius*', *Galilæana* 6 (2009), 61–78.
3. *OG* x 280.
4. *OG* x 298; Amir Alexander, 'Lunar Maps and Coastal Outlines: Thomas Hariot's Mapping of the Moon', *SHPS* 29 (1998), 345–68; *OG* xi 33. But note that Harriot's friend William Lower said, on reading *The Starry Messenger*, 'in the moone I had formerlie observed a strange spottednesse al ouer, but had no conceite that anie parte therof mighte be shadowes': Stevens, *Thomas Hariot*, 116.
5. *OG* x 280; *MC* 55–6, nn. 28, 29.
6. *OG* iii 35.
7. This was presumably the title on the manuscript which went to be licensed, and it was sufficiently unusual as to confuse the authorities, who rendered it 'Astronomica denuntiatio ad astrologos': *OG* xix 227–8. On the date of the licence, which is correct in *OG*, see Antonino Poppi, 'Una implicita ritrattazione di Antonio Favaro sulla licenza di stampa del *Sidereus nuncius*', *Atti e memorie dell'Accademia patavina di scienze, lettere ed arti già Accademia dei Ricovrati: Memorie della classe di scienze morali, lettere ed arti*, 110 (1998), 99–105.
8. See, for example, *OG* xi 179, iv 63, v 192.
9. There has been much discussion of Galileo's title: see, for example, *MC* xxxii–xxxvii.
10. *OG* x 442.
11. *OG* x 499, 504.
12. Lunardi and Sabbatini, eds, *Una casa per memoria*, 36.
13. On the classical origins of the concept of the fact, see Pantin's introduction to Johannes Kepler, *Discussion avec le messager céleste*, ed. Isabelle Pantin (Paris: Les Belles Lettres, 1993), lxix, cx–cxi. Kepler used the word 'factum' in full knowledge of its juridical origins, *OG* x 413–17: 'Et vero, non problema philosophicum, sed quaestio iuridica facti est, an studio Galilaeus orbem deluserit' ('It is not a philosophical question, but a legal question concerning a matter of fact, whether Galileo has deliberately misled the world').
14. The claim that Bacon uses the word 'fact' in the modern sense in English has been made, but appears to be mistaken: Barbara J. Shapiro, *A Culture of Fact: England, 1550–1720* (Ithaca: Cornell University Press, 2000), 109, 117; see the careful formulation in Lorraine Daston,

'Baconian Facts, Academic Civility, and the Prehistory of Objectivity', in Allan Megill, ed., *Rethinking Objectivity* (Durham, N.C.: Duke University Press, 1994), 37–63, at 45.

15. Leonardo Fioravanti, *Il reggimento della peste* (Venice: heirs of Melchior Sessa, 1571), f. 11v; Giovanni Lorenzo d'Anania, *L'universale fabrica del mondo* (Venice: Aniello San Vito, 1576), 105.

16. *OG* ii 188; v 201, 330 (twice), 310, 333, 338, 343; vi 216, 225, 267; vii 276, 291, 317, 375, 395, 399, 413, 480; viii 44, 125, 208, 309, 511; x 432; xi 8; xviii 164. The phrase 'la verità del fatto' is used by those close to Galileo: by Santini, *OG* x 398; by Cesi, xi 458; and by Angelo de Filiis in the preface to *Letters on Sunspots*, v 81.

17. *OG* iii 68, x 275. On Galileo's illustrations, see Ewan A. Whitaker, 'Galileo's Lunar Observations and the Dating of the Composition of the *Sidereus nuncius*', *Journal for the History of Astronomy* 9 (1978), 155–69; Mary G. Winkler and Albert van Helden, 'Representing the Heavens: Galileo and Visual Astronomy', *Isis* 83 (1992), 195–217; Owen Gingerich and Albert van Helden, 'From *Occhiale* to Printed Page: The Making of Galileo's *Sidereus nuncius*', *Journal for the History of Astronomy* 34 (2003), 251–67; Owen Gingerich, 'The Curious Case of the M-L *Sidereus nuncius*', *Galilæana* 6 (2009), 141–65, which provides a critique of Horst Bredekamp, *Galileo der Künstler* (Berlin: Akademie Verlag, 2007).

18. Galileo, *Galileo against the Philosophers*, 104–5.

19. *GIC* 135–217.

20. For an effective refutation of an earlier version of Feyerabend's account of Galileo, see Peter K. Machamer, 'Feyerabend and Galileo: The Interaction of Theories and the Reinterpretation of Experience', *SHPS* 4 (1973), 1–46. The first two editions of *Against Method* (1975, 1988) contain a reply to Machamer as appendix 1. A full-scale critique of Feyerabend (which I do not have space for here) would now have to include a discussion of his influence on *GC*. For another study taking the side of Galileo's opponents, see Roger Ariew, 'The Initial Response to Galileo's Lunar Observations', *SHPS* 32 (2001), 571–81 and 749. Sarpi would have found Ariew's argument specious: see his brisk rebuttal of a similar argument in *Lettere ai gallicani*, 80–1.

21. Paolo Sarpi, *Lettere ai protestanti*, ed. Mario D. Busnelli (2 vols, Bari: Laterza, 1931), i, 122; see also Sarpi's views as reported by Andrea Rey in Philippe Duplessis-Mornay, *Mémoires et correspondance* (12 vols, Paris: Treuttel et Würz, 1824–5), xi, 260–1.

22. *OG* xi 350.

23. *OG* iii 244, 248–9; xi 107. See also Luca Valerio, xi 104. For a contemporary critique of the reliability of the telescope, see Sizi (or Sizzi), *Dianoia*, in *OG* iii 205–50, Italian translation by Clelia Pighetti (Florence: Barbera, 1964).

24. Ewan A. Whitaker, 'Galileo's Lunar Observations', *Science* 208 (1980), 446; Feyerabend's first reply: *Science* 209 (1980), 544; Whitaker, 'Lunar Topography: Galileo's Drawings', *Science* 210 (1980), 136; Feyerabend's second reply: *Science* 211 (1981), 876–7. *Against Method*: 2nd edn, 1988; 3rd edn, 1993; 4th edn, 2010.

25. *OG* xix 116–17.

26. *OG* x 281.

27. *OG* x 283.

28. *OG* x 284.

29. *MC* xc.

30. *SN¹* A3r; *MC* 3; *OG* iii 56. *OG* iii 27; *SN¹* D4r; *MC* 22; *OG* iii 75. *OG* iii 46; *SN¹* G4r; *MC* 47; *OG* iii 95. It seems to me wrong to say that in *The Starry Messenger* Galileo did not 'go so far as to declare himself a follower of Copernicus. That would have been highly inappropriate and counterproductive in a book designed, at least in part, to glorify the House of Medici, and thus to obtain patronage for Galileo': Albert van Helden, 'Galileo, Telescopic Astronomy, and the Copernican System', in René Taton and Curtis Wilson, eds, *Planetary Astronomy from the Renaissance to the Rise of Astrophysics: Part A: Tycho Brahe to Newton* (Cambridge: Cambridge University Press, 1989), 81–105, at 84; for a similar view expressed by Finocchiaro, see ch. 11 n. 15 above. It is precisely Galileo's inappropriate behaviour that needs explaining.

31. *MC* 53 n. 18; 57–8 n. 8; 76 n. 84; 92–3 n. 164.

32. *OG* iii 17, 36; *SN¹* B1v, E2r; *MC* 7, 29; *OG* iii 60, 81.

33. *OG* x 262.
34. Logan Pearsall Smith, *The Life and Letters of Sir Henry Wotton* (2 vols, Oxford: Clarendon Press, 1907), i, 486.
35. *OG* iii 36, 46; *SN¹* E2r, G4r; *MC* 29, 47; *OG* iii 81, 95.
36. *OG* x 68.
37. *OG* x 300.
38. There are three references to the future *Dialogue*: *OG* iii 25, 27, 47; *SN¹* D3r, D4r, G4v; *MC* 20, 21, 48; *OG* iii 73, 75, 96. These are all passages I would date to mid-February.
39. The manuscript, though not the printed text, acknowledges this as a significant difficulty, one which presumably struck Galileo at a late date. *OG* iii 46; *MC* 93 n. 165.
40. The divergence between manuscript and printed text is recorded in *MC* 93 n. 165.
41. *OG* iii 75.
42. *SNᵉ* 57; compare *OG* xi 148.
43. *OG* x 300.

CHAPTER 18: FLORENCE AND BUOYANCY

1. *OG* x 350.
2. *OG* xi 171, xiv 62.
3. *OG* xi 165. See also xi 230 (an echo of Cremonini). Later, in 1640, Galileo was to ask the same question himself: xviii 209.
4. *OG* xi 172.
5. Cesare Cremonini, *Le orazioni*, ed. Antonino Poppi (Padua: Antenore, 1998), 53–69; Cozzi, 'Galileo Galilei, Paolo Sarpi', 143–8.
6. He supplied Galileo with a copy of this correspondence in 1619, when Galileo was once more at odds with the Jesuits: *OG* xii 458.
7. *OG* x 158. On the interpretation of this text, see Favaro, *Galileo Galilei e lo studio di Padova*, i, 77. Nevertheless, we may note that, if his closest ties were with the anticlericals, Galileo still established friendly relations with two Jesuits teaching in Padua, Andreas Eudaemon-Joannes and Giuseppe Biancani: Baldini, *Legem impone subactis*, 371–3.
8. Cf. Martin Hasdale's letter of August 1609 (*OG* x 418), written before he heard that Galileo had changed sides; Muir, *Culture Wars*.
9. *OG* x 107.
10. *OG* ii 513–601, x 171–6, 191–2, xix 222–6.
11. *OG* x 147, 149.
12. *OG* x 153–4.
13. *OG* x 214.
14. *OG* x 221–3; *GC* 120–7.
15. *OG* x 232–4.
16. *OG* x 234.
17. *OG* x 301.
18. *OG* x 233.
19. *OG* x 353, 373.
20. *OG* iv 32 n. 3; Galileo Galilei, *Discourse on Bodies in Water*, trans. Stillman Drake (Urbana: University of Illinois Press, 1960), 215.
21. *OG* iv 345.
22. *OG* iv 64–5.
23. *OG* xii 227.
24. *OG* xix 183.
25. *OG* xii 222–3.
26. *OG* xiv 70.
27. *OG* iv 63–141; Galileo, *Discourse on Bodies in Water*.
28. Galileo's habit of having others publish on his behalf reflected a view that great men should engage in debate only with their equals: cf. Cesi's advice, *OG* xi 409.

29. *OG* iv 21, 112. Steffen Ducheyne, 'Galileo's Interventionist Notion of "Cause" ', *Journal of the History of Ideas* 67 (2006), 443–64.
30. Wootton, 'Accuracy and Galileo', 47.

CHAPTER 19: JESUITS AND THE NEW ASTRONOMY

1. *OG* x 439.
2. *OG* x 120.
3. *OG* x 431. The usually impeccable Favaro misrepresents this letter: *Galileo Galilei e lo studio di Padova*, i, 77.
4. *OG* x 432.
5. Mordechai Feingold, 'Jesuits: Savants', in Feingold, ed., *Jesuit Science and the Republic of Letters* (Cambridge, Mass.: MIT Press, 2003), 1–45, at 18–20; Richard J. Blackwell, *Galileo, Bellarmine, and the Bible* (Notre Dame: University of Notre Dame Press, 1991), 135–64. Acquaviva's letter of 1611 and the even stronger one of 1613 are translated in Richard J. Blackwell, *Behind the Scenes at Galileo's Trial* (Notre Dame: University of Notre Dame Press, 2006), 209–17. Cf. *OG* xi 434, xii 76, xv 254, and below, p. 209.
6. *OG* xi 480. On Grienberger, see Mordechai Feingold, 'The Grounds for Conflict: Grienberger, Grassi, Galileo, and Posterity', in Feingold, ed., *The New Science and Jesuit Science: Seventeenth-Century Perspectives* (Dordrecht: Kluwer, 2003), 121–57.
7. See, for example, *OG* xv 254. A classic and complex case is that of Riccioli, on which see Alfredo Dinis, 'Giovanni Battista Riccioli and the Science of His Time', in Feingold, ed., *Jesuit Science*, 195–224. See also below, pp. 225, 245.
8. *OG* x 287, 317.
9. *OG* x 289.
10. *OG* x 297–8.
11. *OG* x 299, 302, 307, xvii 174, 181, 220. At that time the grand duke had five telescopes made by Galileo: *OG* xvii 16.
12. *OG* x 289, 318.
13. *OG* x 298. A Galilean telescope reached France only in September: *OG* x 430.
14. *OG* x 378; *GIC* 86–7. It was presumably the fact that Santini lacked academic and social status that prevented Galileo from publicising his achievement. He does not mention him to Kepler on 19 August when listing those who had seen the moons: *OG* x 422.
15. See, for example, *OG* x 430, 442. Galileo's wording on the title page of *The Starry Messenger* was ambiguous: 'perspicilli nuper a se reperti'. But the description in the body of the text is perfectly clear and accurate.
16. *OG* x 423, xi 100, 117, 165, 320, xv 12. On Cremonini and the microscope, see *OG* iii 164; Spini, *Galileo, Campanella*, 84. Some continued to deny the relevance of telescopic observations as late as 1620 (*OG* xiii 16), although Welser thought they might be an extinct species as early as the end of 1613 (xi 609).
17. *OG* x 364.
18. *OG* iii 97–126; Johannes Kepler, *Kepler's Conversation with Galileo's Sidereal Messenger*, trans. Edward Rosen (New York: Johnson Reprint Corporation, 1965). The best edition is now the bilingual text, Kepler, *Discussion*.
19. Campanella too immediately thought in terms of other inhabited worlds: *OG* xi 21–2. Because Kepler's *Conversation* appeared in three editions, and was more widely disseminated than Galileo's *Starry Messenger*, the idea that the new discoveries supported the notion of other inhabited worlds spread rapidly. William Lower, reading *The Starry Messenger*, also saw at once that the discovery of an unquantifiable number of new stars provided support for Bruno's argument that the universe was infinite: Stevens, *Thomas Hariot*, 116. For Kepler's references to Bruno as Galileo's unacknowledged source, see *OG* x 315, 321, 333–4, 338–9; Kepler, *Discussion*, 7, 23–5, 30–1; Kepler, *Kepler's Conversation*, 11, 34–9, 44–5.
20. *OG* x 316.

21. As he himself said: see the 'Narratio', *OG* iii 179–90, at 183–4; Kepler, *Discussion*, 34–41, at 35–6.
22. *OG* x 343, 390.
23. *OG* x 358–9, 365, 376, 384, 386–7, 390, 398–401, 408, 411–12, 414, 417–19, 428–9, 436, 440, 446, 450, 457–8.
24. *OG* x 380. See also Rey, in Duplessis-Mornay, *Mémoires et correspondance*, xi, 260–1.
25. Sarpi was clearly very irritated by the failure to provide more information: *OG* x 290 (full text in Sarpi, *Lettere ai gallicani*, 73–4).
26. It is striking that the first person of modest rank to own a Galileo telescope appears to have been Paolo Gualdo, who had been Pinelli's close companion. Gualdo, a priest, belonged to the pro-papal wing of Venetian society; he was also unlikely to use the instrument to make significant telescopic discoveries. On Gualdo, see G. Ronconi, 'Paolo Gualdo e Galileo', in Santinello, ed., *Galileo e la cultura padovana*, 375–88.
27. *OG* x 420, 427.
28. *OG* x 414.
29. *OG* x 435. In fact Harriot in England and Peiresc in France had also seen the moons of Jupiter, but Galileo did not know this: *GIC* 88. On Peiresc's astronomy, see Jane T. Tobert, 'Peiresc and Censorship: The Inquisition and the New Science, 1610–1637', *Catholic Historical Review* 89 (2003), 24–38; P. N. Miller, 'Description Terminable and Interminable', in Gianna Pomata and Nancy G. Siraisi, eds, *Historia* (Cambridge, Mass.: MIT Press, 2005), 355–97, at 373–80.
30. *OG* x 398, 446.
31. *OG* x 410, 422, 439.
32. *OG* x 451, 435, 437.
33. *OG* iii 185, x 427–8.
34. *OG* x 431, 433.
35. *OG* x 451, 478.
36. *OG* x 480, 484–5, xi 34. Several people had seen the moons of Jupiter in Rome by mid-November: *OG* x 475.
37. *OG* x 484–5.
38. He will have been thinking of Grienberger's letter of January 1611 (*OG* xi 31–5), which gave a detailed account of the confirmation of Galileo's discoveries by the Jesuits in Rome.
39. *OG* xi 41; see also 45.
40. *OG* xi 127.
41. *OG* xi 87–8.
42. *OG* xi 92–3: in December he had been in no doubt that the moon's surface was rough not smooth (x 485).
43. *OG* x 410, 426.
44. *OG* x 474, xix 229.
45. *OG* xi 11–12.
46. *OG* x 440–1.
47. *OG* xi 78.
48. Neil Thomason, 'Sherlock Holmes, Galileo, and the Missing History of Science', *PSA 1994: Proceedings of the Biennial Meeting of the Philosophy of Science Association*, i, 323–33.
49. Alan Chalmers, 'Galileo's Telescopic Observations of Venus and Mars', *British Journal for the Philosophy of Science* 36 (1985), 175–84, maintains that this argument was mistaken (see above, p. 53, on the geometrical equivalence of the three systems), but the mistake was not apparent to Galileo or his contemporaries.
50. Westfall, 'Science and Patronage'; the charge was made in the nineteenth century by Raffaello Caverni (1837–1900). On Caverni, see Giuseppe Castagnetti and Michele Camerota, 'Raffaello Caverni and His *History of the Experimental Method in Italy*', *Science in Context* 13 (2000), 327–39. Ironically, Caverni himself was guilty of fabricating texts supposedly by Galileo: Favaro, *Adversaria galilæiana*, 141–54. For a discussion of two cases where Galileo really did take someone else's ideas without acknowledgment, see below, pp. 150, 208.

51. Paolo Palmieri, 'Galileo Did Not Steal the Discovery of Venus' Phases: A Counterargument to Westfall', in *LCF* 433–44; and Palmieri, 'Galileo and the Discovery of the Phases of Venus', *Journal for the History of Astronomy* 32 (2001), 109–29.
52. *OG* x 481, 503–4.
53. *OG* xi 24, xii 124; *Comets* 153, 154. For Campanella, on the other hand, Galileo was another Columbus; for Lower another Magellan.
54. *OG* x 425.

Chapter 20: Sunspots

1. Albert van Helden, 'Galileo and Scheiner on Sunspots: A Case Study in the Visual Language of Astronomy', *Proceedings of the American Philosophical Society* 140 (1996), 358–96, at 360–1.
2. *A&C* 658; Reeves, *Painting the Heavens*, 80.
3. *OG* xi 208–9, 305, 335.
4. On Scheiner, see Blackwell, *Behind the Scenes*, 65–91.
5. *OG* xi 537.
6. Jochen Büttner, Peter Damerow and Jürgen Renn, 'Traces of an Invisible Giant: Shared Knowledge in Galileo's Unpublished Treatises', in *LCF* 183–201, at 184; and Matthias Schemmel, 'England's Forgotten Galileo', in *LCF* 269–80.
7. For Galileo's own impatience with those who claimed to be familiar with Copernicus but were not, see *OG* xi 153, xiv 66. Recently Koestler's charge that Galileo evidently did not study Copernicus with care has been revived on the basis of the scanty annotations in his copy of *On the Revolutions*: Owen Gingerich, *The Book Nobody Read: Chasing the Revolutions of Nicolaus Copernicus* (New York: Penguin Books, 2005), 200. It needs to be remembered that Galileo owned two copies, only one of which survives: Favaro, 'La libreria di Galileo Galilei', 92, 93. I prefer Edward Rosen's judgement that Copernicus was 'one of Galileo's constant companions': 'Galileo the Copernican', in Carlo Maccagni, ed., *Saggi su Galileo Galilei* (Florence: Barbèra, 1967), 181–92.
8. *OG* xiv 299. Note, however, that the testimony of the Jesuit Adam Tanner, whose importance was stressed by Galileo and by Favaro, has now been shown to be based on a misattribution: Massimo Bucciantini, 'Novità celesti e teologia', in *LCF* 795–808, at 807–8. For an up-to-date survey of the evidence, see Michele Camerota, *Galileo Galilei e la cultura scientifica nell'età della controriforma* (Rome: Salerno Editrice, 2004), 241–5.
9. *OG* xi 480.
10. See, for example, *OG* xi 230.
11. *OG* xi 46. Sarpi, keeping his distance, did not reply in person, but had Micanzio reply for him: *OG* xi 57–8.
12. *OG* x 418.
13. *OG* xi 14, 45, 98–9.
14. Cf. *OG* xi 395.
15. *OG* v 228.
16. *OG* xvi 463–9; xvii 19, 60, 96–105.
17. Cf. *OG* xii 390 for an early version of the telescope clamped to the body.
18. *OG* vii 197–8; *Dialogue*, 171–2.
19. *OG* xii 361.

Chapter 21: The Catholic scientist

1. *OG* v 232.
2. *OG* v 96, x 423, xi 113, and Kepler at xi 77.
3. Galileo invokes the Aristotelian view of miracles at *Dialogue* 421; *OG* vi 447.
4. Cf. *OG* xi 147–8.
5. *OG* xi 354–5, 376.

6. *OG* xi 465.
7. *OG* xi 437, 483.
8. *OG* xi 439, 450, 458.
9. *OG* v 258.
10. Stillman Drake and Charles T. Kowal, 'Galileo's Sighting of Neptune' [1980], in Drake, *Essays on Galileo*, i, 430–41.

CHAPTER 22: COPERNICUS CONDEMNED

1. *OG* xi 605–6; *GA* 47–8: 'last Thursday', says Castelli. For earlier religious opposition to Galileo's Copernicanism, see *OG* xi 241–2, 427. Galileo laughed off these early attacks, some of which came from Lorini: *OG* xi 461.
2. Lodovico Cardi had warned Galileo as early as December 1611 that there were plans to attack him from the pulpit: *OG* xi 241–2. That such attacks actually took place has been confirmed by Luigi Guerrini, *Galileo e la polemica anticopernicana a Firenze* (Florence: Edizioni Polistampa, 2009), although, as the evidence is drawn from sermons published some time after they were delivered, they are difficult to date exactly.
3. *OG* xix 275–6, 293, 297–8.
4. *OG* v 291–5 (*GA* 55–8), xii 151.
5. For the two versions, see *OG* v 281–8 (*GA* 49–54), xix 299–305.
6. *OG* xii 154 (Favaro's note here seems to me a sleight of hand).
7. *OG* v 291; *GA* 55.
8. Cf. Mauro Pesce, 'Le redazioni originali della lettera "copernicana" di G. Galilei a B. Castelli', *Filologia e critica* 17 (1992), 394–417, and Pesce's introduction to Galileo Galilei, *Lettera a Cristina di Lorena* (Genoa: Marietti, 2000), 20–1, though Pesce does not note the key phrase in Castelli's letter of 12 March. Others are not convinced: Maurice Clavelin, *Galilée copernicien* (Paris: Albin Michel, 2004), 490 n.; Massimo Bucciantini, *Contro Galileo: Alle origini dell'affaire* (Florence: Olschki, 1995), 34–5.
9. *OG* xix 305; *DV* 12; *GA* 135–6.
10. *OG* iv 248, v 190. He does so again at *OG* xviii 295. See the origin of this in copying painting versus painting from nature (*OG* iii 396) and in attacking Aristotelianism as book-based learning (see above, pp. 27, 132). On the image of the book of nature in Galileo, see *GIC* 219–59.
11. *OG* v 283 (*GA* 50), 316; see also, for example, xv 24 (*GA* 224), vi 337; *Comets* 298; Philippe Hamou, ' "La nature est inexorable": Pour une reconsidération de la contribution de Galilée au problème de la connaissance', *Galilæana* 5 (2008), 149–77; *GIC* 247–52.
12. *OG* v 310–15; *GA* 88–92. See also *OG* v 300 (*GA* 62), xii 184, 230, 244 (*GA* 151), 251.
13. *OG* v 292–3, 295; *GA* 56, 58.
14. *OG* xii 244; see also 230, 251.
15. *GA* 135; *DV* 13–14; *OG* xix 298.
16. *DV* 21–9, 174–5; *OG* xix 276–7, 307–11; *GA* 136–41.
17. They did indeed call themselves this: *OG* xiv 387.
18. *OG* xii 173, 174.
19. *DV* 31–3; *OG* xix 313–15.
20. *DV* 36–8; *OG* xix 316–17; *GA* 141–3.
21. *DV* 39–42; *OG* xix 318–20; *GA* 143–6.
22. *OG* xix 277–8.
23. See in particular the significant new evidence in Germana Ernst and Eugenio Canone, 'Una lettera ritrovata: Campanella a Peiresc, 19 giugno 1636', *Rivista di storia della filosofia* 49 (1994), 353–66; the contextualisation of this text in Federica Favino, 'A proposito dell'atomismo di Galileo: Da una lettera di Tommaso Campanella ad uno scritto di Giovanni Ciampoli', *Bruniana e campanelliana* 3 (1997), 265–82; and in general Pietro Redondi, 'Vent'anni dopo', in Redondi, *Galileo eretico* (3rd edn, Turin: Einaudi, 2004), 467–85, at 467–73.
24. E.g. *OG* xiv 164–5.

25. *OG* xii 345.
26. *OG* xiii 431.
27. *OG* xii 161.
28. *OG* xi 35.
29. *OG* xii 392, xvii 206 (he went with Lorenzo Ceccarelli), xix 424. He talked of going again in 1628 after an illness (*OG* xiii 408) and also it seems in 1634 (xvi 93). Spini, *Galileo, Campanella*, 39 n. 27. Ciampoli, who seems to have been a Lucretian rather than a Christian, wrote poems in praise of the Santa Casa.
30. Olaf Pedersen, 'Galileo's Religion', in G. V. Coyne, M. Heller and J. Życiński, eds, *The Galileo Affair: A Meeting of Faith and Science* (Vatican City: Specola Vaticana, 1985), 75–102, at 91.
31. *OG* xiii 360, 451–2. Colonna evidently assumed, rightly or wrongly, that the treatise included a continuation of the discussion of the fiery furnace in *The Assayer*: *OG* vi 366; *Comets* 329–30; *postilla* no. 190, *OG* vi 174. Of course one could read the *Letter to the Grand Duchess Christina* as a treatise on miracles, but it seems unlikely that Baliani and Colonna had only just heard rumours of a text written more than a decade earlier.
32. *OG* xii 478. See also Galileo's account of the views of Riccardi: *OG* xiii 183.
33. *OG* iii 340. It is interesting that Galileo uses the word 'law' only in the context of divine legislation – here, and at *OG* vi 538. Compare Copernicus, as described in Jane E. Ruby, 'The Origins of Scientific "Law"', *Journal of the History of Ideas* 47 (1986), 341–59. Galileo certainly has the concept of a law of nature, but he avoids the language.
34. See above, p. 138.
35. David Wootton, *Paolo Sarpi: Between Renaissance and Enlightenment* (Cambridge: Cambridge University Press, 1983), 15.
36. *OG* xi 26, xvii 352. Galileo was already implicitly an atomist when he wrote *On Motion*: see Camerota, 'Galileo, Lucrezio', 141–3.
37. *GA* 146; *DV* 42–3; *OG* xix 320–1.
38. *OG* xii 129.
39. *OG* xii 151 (*GA* 58–9), 160.
40. *OG* xii 171–3 (*GA* 67–9); *OG* xix 339 (*DV* 68–9; *GA* 258–9). A measure of the importance Galileo attributed to this text is the fact that he carried it with him to his first interrogation in 1633.
41. Blackwell, *Galileo, Bellarmine, and the Bible*, 254. Translations of Foscarini's pamphlet and Latin defence (neither of which appear in *OG*) are to be found in appendices 6 and 7.
42. *GA* 68 (revised).
43. Galileo's reply implies a good grasp of probability theory: see Mario Barra, 'Galileo Galilei e la probabilità', in *LCF* 101–18. This should reassure those who think, following Ian Hacking, that probability theory and the modern concept of evidence are inseparable.
44. *GA* 68.
45. *OG* xii 151; *GA* 59.
46. Baldini, *Legem impone subactis*, 285–346. See also *OG* xiii 429–30.
47. *OG* v 181, 189–90, xii 173, 181, 190.
48. He later produced a detailed reply to Bellarmine: *OG* v 364–70.
49. *OG* v 297–305 (*GA* 60–7), xii 173, 175–6.
50. *OG* xii 183–5.
51. We may note that Galileo was suffering from illness and anxiety soon after Caccini's sermon: *OG* xii 128. Galenic medicine stressed the interplay between mind and body, so that while 'psychosomatic' and 'hysterical' may be anachronistic terms, the idea of mind influencing body is not (see Cavalieri to Galileo, *OG* xvii 243). See *OG* xi 248 for Galileo's description of the way illness makes him depressed and being depressed makes him ill.
52. *OG* xii 205, v 395. Not in the embassy itself, *pace GA* 301: a wise decision as there was ill feeling between Galileo and the ambassador, dating back to 1611 (*OG* xii 207). A similar confusion between the ambassadorial residence and the Villa Medici (the latter identical rather to the 'palazzo di Sua Altezza, posto alla Trinità de Monti') appears in the work of the best scholars (Fantoli, Camerota, Pagano) and in Dan Hofstadter, *The Earth Moves* (New

York: Norton, 2009), 151. The error goes back to Viviani's biography (*OG* xix 617), but was pointed out to Viviani by Michelangelo Ricci: Favaro, 'Documenti inediti', 165.

53. *OG* xii 206–7; Guicciardini was right about the Inquisition's concerns dating back to 1611: *OG* xix 275.

54. *OG* xii 242.

55. *OG* xii 209, 229.

56. *OG* xii 212.

57. *OG* xii 229.

58. *OG* xii 220.

59. *OG* x 228.

60. *OG* xii 238, 244, 251. Two days after he learnt of the condemnation of Copernicanism Galileo hurried to put into writing a conversation he had had with Cardinal Muti about the possibility of other inhabited worlds: *OG* xii 240–1.

61. *OG* xii 126–7. On the question whether Galileo had access to Foscarini's *Defence*, see Ernan McMullin, 'Galileo's Theological Venture', in *C&G* 88–116, at 105–6.

62. *OG* xi 403, 429.

63. Bernard R. Goldstein, 'Galileo's Account of Astronomical Miracles in the Bible: A Confusion of Sources', *Nuncius* 5 (1990), 3–16.

64. Favaro, 'La libreria di Galileo Galilei', 239–41, lists the theological works Galileo did own.

65. *OG* xi 355, xii 216.

66. *OG* xii 233.

67. Cf. *OG* xv 251–2.

68. *OG* xii 450. I do not think we can conclude from this letter that Galileo had direct knowledge of Bacon on the tides (see n. 69 below), although there was a proposal to elect Bacon to the Lincean Academy in 1625: see Favino, 'Scetticismo ed empirismo', 185–7. Bacon refers to Galileo's views on tides in the *Novum Organum* (1620), Galileo having sent him a copy of his 1616 essay (see above, p. 153).

69. *OG* xiv 343. Had he been familiar with Bacon's *On the Ebb and Flow of the Sea* (*The Philosophical Works of Francis Bacon*, ed. James Spedding, 5 vols, London: Longman, 1861, iii, 47–61) he would already have known this.

70. *GA* 132–3; *OG* v 393–4. Galileo had rejected this view in 1611: *OG* iii 271.

71. *OG* xii 215–20.

72. *Dialogue* 440–1; *OG* v 394, vii 466; Galileo, *Dialogo*, ii, 871.

73. *OG* xii 230.

74. *OG* xii 242.

75. *DV* 46–7; *OG* xix 322–3; *GA* 148–50.

76. *DV* 45–6; *OG* xix 321; *GA* 147–8.

77. *DV* 177; *OG* xix 278; *GA* 148.

78. *GA* 153; *DV* 78–9; *OG* xix 348. For the rumours, see *OG* xii 250, 254, 257, 265.

79. For careful analysis of the phrase 'successive ac incontenti', which may mean no more than 'without further delay', see Thomas F. Mayer, 'The Roman Inquisition's Precept to Galileo (1616)', *British Journal for the History of Science* (online pre-print, 2009), 14–19 (pre-print pagination).

80. Note that the Venetian ambassador reported that Galileo 'sia stato ammunito rigorosamente': *OG* xx 570. Camerota, *Galileo Galilei*, 319, however, following Francesco Beretta, 'Le procès de Galilée et les Archives du Saint-Office', *Revue des sciences philosophiques et théologiques* 83 (1999), 441–90, at 476–7, continues to believe in the possibility of a forgery. For a robust refutation of this view, see Mayer, 'The Roman Inquisition's Precept', pre-print 7–11.

81. See Annibale Fantoli, 'The Disputed Injunction and Its Role in Galileo's Trial', in *C&G* 117–49, and Mayer, 'The Roman Inquisition's Precept'.

82. Pagano is quite wrong to say (*DV* cxxiv, clxvii, clxxxvii, cxcvi) that Galileo had received a written warning from Bellarmine prior to the decree of the Congregation of the Index on 5 March, that he had a copy amongst his papers, and that he produced it in 1633. This is a remarkable lapse.

83. 'Non credo che non mi debba esser prestato fede che io nel corso di 14 o 16 anni ne habbia haver persa ogni memoria': *OG* xix 346; *DV* 136.
84. *OG* xii 151; *GA* 58; *DV* 172–3; *OG* xix 275–6.
85. *OG* v 374, xii 287, 391, 450.
86. *OG* xii 242.
87. There is only one reference to Guicciardini in *GC*: on 104 n. 4 we are given his salary in 1624. In that year Guicciardini had the highest salary at court, being paid the same as Galileo.
88. *OG* xii 241–3.
89. *OG* xii 251: 'Absolutely the key to my reputation is the affection shown by Their Serene Highnesses.'
90. *OG* xiv 96–7.
91. *OG* xii 237.
92. *OG* xii 251.
93. Translation from William R. Shea and Mariano Artigas, *Galileo in Rome: The Rise and Fall of a Troublesome Genius* (Oxford: Oxford University Press, 2003), 90; *OG* xii 248; *GA* 152.
94. Translation from Shea and Artigas, *Galileo in Rome*, 92–3; *OG* xii 259.
95. *OG* xii 261.
96. *OG* xviii 421–2.
97. *OG* xviii 421.
98. *OG* xii 260–1; see also 294–5.
99. And, it seems, in later life he had a preference for republics: *OG* xvii 104–5.
100. *OG* xiii 175 (a passage not quoted in *GC*); see also 178–9.
101. In 1640 he described himself as 'homo rozzo e cortigiano poco accorto': *OG* xviii 233.
102. After Castelli had moved to Rome, we find Galileo once again teaching the grand duke in 1625: *OG* xiii 289, 306.
103. *OG* xiii 64.
104. *OG* xiii 56, 91, xix 444–5.
105. *OG* xiii 61.
106. *OG* xiii 70, 96, 98, 102.
107. See, for example, *OG* xiii 97–8.

CHAPTER 23: COMETS

1. Christopher M. Graney, 'But Still, It Moves: Tides, Stellar Parallax, and Galileo's Commitment to the Copernican Theory', *Physics in Perspective* 10 (2008), 258–68.
2. Grassi, 'On the Three Comets of the Year MDCXVIII', in *Comets* 6. See also the *Assemblea celeste* (1619), reprinted in Ottavio Besomi and Michele Camerota, *Galileo e il Parnaso Tychonico* (Florence: Olschki, 2000), 160–233, at 179.
3. *OG* xii 422.
4. *OG* xii 421, 422, 424–5, 428, 430.
5. *OG* xii 423.
6. *OG* xii 428; see also xiii 48; and xiii 67, for an offer to print in Antwerp.
7. Galileo implies (at least as I read him; see also *GH* 31) that he never saw the comets; Viviani (obviously aware that Galileo's account is somewhat implausible) says he barely saw them: *OG* xix 615. Cesi was slow to observe them, but seems to have observed the last of them.
8. Galileo, *Il saggiatore*, 111; *OG* xii 415, 417, 420, 422. For references to Galileo's illness, see *OG* xii 421, 435.
9. Tycho's book on the comet is available in French translation: *Sur des phénomènes plus récents du monde éthéré*, trans. Jean Peyroux (Paris: A. Blanchard, 1984).
10. For a succinct exposition of the case for Tycho against Copernicus, see Agucchi's letter, *OG* xi 532–5. Kepler, who was above all an astronomer, was always keen to defend Tycho's reputation, despite the fact that he was in favour of Copernicanism and Tycho was opposed (for example, *OG* xiii 299; 'Appendix to the Hyperaspistes', in *Comets* 337–55, at

342–3); for Galileo the crucial question was whether the earth moved, and consequently Tycho's system was merely a minor revision to Ptolemy's (see below p. 179, and, for example, *Comets* 184).

11. *Comets* 173. There is a tendency to read Grassi as a much more Aristotelian thinker than he actually is: see the discussion in Sharratt, *Galileo*, 135–6.
12. *Comets* 6.
13. The best discussion of the relationship between Galileo and Grassi is Feingold, 'The Grounds for Conflict'.
14. *Comets* 27.
15. See, for example, *OG* xi 107.
16. *Comets* 49.
17. *Comets* 36–7.
18. Having made comets terrestrial, Galileo also attributed to them a form of movement appropriate, in the view of Aristotelians, to a terrestrial body, claiming that they moved straight upwards in a vertical line – a claim that could scarcely be reconciled with their apparent path through the sky. The central problems with Galileo's argument were immediately identified by the ever astute Baliani: *OG* xii 476–8. We may note that Galileo repeatedly steps back from his own arguments in *The Assayer*: *Comets*, 231–2, 248, 261–2.
19. *OG* xii 485: 'pur io vedo ultimamente che V. S. vole essercitar l'ingegni speculativi'.
20. For his late views, see *OG* xviii 294–5. For literary theatre, see *OG* xiii 46, 69, 74. There is now an extensive literature that reads Galileo's books as rhetorical performances rather than logical arguments: see, for example, Jean Dietz Moss, *Novelties in the Heavens* (Chicago: University of Chicago Press, 1993). This seems to me a sound approach to the debate on comets, but less convincing when applied to other texts, except when handled with great care, as by Brian Vickers, 'Epideictic Rhetoric in Galileo's *Dialogo*', *Annali dell'Istituto e Museo di storia della scienza di Firenze* 8 (1983), 69–102.
21. *Comets* 314–26. Feyerabend regarded the diffraction explanation as incomplete. For the general acceptance of Grassi's argument, see, for example, the letter of Francesco Stelluti, *OG* xiii 430–1.
22. Galileo Galilei, *Discorso delle comete*, ed. Ottavio Besomi and Mario Helbing (Rome: Antenore, 2002), 15 (citing Tycho: *Opera omnia*, ed. J. L. E. Dreyer (15 vols, Copenhagen: Libraria Gyldendaliana, 1913–29; repr. Amsterdam: Swets & Zeitlinger, 1972), ii, 293, 295); *OG* xii 443; Bucciantini, *Galileo e Keplero*, 263–4.
23. *OG* xii 466. Ciampoli later advised Galileo to make his peace, if not with Grassi then with the Jesuits (*OG* xiii 46); Cesarini, however, writing on behalf of the Linceans, insisted that Galileo should reply *in propria persona* to *The Astronomical Balance* (xiii 68–9; see also 74, 76), thus extending the conflict.
24. *Comets* 179; *OG* vi 228.
25. *Comets* 70; *OG* vi 115.
26. *Comets* 81; *OG* vi 127. For Galileo's response, see *OG* vi 257; *Comets* 211.
27. *OG* vi 217–19; *Comets* 168–9.
28. *OG* xiii 314.
29. *OG* xiii 154, 161.
30. *OG* xiii 205–6; Galileo's report, 209; see also 199, 202–3, 210, 232–3, 236.
31. *OG* xiii 307.
32. *OG* xiii 313.
33. *OG* vi 227; *Comets* 178.
34. *OG* vi 186–7; *Comets* 137.
35. *OG* xiv 367, xv 88, 115, 183, 254. For an assessment of the role of the Jesuits in Galileo's trial, see Fantoli, *Galileo*, 341–3.
36. *OG* xvi 117.
37. *OG* xix 616.
38. *Comets* 189, *OG* vi 236–7; see also *postilla* no. 25 (*OG* vi 119–20).

39. *Comets* 252; *OG* vi 296.
40. *Comets* 300–1; *OG* vi 340.
41. *OG* xi 327, 335.
42. Elsewhere (*OG* iv 65) Galileo quotes the second-century Platonist Alcinous as saying 'philosophising wants to be free'.
43. *Comets* 118–19; *OG* vi 163–5.
44. *Comets* 147; *OG* vi 194.
45. *Comets* 301; *OG* vi 340.
46. See also *OG* xviii 423–5 and, stressing the innovative character of Grassi's use of experimental evidence, Peter Dear, 'Jesuit Mathematical Science and the Reconstitution of Experience in the Early Seventeenth Century', *SHPS* 18 (1987), 133–75, at 169–72. For Grassi's later interest in (and scepticism with regard to) experiments to demonstrate a vacuum, see his correspondence with Baliani, in Serge Moscovici, *L'expérience du mouvement: Jean-Baptiste Baliani, disciple et critique de Galilée* (Paris: Hermann, 1967), 230–63; in 1653 he was prevented from publishing a work on colour because it was too innovative (253).
47. Sharratt, *Galileo*, 140; *OG* vi 232 (Sharratt's translation).
48. Biagioli gives a rather different account of the ambiguities at the heart of Galileo's argument: *GIC* 241–58.
49. *DV* 180–4; *GH* 335–40; see also *OG* xiii 186 (which suggests that the most likely date for this document is 1624) and xiii 393–4 (where Riccardi dismisses Grassi's charges against Galileo).
50. *OG* xiv 124–5.
51. *OG* xiv 127–30.
52. *OG* xiv 158.
53. *OG* xiii 104, 107, 116.
54. *EDG* 248. Garcia assumes that this visit of Diodati's to Italy must be identical with one he is known to have made in 1626, but, as he recognises, the two visits seem quite different in character and itinerary. In 1635 Diodati referred in print to a visit made to Galileo fifteen years earlier: *EDG* 293–4.
55. *OG* xiii 48; *EDG* 243–5.
56. *OG* xiii 53.
57. Tommaso Campanella, *Apologia pro Galileo*, ed. Michel-Pierre Lerner (Pisa: Scuola Normale Superiore, 2006), ix.

CHAPTER 24: THE DEATH OF GIANFRANCESCO SAGREDO

1. On Sagredo, see Nick Wilding, 'Galileo's Idol: Gianfrancesco Sagredo Unveiled', *Galilæana* 3 (2006), 229–45.
2. *OG* xii 199.
3. *OG* xii 415–17.
4. *OG* xii 200, 446.
5. *OG* xi 522.
6. *OG* xii 168.
7. *OG* xi 356; cf. *OG* xi 363–5.
8. *OG* xi 379.
9. *OG* xix 590; Jean Tarde, *À la rencontre de Galilée: Deux voyages en Italie* (Geneva: Slatkine, 1984), 63, reporting a meeting with Galileo in 1614. Galileo denied that Kepler had had any significant influence on his work: *OG* xvi 163.
10. *OG* xii 257–8.
11. *OG* xiii 27.
12. *OG* xvi 414.
13. *OG* xi 552, xii 51.
14. *OG* xi 448. For a similar complaint of ingratitude from Paolo Gualdo, who had received no thanks for his gift of melon seeds, see *OG* xiii 27–8.

15. *OG* xii 418–19.
16. *OG* xvi 414.

CHAPTER 25: URBAN VIII

1. *OG* xiii 121.
2. *OG* xiii 133.
3. *OG* xiii 141–8. The story of Grassi's purchase of *The Assayer* is misrepresented in Adrian Johns, *The Nature of the Book* (Chicago: University of Chicago Press, 1998), 25–6: there is no evidence of 'a plan'; Sarsi did not attack the bookseller; and Galileo was already coming to Rome. The claim that 'access to the bookshop . . . had transformed the dispute' is wrong; what had transformed the dispute was papal patronage.
4. Translation from Shea and Artigas, *Galileo in Rome*, 105; *OG* xiii 135.
5. *OG* xiii 127, 139, 140, 144, 160–1, 164, 165, 166, 168.
6. Francesco Stelluti, letter of 8 January 1633, in Lino Conti, 'Francesco Stelluti, il copernicanesimo dei Lincei e la teoria galileiana delle maree', in Carlo Vinti, ed., *Galileo e Copernico* (Assisi: Porziuncola, 1990), 141–236, at 229–36.
7. *OG* xiii 175, 179, 183.
8. *OG* xiii 182.
9. *OG* xiii 183–4.
10. *OG* xiii 145.
11. *OG* v 397–412 (Ingoli's *Disputatio*), vi 501–61 (Galileo's letter), xiii 186–8, 200, 203, 209. Ingoli's *Replicationes* to Kepler (1618) are printed in Bucciantini, *Contro Galileo*, 177–205.
12. *GA* 156; *OG* vi 511–12.
13. Howard Margolis, 'Tycho's System and Galileo's *Dialogue*', *SHPS* 22 (1991), 259–75, at p. 264.
14. *GA* 179; *OG* vi 539.
15. This is misunderstood by Shea and Artigas, *Galileo in Rome*, 117; and see the comparable passage in *Dialogue* 455. See also *OG* xvii 292 (a passage that the recipient thought would need to be removed prior to publication but which Micanzio (317, 329) thought could be published in Venice, which would not have been the case if it were an explicitly Copernican argument); xviii 293–4; and xi 11–12, the earliest occurrence of this form of expression.
16. *OG* vi 529–30; *GA* 171–2. In the *Dialogue* the discussion is reduced to little more than a sentence: *OG* vii 347; *Dialogue*, 319.
17. *OG* xiii 220–1.
18. *OG* xiii 218–21; in December he sent a copy to Cesare Marsili in Bologna: xiii 235.
19. *OG* xiii 230, 261.
20. *OG* xiii 267.
21. *OG* xiii 265–6 (*GA* 204–5), 269.
22. *OG* xiii 236, 249, 253, 254, 260, 264.
23. *OG* xiii 328.
24. *OG* xiii 301, 305; see also his difficulty in reading Liceti: xiii 93.
25. *OG* xiii 302, 308. Galileo wrote with delightful irony of the extraordinary fluency of Fortunio Liceti, *OG* xviii 107.
26. *OG* xiii 365.
27. *OG* xiii 295.
28. *OG* xiii 400.

CHAPTER 26: FAMILY TIES

1. *OG* xiii 405 (although they would agree to bury their differences, yet again, in March 1628).
2. *OG* xv 93.
3. *OG* xiv 174–5; see also xiii 415.

4. It seems highly unlikely that Maria Virginia is (as Favaro thought) the child called Virginia who was living with Galileo around the time that Anna was pregnant with Benedetto. Galileo was trying to make arrangements to board this Virginia in a nunnery, perhaps to keep her at a distance from the plague, but our Virginia would have been too young to be sent into a nunnery.
5. *OG* xix 506–10.
6. *OG* xv 195, 201.
7. *OG* xix 510–14. Galileo's claim that he decided to fund Virginia in fulfilment of a religious vow (*OG* xx 612–13) must be discounted as a misdirection.
8. *OG* xviii 126–7.
9. *OG* xviii 112.
10. *OG* xviii 148, 211, 221, xix 515–19; see also xx 631.
11. *OG* xviii 130.
12. *OG* xviii 313.
13. Favaro, *Scampoli galileiani*, ii, 460–5.
14. *OG* xiii 462.
15. *OG* x 157, 192–4.
16. See, for example, *OG* xiii 401, 406. A friend insisted he was capable of economy: *OG* xi 473.
17. *OG* xiii 346–8, 353–5, 371–2.
18. *OG* xiii 409, 417, xiv 257.
19. *OG* xiii 394–5.
20. *OG* xiii 409, 416.
21. *OG* xiv 257.
22. *OG* xiii 405, 438.
23. *OG* xiii 415, xiv 258.
24. *OG* xiv 177–9, 209–10.
25. *OG* xiv 257–8, 310, xv 369, xix 475.
26. The claim that she died in Florence, made by Righini Bonelli and William Shea, *Galileo's Florentine Residences*, 50, and implicitly by Locovico Geymonat, *Galileo Galilei* (New York: McGraw Hill, 1965), 123, is perhaps based on what I take to be a misinterpretation of Alberto's letter of August 1636 (*OG* xvi 459), though it has the advantage of making Galileo truthful when he tells the pope immediately after his condemnation in 1633 that he is expecting the arrival of his sister(-in-law) and her eight children (xix 362). It seems more likely that Galileo was simply trying to evoke Urban's charitable sentiments, as there is no other evidence that Chiara was planning to return to live with her brother-in-law: Favaro thought this a mere pretext (Favaro, *Scampoli galileiani*, i, 122). Galileo would have been in touch with Alberto already had he had to inform him of his mother's death, and his letter to Micanzio (*OG* xvi 476) implies that he has understood her to have died during the sack of Munich. In fact Alberto's letter implies she had died earlier. Favaro's 'albero genealogico' gives 1634 as the date of Chiara's death, on the grounds that this is the last year in which she was paid her pension (*Scampoli galileiani*, i, 118).
27. *OG* xvi 436, 439–40, 441–2, 475–6.
28. *OG* xvii 187, 216, 218, 221, 323, 326–7, 376, 384, 392.
29. *OG* xix 520–35.
30. *OG* xviii 119–20.
31. *OG* xviii 226, 265.
32. *OG* xvii 174, 176, 180–1.
33. *OG* xiii 390–1, 403.
34. *OG* xiii 422–3, 427–8.
35. An anxiety about Chiara Galilei being struck runs through her husband's correspondence: *OG* xiii 402, 409, 418. Vincenzo had certainly done something to his mother: *OG* xiii 443.
36. *OG* xiii 430–5, 437–8, 443–4, 453.
37. *OG* xiv 210.
38. *OG* xviii 266.

39. *OG* xiii 156.
40. Those who corresponded with Galileo often expressed anxiety that their letters might be being monitored: *OG* xi 359 (an irrational fear perhaps), xii 430, xiii 11.
41. John M. Headley, *Tommaso Campanella and the Transformation of the World* (Princeton: Princeton University Press, 1997).
42. Stillman Drake, 'Galileo Gleanings III: A Kind Word for Sizzi' [1958], in Drake, *Essays on Galileo*, i, 442–57.

CHAPTER 27: PERMISSION TO PUBLISH

1. He had made the same resolution after his illness in 1628, but then he had had his brother's family underfoot: *OG* xiii 419.
2. *OG* xiii 419; see also 448.
3. *OG* xiv 49.
4. *OG* xiv 77–8.
5. *OG* xiv 59.
6. *OG* xiv 64.
7. *OG* xiv 70.
8. *OG* xiv 60.
9. *OG* xiv 52–5.
10. *OG* xiv 73–6. This exchange is seriously misrepresented in Shea and Artigas, *Galileo in Rome*, 132.
11. *OG* xiv 79, 80, 83, 85.
12. *OG* xix 487–90.
13. Favino, 'A proposito dell'atomismo di Galileo', 267–72.
14. *OG* xiv 130.
15. *OG* xiv 216 (*GA* 207).
16. *OG* xiv 78.
17. George S. Lechner, 'Tommaso Campanella and Andrea Sacchi's Fresco of Divina Sapienza in the Palazzo Barberini', *Art Bulletin* 58 (1976), 97–108, followed in 59 (1977) by an exchange of letters between Lechner and Ann Sutherland Harris; Francesco Grillo, *Tommaso Campanella nell'arte di Andrea Sacchi e Nicola Poussin* (Cosenza: Edizioni Pellegrini, 1979); John Beldon Scott, *Images of Nepotism: The Painted Ceilings of Palazzo Barberini* (Princeton: Princeton University Press, 1991), 38–94; Scott, 'Galileo and Urban VIII: Science and Allegory at Palazzo Barberini', in Lorenza Mochi Onori, Sebastian Schütze and Francesco Solinas, eds, *I Barberini e la cultura Europea del Seicento* (Rome: De Luca Editori, 2007), 127–36. The interpretation of the fresco as Copernican, under the influence of Campanella, hardly squares with Campanella's insistence to Urban that he was no Copernican: *OG* xx 604.
18. On Urban and Campanella, see D. P. Walker, *Spiritual and Demonic Magic from Ficino to Campanella* (London: Warburg Institute, 1958), 203–36; Headley, *Tommaso Campanella*, 106–14; Germana Ernst's introduction to Tommaso Campanella, *Opuscoli astrologici* (Milan: Rizzoli, 2003).
19. See Campanella's letter to Ferdinando II in 1638: 'Vederà in questo libro V. A. che in alcune cose io non accordo con l'ammirabile Galileo' (*OG* xvii 352).
20. *OG* xiv 415. The dating of the composition of Campanella's *Defence* has been much discussed: see Lerner's introduction to Campanella, *Apologia pro Galileo*, ix–xxx. The most recent English text is Campanella, *A Defense of Galileo*, trans. Richard J. Blackwell (Notre Dame: University of Notre Dame Press, 1994).
21. Shea and Artigas, *Galileo in Rome*, 134; see his letter to Galileo, *OG* xi 21–2.
22. *OG* xx 604. Michael H. Shank, 'Setting the Stage: Galileo in Tuscany, the Veneto, and Rome', in *C&G* 57–87, at 85–6.
23. *OG* xvi 182–3.
24. *OG* xiii 174.
25. Ingrid D. Rowland, *The Scarith of Scornello: A Tale of Renaissance Forgery* (Chicago: University of Chicago Press, 2004), 59.

26. *OG* xiv 381.
27. *OG* xiv 113.
28. *OG* xiv 88.
29. Francesco Beretta, 'Galileo, Urban VIII, and the Prosecution of Natural Philosophers', in *C&G* 234–61, at 250–1, and Beretta, 'Urbain VIII Barberini protagoniste de la condemnation de Galilée', in *LCF* 549–73, at 558–64; also Luca Bianchi, 'Agostino Oreggi, qualificatore del *Dialogo*, e i limiti della conoscenza scientifica', in *LCF* 575–84. In Urban's thinking, the issue of divine omnipotence was inseparable from the question of astrological determinism: Shank, 'Setting the Stage', 77–9.
30. *OG* xiii 295.
31. *OG* vii 157–8; *Dialogue* 131; Redondi, 'La nave di Bruno', 329–42.
32. Vickers, 'Epideictic Rhetoric', 83–4, 96–7.
33. *DV* 80 (n. 214), 95 (n. 283).
34. *DV* 58.
35. *OG* xiv 120, 121, 216.
36. *OG* xiv 132–3, xix 465–8.
37. *OG* xiv 126–7.
38. *OG* xiv 135. By concentrating on 1632 rather than 1630, David Marshall Miller, 'The Thirty Years War and the Galileo Affair', *History of Science* 46 (2008), 49–74, seems to me to miss the key turning point.
39. David Parrott, 'The Mantuan Succession, 1627–31,' *English Historical Review* 112 (1997), 20–65. Thomas Salusbury believed, surely correctly, that the trial of Galileo was linked to Italy's civil wars: Nick Wilding, 'The Return of Thomas Salusbury's *Life of Galileo* (1664)', *British Journal for the History of Science* 41 (2008), 1–25, at 19.
40. Spini, *Galileo, Campanella*, 71. For a clear expression of the conventional view that Venice was pro-French and Florence pro-Spanish, see the letter from the French ambassador in Rome to Richelieu, early 1639, in *EDG* 312–13.
41. *OG* xiv 90.
42. *OG* xiv 82; cf. 80.
43. The trial of Galileo of course exacerbated the tensions between Rome and Florence: see Rowland, *The Scarith of Scornello*, and below, pp. 242–4.
44. *OG* xiv 111.
45. On the Morandi case, see Germana Ernst, 'Astrology, Religion, and Politics in Counter-Reformation Rome', in S. Pumfrey, P. L. Rossi and M. Slawinski, eds, *Science, Culture and Popular Belief in Renaissance Europe* (Manchester: Manchester University Press, 1991) 249–73; Brendan Dooley, *Morandi's Last Prophecy and the End of Renaissance Politics* (Princeton: Princeton University Press, 2002). The introduction to *DV*, cxi–cxii, confuses the chronology.
46. *OG* xiv 134.
47. See Morandi's letter of 1613 (*OG* xi 530), which plays almost blasphemously with the biblical theme of the lost sheep.
48. *OG* xix 24.
49. According to Campanella, Visconti had in fact agreed with predictions of the pope's imminent demise: see Ernst in Campanella, *Opuscoli astrologici*, 20–1.
50. *OG* xiv 130.
51. *OG* xiv 152.
52. *OG* xiv 160.
53. *OG* xiv 134.
54. *OG* xiv 161, 171, 312.

CHAPTER 28: ALESSANDRA BUONAMICI

1. *OG* xiv 126. My account of the relationship between Galileo and Alessandra may be compared with that of Geymonat, *Galileo Galilei*, 190–2. Galileo seems to have had a liking for independent women: he also corresponded with the artist Artemisia Gentileschi: *OG* xvi

318–19; English translation in Mary D. Garrard, *Artemisia Gentileschi* (Princeton: Princeton University Press, 1989), 383–4.
2. *OG* xiv 130–2.
3. *OG* xviii 312.
4. *OG* xviii 194–5.
5. *OG* xviii 312–13.
6. *OG* xviii 319–20.

CHAPTER 29: A RIVER FLOODS

1. *OG* vi 619–47; substantial excerpt in Stillman Drake, *Galileo at Work: His Scientific Biography* (Chicago: University of Chicago Press, 1978), 321–9. On Galileo's involvement in flood control, see Richard S. Westfall, 'Floods along the Bisenzio: Science and Technology in the Age of Galileo', *Technology and Culture* 30 (1989), 879–907.
2. *OG* xiv 176–206.
3. *OG* xiii 348–9.
4. Discussions of such an experiment took place at the time (Westfall, 'Floods along the Bisenzio', 894–5); it is not clear if the experiment was actually performed, or if (as seems more likely) it was merely a thought experiment.

CHAPTER 30: PUBLICATION

1. *OG* xiv 148, 150–1, 156–7.
2. *OG* xiv 167, 190.
3. *OG* xiv 224, 250.
4. *OG* xiv 217–18 (*GA* 208), 224.
5. *OG* xiv 253, 407.
6. *OG* xiv 259; *GA* 210.
7. *OG* xiv 242, 340.
8. *OG* xiv 218–19, 254 (*GA* 209).
9. *OG* xiv 251.
10. *OG* xiv 258–60; *GA* 210–11.
11. *OG* xiv 266–7, 284–5.
12. *OG* xiv 281, 289.

CHAPTER 31: THE *DIALOGUE*

1. *OG* xiv 387–8; Blackwell, *Behind the Scenes*, 220.
2. Drake, 'A Kind Word for Sizzi'.
3. *OG* xv 47; Galileo, *Dialogo*, ii, 40, 720–34.
4. *OG* vii 26, *Dialogue* 2; *OG* xiv, 294–5.
5. *OG* xi 491–3, 530; Drake, 'A Kind Word for Sizzi'. Recent scholarship, however, favours Scheiner's view that Galileo had lifted his discovery without acknowledgement: see, for example, Camerota, *Galileo Galilei*, 434.
6. *OG* xiii 300, xv 254; see also Peiresc's view (xv 254) and Descartes's view (xvi 56).
7. The most recent contributions to the debate on the validity of Galileo's argument are David Topper, 'Colluding with Galileo', *Journal for the History of Astronomy* 34 (2003), 75–7, and Owen Gingerich, 'The Galileo Sunspot Controversy', ibid., 77–8.
8. *OG* xv 279. So Fulgenzio Micanzio found that reading the *Dialogue* forced him to meditate on the greatness of God, maker of the universe: *OG* xvi 162, xvii 15–16. But his conclusion, that God made the world, though not to be understood by us, is one Galileo may well have shared: *OG* xvii 28.
9. Compare Campanella's letter to Ferdinando II de' Medici: *OG* xvii 352–3.
10. *Dialogue* 368.

11. *Dialogue* 320, 369.
12. *Dialogue* 61.
13. *Dialogue* 62.
14. *Dialogue* 371.

CHAPTER 32: MARIA CELESTE AND ARCETRI

1. Thomas Hobbes, *The Correspondence of Thomas Hobbes*, ed. Noel Malcolm (2 vols, Oxford: Oxford University Press, 1994) 19; *OG* xx 606–7.
2. *OG* xx 586–8.
3. For a bilingual edition, see *To Father: The Letters of Sister Maria Celeste to Galileo, 1623–1633*, trans. Dava Sobel (London: Fourth Estate, 2001).
4. To place Maria Celeste's life in context, see Sharon T. Strocchia, *Nuns and Nunneries in Renuissance Florence* (Baltimore, Md: Johns Hopkins University Press, 2009).
5. *OG* xiv 39, 56.
6. *OG* xiv 291.
7. *OG* xv 113.
8. *OG* xv 318.
9. *OG* xv 270.
10. *OG* xv 308.
11. *OG* xv 292–3.
12. *OG* xv 247, 258.
13. *OG* xv 363.

CHAPTER 33: TRIAL

1. *OG* xiv 351, 357, 368, 379, xv 103.
2. *OG* xiv 370; see also xv 88.
3. *OG* xiv 379, 384 (*GA* 230), xv 68 (*GA* 247), xvi 171, 363, 449–50, 455. Galileo was of the view that his failure to give due weight to Urban's argument was 'il primo motore di tutti i miei travagli'. He had in fact given a similar argument earlier to Salviati and Sagredo: *Dialogue* 101–4. Jules Speller, *Galileo's Inquisition Trial Revisited* (Frankfurt: Peter Lang, 2008), sees this failure as the fundamental charge against Galileo (155–60, 375–96), and it may have been so in Urban's own mind, but (unlike the other concealed charge, that of denying transubstantiation) it was not a charge that could easily be pressed before the tribunal of the Inquisition, or every Aristotelian philosopher would be in danger of being convicted of heresy.
4. *OG* xiv 379.
5. *OG* xv 103–4; *GA* 252.
6. *OG* xiv 383–4 (*GA* 229–30), 429, xv 56 (*GA* 245), 68 (*GA* 247).
7. *OG* xiv 352, 358. Federica Favino, ' "Quel petardo di mia fortuna": Reconsiderando la "caduta" di Giovan Battista Ciampoli', in *LCF* 863–82.
8. I see no evidence, however, to support the argument (e.g. *GC* 313–52; *GH* 227–71) that the trial of Galileo was the result of Urban making concessions to Spain. See (in reply to *GH*) Richard S. Westfall, *Essays on the Trial of Galileo* (Vatican City: Vatican Observatory Publications, 1989), 92–7. New evidence adduced by Franceso Beretta (*LCF* 572), does little, as he recognises (568), to stengthen the Spanish pressure thesis.
9. *OG* xiv 416, 430.
10. On Ciampoli, see Federica Favino's work, including 'Un caso di censura postuma: La "filosofia naturale" di Giovanni Ciampoli', in *I primi Lincei e il Sant'Uffizio* (Rome: Bardi, 2005), 141–56.
11. See, for example, *OG* xiv 384 (*GA* 230), 392, xv 55 (*GA* 245), 95.
12. *OG* xiv 381, 384 (*GA* 230–1), 388 (*GA* 232); see also xv 45 (*GA* 243).
13. *OG* xiv 383 (*GA* 229), xv 68 (*GA* 247–8).

14. *OG* xiv 379–80, 382, 383 (*GA* 229), 389 (*GA* 233).
15. *OG* xiv 389 (*GA* 233), 391–3, xv 56 (*GA* 245).
16. *OG* xv 56; *GA* 246.
17. *DV* 49–57 (which is evidently the special commission report, and can be securely dated by the reference to the instructions to Florence to supply Galileo's manuscript: 51); *OG* xix 324–7; *GA* 218–22.
18. *OG* xiv 406–10.
19. *OG* xiv 427–9 (*GA* 238–40), 431–2, 438–9 (*GA* 240).
20. *OG* xiv 443–4 (*GA* 241), xv 21.
21. *OG* xv 36.
22. *OG* xv 21, 37, 56–7.
23. *OG* xv 22.
24. *OG* xv 44, 51.
25. *OG* xv 44, 51, 55, 95.
26. *OG* xv 51.
27. *OG* xv 85; *GA* 249.
28. *OG* xv 28 (*GA* 241), 68 (*GA* 247–8).
29. *OG* xv 85; *GA* 249.
30. *DV* 191–2; Francesco Beretta, 'Un nuovo documento sul processo di Galileo Galilei: La lettera di Vincenzo Maculano del 22 aprile 1633', *Nuncius* 16 (2001), 629–41. There is a charming document which reports that Niccolini and his wife were allowed to visit Galileo in the evenings and entertain him; but this is surely a forgery: *OG* xv 104.
31. *DV* 69–70, 76–8; *OG* xix 340, 345–7; *GA* 259–60, 279–81. Mayer argues that Galileo 'destroyed himself' in the course of answering questions relating to Seghezzi's injunction ('The Roman Inquisition's Precept', pre-print 11–13), but this seems to me an exaggeration: Galileo's line was that if there was an injunction he had no reason to think he was in breach of it.
32. An error in one of Favaro's footnotes (*OG* xiv 401) has caused a confusion between Maculano and his predecessor, Ippolito Maria Lanci, a friend of Castelli's who seems to have been sympathetic to Copernicanism: see, for example, Hofstadter, *The Earth Moves*, 178–9, following the second edition of Fantoli, *Galileo*, but then see the correction in the third edition, 290, 530.
33. *DV* 233–4; *OG* xv 106–7; *GA* 276.
34. *DV* 184–6; Pagano here dates this document to 1628–31, but in the introduction (lxxxviii) he follows Camerota (and Cerbu) in dating it to 1632. For an English translation, see Mariano Artigas, Rafael Martínez and William R. Shea, 'New Light on the Galileo Affair', in *C&G* 213–33, at 228–30.
35. Artigas, Martínez and Shea, 'New Light', 220–1.
36. *DV* 81, 83; *OG* xix 349, 351; *GA* 263, 265.
37. On this reading Inchofer's paper may well be independent of the earlier, anonymous denunciation of *The Assayer* which appears alongside it in the Congregation's file – although Inchofer may have been shown the earlier denunciation and used it as a guide.
38. For Scheiner's view, see Blackwell, *Behind the Scenes*, 46–7; for Grassi, see below, p. 225.
39. My argument here is a watered-down version of the argument of *GH*, in which the early charges of heresy made against *The Assayer* were first discussed, but which was written long before the discovery of Inchofer's denunciation. For early responses to the new evidence, see Artigas, Martínez and Shea, 'New Light', and (interpreting its significance very differently) Thomas Cerbu, 'Melchior Inchofer, "un homme fin et rusé"', in *LCF* 587–611, at 593–8. Note that where Richard Blackwell has argued that a plea-bargain was reached at the meeting between Maculano and Galileo, but that the bargain later broke down (*Behind the Scenes*, 1–27; see also Christopher F. Black, *The Italian Inquisition* New Haven, Conn. and London: Yale University Press, 2009, 191–2), I am arguing that a bargain was reached and that it held: Blackwell fails to discuss the relevance of the Inchofer denunciation to any bargaining.
40. *OG* xv 106–7; *GA* 276–7.
41. *DV* 72–4; *OG* xix 342–4; *GA* 277–9.
42. *OG* xv 178.

43. *OG* xv 109 (*GA* 252), 112 (*GA* 253), 124 (*GA* 253), 142.
44. *OG* xv 132 (*GA* 253–4), 160.
45. *OG* xv 123.
46. *OG* xv 123.
47. *OG* xv 140 (*GA* 254), 159.
48. *DV* 101–2; *OG* xix 360–2; *GA* 286–7.
49. *OG* xv 164, xix 402–7 (*GA* 287–93).
50. *OG* xv 164, 165, 171.
51. Translation in Blackwell, *Behind the Scenes*, 105–206.
52. In 1911 a picture of Galileo in prison, supposedly painted in the 1640s and bearing the famous phrase, was discovered (J. John Fahie, *Memorials of Galileo Galilei 1564–1642: Portraits and Paintings, Medals and Medallions, Busts and Statues, Monuments and Mural Inscriptions* (Lemington, privately printed, 1929). 72–5). One needs to be credulous, in my view, to think that this might be anything but a forgery.
53. *OG* xix 411.
54. *DV* 104–45; *OG* xv 169, xix 363–93.
55. *DV* 120; *OG* xv 171, xix 374.
56. *OG* xv 273.
57. My reading here seems to be the opposite of that of Redondi, who says (*GH* 262) 'even though [my italics] Father Grassi had never thought to denounce the Copernican doctrine, he was sent away from Rome'. Thus I see no need for the puzzlement expressed by Westfall, *Essays*, 95.
58. *DV* 104, 195; *OG* xix 362–3.
59. *OG* xv 186.
60. *OG* xv 354, 363.
61. *OG* xv 288.
62. *DV* 146–7, 200; *OG* xix 393.
63. *OG* xv 344–5, 350.
64. *OG* xvii 346–7, 349–50.
65. *OG* xvii 321.
66. *OG* xvi 531.
67. *OG* xiv 104 ('stupida et inhumana'), xv 202, 208.
68. *OG* xv 356.
69. Already in 1612: *OG* xi 359–60; after the condemnation of Copernicus: xii 430; when Galileo was in Rome: xv 72, 73, 77–8; post condemnation: xvii 506, xviii 313–14, 343. Thus we find Galileo writing that there is much that he would like to say, but that he dare not put the words on paper: *OG* xvi 475. Peiresc gave Gassendi instructions on how to write to Galileo, given that the post might be intercepted: *OG* xvi 14–15. At least one letter was in fact intercepted: *OG* xvi 117. Peiresc was also afraid that Galileo's private papers might be confiscated: Stéphane Garcia, 'Peiresc, Bernegger et Diodati: Cinq lettres inédites en rapport avec Galilée', *Galilæana* 6 (2009), 219–33, at 231.
70. See, for example, *OG* xvi 34, 64, 71, 90, 96, 136, 139, 174, 206, 217, 298, 411, 448, 453, xvii 144, 149, 170, 232, 248. It may be these precautions which gave rise to the belief that it was illegal to write to Galileo: *OG* xvi 77, 170. Packets of books and bulky objects were sent within bolts of cloth, or in the diplomatic pouch, or to be delivered by hand: *OG* xvi 191, 192, 217, 237, 298, xvii 363.
71. *OG* xvi 448, 453, 462. Favaro half-heartedly tries to rescue Galileo from mendacity here (*A&C* 1387), but the enterprise is futile: Galileo had known all along that the plan was to publish *all* his works and had not objected. If one takes Favaro's line, how far must one go? Must one conclude that Galileo is telling the truth when he roundly declares that the Copernican system is false (*OG* xviii 314)? For a clear example of Fulgenzio's writing in code, see the letter in which he tells Galileo that the Copernican system, which of course he knows from Galileo's letters that Galileo regards as entirely false, is now being adopted by all the leading intellectuals (*OG* xvii 385) – information that Galileo immediately passes on to the grand duke. Galileo, in his reply, carefully avoids referring directly to Copernicanism (*OG* xvii 390–1).

72. *OG* xvii 15.
73. *OG* xvii 312–13.
74. R. H. Naylor, 'Galileo's Physics for a Rotating Earth', in *LCF* 337–55, at 350–3; also Naylor, 'Galileo's Tidal Theory', *Isis* 98 (2007), 1–22; *OG* xvii 215, 219, 270, 287. There are a series of interpretative cruces which depend on deciding whether Galileo is to be understood as speaking his mind: see above, pp. 160, 208, ch. 33 n. 71, and below, ch. 34 nn. 1 and 5.
75. *OG* xvi 63–4. Galileo was of course forbidden to say anything that might be construed as favourable to Copernicanism, and he was also forbidden to talk to any more than one or two people at a time: *DV* 200, 233 (*OG* xv 345). Consequently when the grand duke visited him on his return to Arcetri he came with only one servant: *OG* xvi 59.
76. *OG* xvi 522. On Magiotti, see Maurizio Torrini et al., *Il diavolo e il diavoletto* (Montevarchi: Accademia valdarnese del Poggio, 1997).
77. *OG* xvii 50, 64, 80–1.
78. C. de Waard, *L'expérience barométrique, ses antécédents et ses explications* (Thouars: Imprimerie Nouvelle, 1936), 119–37, 178–82.

CHAPTER 34: THE *TWO NEW SCIENCES*

1. *OG* xv 236. This is a major interpretative crux, as Galileo had, in his abjuration, promised that he would never in future say or write anything that might give the impression that he was a Copernican (*OG* xix 407; *GA* 292). Pedersen ('Galileo's Religion', 81) is not alone in maintaining that 'during the rest of his life Galileo kept this promise', although he also acknowledges that 'it would be easy to construe Galileo's statements on his astronomical ideas in and after 1633 as flagrant examples of conscious dissimulation, and to excuse them on account of the forced conditions in which he was now placed'. Favaro insisted that Galileo had never requested the Latin translation of the *Dialogue* (*A&C* 1358–64), and Garcia's review of the evidence does not dispute this claim (*EDG* 277–87, 292). We may note that Galileo pretended not to have been involved in the publication of the *Letter to the Grand Duchess Christina* and the *Two New Sciences*, but that in each case, as we shall see, he was. Stéphane Garcia, 'Galileo's Relapse: On the Publication of the Letter to the Grand Duchess Christina (1636)', in *C&G* 265–78 and *EDG* 293–9; Maurizio Torrini, 'Galileo e la repubblica degli scienziati', in *LCF* 783–94, at 788–9.
2. *OG* xvi 209, 272–3, 293–4, 300–2, 386, 406. See also Galileo, *A Long-Lost Letter*, which is not in *OG*. The 'great prince' referred to in this letter, of 12 May 1635, is presumably Prince Mattias de' Medici.
3. Garcia, 'Galileo's Relapse'. Although Galileo had (as I believe) authorised all three Elsevier editions, and supplied the text of the letter to Christina, precautions were taken so that he could deny responsibility: see, for example, *OG* xvi 434. The turn to a Protestant publisher was made easier by a shift in Florentine foreign policy: Spini, *Galileo, Campanella*, 71–3.
4. *OG* xvii 59, 71, 75–6.
5. This is another interpretative crux. Favaro insists that Noailles had been given a copy (*A&C* 1340), but he relies on a passage in Viviani's handwriting which is crossed out, and which is at odds with the rest of the letter (as recorded in an incomplete and possibly censored copy made by Viviani) to which it is supposed to belong (*OG* xvi 524), for the rest of the letter shows that Galileo was still writing the *Two New Sciences* after his meeting with Noailles. To my mind this is a clear case of Viviani's drafting a revision to the historical record, and we can see that he tried out several versions. Stillman Drake, in *TNS* xi, follows Favaro. Pierre Costabel and Michel-Pierre Lerner, in Marin Mersenne, *Les nouvelles pensées de Galilée* (2 vols, Paris: Vrin, 1973), 19, disagree, but it is apparent both that they have failed to locate Favaro's reference, and that they have misunderstood *OG* xvii 174. It is absolutely clear, however, that Galileo was not telling the truth when he claimed that he did not know how the Elseviers had obtained a copy (*OG* viii 365), as he had supplied them with one that had not reached them through

Noailles; it is also clear from Noailles's letters that he had no knowledge of the book, except from Diodati's report, before its publication: *OG* xvii 246, 357–8. Thus it is evident that the story of a copy being given to Noailles is, in all its versions, a ruse. Benjamin Engelcke had provided a similar cover story for the translation of the *Dialogue*: *A&C* 606–16.

6. *DV* 204.
7. *Dialogue* 207–8; *OG* vii 233–4.
8. Cf. *OG* vii 61–2; *Dialogue* 37.
9. In July 1624 Antonio Santini wrote to Galileo asking for a microscope (*OG* xiii 190); in September 1624 he sent one to Cesi (xiii 208–9); and in April 1625 Giovanni Faber named the new instrument a microscope (xiii 264). Yet Galileo described to Jean Tarde as early as 1614 how a long telescope (he had not yet devised the table-top microscope) made it possible to understand how flies could hang upside down from a smooth surface: *À la rencontre de Galilée*, 64; *OG* xix 590.
10. *OG* xviii 12–13 (cf. xvii 90–1). See Baliani's reply (*OG* xviii 68–71), and Moscovici, *L'expérience du mouvement*.
11. Preface to Michelangelo Galilei, *Il primo libro d'intavolatura di liuto* (Munich: privately printed, 1620). Michelangelo's music is still played today.
12. See the criticisms of the engineer Antoine de Ville in a letter to Galileo: *OG* xvi 221–8. On this correspondence, see Hélène Vérin, 'Galilée et Antoine de Ville: Un courier sur l'idée de matière', in *LCF* 307–21.
13. *TNS* 14, *OG* vii 52–3.

CHAPTER 35: VINCENZO, SON OF GALILEO

1. *OG* xi 417. On Marina Gamba's death, see Bellinati, 'Integrazioni e Correzioni', 278–80.
2. *OG* xiii 424–6, xix 220, 425–30.
3. *OG* xiii 155–6, 228–9.
4. *OG* xiii 296, 392.
5. *OG* xiii 351, 358, xix 460–5.
6. *OG* xiii 405, 438.
7. *OG* xiii 465, xix 430–1.
8. *OG* xiv 214, 220, 222, 246.
9. *OG* xix 432–3, 476–86.
10. *OG* xv 329–30, 334, 340, 346, 347, 348, xix 433–5.
11. *OG* xix 436–7.
12. *OG* xv 199, xix 491–8.
13. *OG* xv 285.
14. *OG* xv 119.
15. *OG* xv 110.
16. *OG* xv 209–11, 238.
17. *OG* xix 394. Thus we find Galileo describing himself as being held in prison, which was technically correct: *OG* xvi 458–9, xvii 297.
18. *OG* xvi 171, 183, 339, 449–50, 456, 461, 480, 500–1, 507, xvii 26–7.
19. *OG* xvi 74, 80, 116.
20. *OG* xvi 84–5.
21. *OG* xviii 291.
22. *OG* xvii 290, xix 552–7.
23. *OG* xvii 370.
24. *OG* xvii 126–7, 174–5, 212, 214, 295–6.
25. *Dialogue* 65–7.
26. *OG* xvii 215.
27. *OG* xvii 219. My account differs from that of Naylor, 'Galileo's Physics for a Rotating Earth', 350–2. But he is perceptive when he says, 'There is nonetheless an almost haunting degree of hesitancy about Galileo's posing of the question to Michanzio [*sic*]. It is as though Michanzio

could answer where Galileo dare not.' Exactly: Micanzio could argue for Copernicanism, while Galileo dared not.

28. *OG* xvii 328.

CHAPTER 36: GALILEO'S (UN)BELIEF

1. *Naudæana et Patiniana*, n. 116.
2. David Wootton, *Bad Medicine: Doctors Doing Harm Since Hippocrates* (Oxford: Oxford University Press, 2006), 126–9.
3. Here I part company with *GH* 304–5, where he is described, strangely, as 'the new spokesman for Aristotelianism in Tuscany'. We may note that de Monconys met Nardi in his tour of Florentine *galileisti*: see below, p. 244.
4. *OG* ix 29–57; Wootton, 'Accuracy and Galileo', 51–4.
5. *OG* xviii 304, 305, 310, 339, 366, 369.
6. *OG* xviii 40–2.
7. *OG* xvi 476.
8. Karen Liebreich, *Fallen Order: A History* (London: Atlantic Books, 2004), 152–7; Spini, *Galileo, Campanella*, 66–8; Leodegario Picanyol, *Le scuole pie e Galileo Galilei* (Rome: PP. Scolopi di S. Pantaleo, 1942); Michele Cioni, *I documenti galileiani del S. Uffizio di Firenze* [1908] (Florence: Giampiero Pagnini, 1996), 55–60.
9. *OG* xvii 399–400, xviii 372.
10. *OG* xvi 76.
11. Michelini's *Treatise on the Management of Rivers* (1664) is not regarded as a pioneering study: see Cesare S. Maffioli, *Out of Galileo: The Science of Waters, 1628–1718* (Rotterdam: Erasmus Publishing, 1994), 98–104. Note that Michelini held the title of philosopher and mathematician to the grand duke – contrary to Mario Biagioli, 'Scientific Revolution, Social Bricolage, and Etiquette', in Roy Porter and Miluláš Teich, eds, *The Scientific Revolution in National Context* (Cambridge: Cambridge University Press, 1992), 11–54, at 22, who states that no one obtained this title after Galileo.
12. *OG* xvi 147.
13. *OG* xviii 130.
14. *OG* xvii 407.
15. Spini, writing in 1996 (*Galileo, Campanella*, 66–7), described Sozzi's accusation as 'un documento molto importante', and was puzzled that more attention has not been paid to it. Yet it continues to be ignored: no mention in Redondi's 'Vent'anni dopo' (2004) or in what is now the standard biography of Galileo, Camerota's *Galileo Galilei*. A partial explanation is provided by the fact that Favaro's lengthy and extremely hostile review of Cioni's *Documenti galileiani* (1909, reprinted in the new edition of the *Documenti*) made absolutely no mention of Sozzi's accusations and effectively discouraged scholars from using Cioni's volume.
16. Balthasar de Monconys, *Les voyages*, ed. Charles Henry (Paris: A. Hermann, 1887), 36.
17. Lorenzo Magalotti, *Delle lettere familiari* (2 vols, Florence: SAR, 1769), i, 49–51.
18. [Angelo Fabroni], 'La vita del conte Lorenzo Magalotti', in Magalotti, *Lettere*, i, xi–lxi, at xxi, xxvi–xxvii, xxxiv–xl, xlix; Anna Maria Crinò, 'An Unpublished Letter on the Theme of Religion from Count Lorenzo Magalotti to the Honourable Robert Boyle in 1672', *Journal of the Warburg and Courtauld Institutes* 45 (1982), 271–8.
19. Magalotti, *Lettere*, i, 51–4.
20. *GA* 60–7; *OG* v 297–305.
21. See the notes and commentary in Galileo Galilei, *Scienza e religione: Scritti copernicani*, ed. Massimo Bucciantini and Michele Camerota (Rome: Donzelli, 2009), 17–32.
22. For a snapshot of Plato's theology, see William C. Greene, ' "God" in Plato's Theology', *Classical Weekly* 35 (1942), 220.
23. *OG* xviii 56–8. There is still a certain pious excitement in Castelli's next letter (62–6).
24. *OG* xiii 40.
25. *OG* xviii 81, 126, 130, 131; for an earlier example, see xvii 271.

26. Note his excitement in July 1638 because Galileo had written to him saying, 'If it pleases God then it must please us' (*OG* xvii 355, 361). Why the excitement, if Galileo was a good Christian? Clearly this was an unexpected sentiment.

27. *OG* x 169–71.

28. J. M. Rist, *Epicurus: An Introduction* (Cambridge: Cambridge University Press, 1972), 140–63.

29. *OG* iii 340.

30. Thomas Kjeller Johansen, *Plato's Natural Philosophy* (Cambridge: Cambridge University Press, 2004), 4.

31. Giorgio Vasari, *The Lives of the Artists*, trans. Julia Conway Bondanella and Peter Bondanella (Oxford: Oxford University Press, 1991), 226.

32. Sarpi, *Lettere ai gallicani*, 133.

33. Giorgio Spini, *Ricerca dei libertini* (2nd rev. edn, Florence: La Nuova Italia, 1983), 161–6; Antonio Rocco, *L'Alcibiade fanciullo a scola* (Rome: Salerno Editrice, [1988]).

34. I have used Melchior Inchofer, *La monarchie des solipses* (Amsterdam: Herman Uytwerf, 1722). On the question of authorship, see Cerbu, 'Melchior Inchofer', 600–11.

35. My account of Galileo's irreligion goes beyond previous arguments. The orthodox line, maintained by Favaro and most modern scholars, is that Galileo was a good Catholic: see, for example, Pedersen, 'Galileo's Religion'. (It is hard to imagine what Pedersen was thinking when he stated that 'no Catholic questioned his orthodoxy or loyalty to the Church until 1633' (92), since, even on the basis of the information available when his paper was given in 1984, he certainly knew of Caccini's and Lorini's denunciations. 'They let him receive the sacraments' (99) is surely a mistranslation of Simone Contarini's comment of February 1616 'il fanno frequentar i Sagramenti' (*OG* xx 570): rather it would seem that Contarini assumed Galileo would frequent the sacraments only under compulsion. See above, p. 185, for Galileo's claim to have taken a vow to rescue a young woman from moral danger – this is not, as Pederson believes (98), an indication of piety.) It was in the face of this consensus that *GH* provoked such an uproar, but it should be emphasised that Redondi has always regarded Galileo as a Christian atomist: see Pietro Redondi, 'From Galileo to Augustine', in Machamer, ed., *Cambridge Companion*, 175–210. Nuanced accounts are to be found in Giorgio Spini, 'The Rationale of Galileo's Religiousness', in Carlo L. Golino, ed., *Galileo Reappraised* (Berkeley: University of California Press, 1966), 44–66, and Spini, 'Postilla Galileiana', in Spini, *Ricerca dei libertini*, 389–92; Isabelle Pantin, ' "Dissiper les ténèbres qui restent encore à percer": Galilée, l'église conquérante et la république des philosophes', in Alain Mothu, ed., *Révolution scientifique et libertinage* (Turnhout: Brepols, 2000), 11–34. On unbelief in the world of Galileo, see Muir, *Culture Wars*; Nicholas Davidson, 'Unbelief and Atheism in Italy: 1500–1700', in Michael Hunter and David Wootton, eds, *Atheism from the Reformation to the Enlightenment* (Oxford: Oxford University Press, 1992), 55–86; and two studies of Sarpi: Wootton, *Paolo Sarpi*, and Vittorio Frajese, *Sarpi scettico: Stato e chiesa a Venezia tra Cinque e Seicento* (Bologna: Il Mulino, 1994). It should perhaps be stressed that the claim that Galileo was not a Christian is entirely compatible with the claim that the theology of the *Letter to the Grand Duchess Christina* is orthodox (Carroll, 'Galileo Galilei and the Myth of Heterodoxy', 123–43).

36. See, for example, Stephen Gaukroger, *The Emergence of a Scientific Culture* (Oxford: Oxford University Press, 2006), 3.

CHAPTER 37: THE COSMOGRAPHY OF THE SELF

1. He also invented binoculars: *OG* xiii 372.

2. One can see Galileo trying to find the language for this distinction at *OG* v 234–5, echoed in Milton, *Paradise Lost* (rev. edn, London: S. Simmons, 1674) bk viii, ll. 15–38.

3. *Oxford English Dictionary*; *EDG* 330. On the origin and significance of Galileo's phrase, see Michel-Pierre Lerner, *Le monde des sphères* (2nd edn, 2 vols, Paris: Les Belles Lettres, 2008), ii, 195–217.

4. *Dialogue* 367–71; *OG* vii 394–8.
5. *OG* xviii 106.
6. *OG* xvii 15–16, xiii 276, xvi 524.
7. *OG* xii 391, xvii 276: a reference to Galileo's 'profonda immaginatione'.
8. *Dialogue* 328, 334–5, 339; *OG* vii 355, 362–3, 367.
9. *OG* xvi 162.
10. *OG* xii 243.
11. Translation from Liebreich, *Fallen Order*, 106; *OG* xvii 247.
12. *Dialogue* 102–4; *OG* vii 128–30, xix 327.
13. *OG* x 212–13.
14. *OG* x 210–12.
15. *OG* xvi 346–7.
16. John Wedderburn reports the use of the telescope as a microscope in 1609: *OG* iii 164.
17. *OG* xix 563. Galileo died leaving a substantial debt to the heirs of the portrait painter Tiberio Titi (d. 1637), who was the official portrait painter to the Medici family: *OG* xix 567. Since no portrait of Galileo by Titi survives, we can surmise that Titi had been commissioned to produce the portraits of his friends, copying them from originals by lesser artists, in order to produce a set uniform in size and style. For a discussion of the importance of Galileo's artistic taste for an understanding of his intellectual life, see Erwin Panofsky, 'Galileo as a Critic of the Arts', *Isis* 47 (1956), 3–15.
18. *OG* xiv 70.
19. *OG* xv 186.
20. *OG* xiii 400, 426–7.
21. *OG* v 235.
22. *OG* xii 253, 240; cf. John Donne, *Ignatius His Conclave* (London: Richard More, 1611), 13–14.
23. See, for example, *OG* xii 360.
24. There is a debate as to whether Milton's claim to have met Galileo is true: See George F. Butler, 'Milton's Meeting with Galileo: A Reconsideration', *Milton Quarterly* 39 (2005), 132–9. The crux of the matter is that there is no contemporary evidence for such a meeting. Since a meeting between Galileo, who was a prisoner to the Inquisition under house arrest, and a Protestant would have been suspect, it is not surprising that there is no contemporary evidence.
25. Milton, *Paradise Lost*, bk viii, ll. 15–27.

CODA: GALILEO, HISTORY AND THE HISTORIANS

1. *OG* xix 398–9. Paolo Galluzzi, 'The Sepulchers of Galileo: The "Living" Remains of a Hero of Science', in Machamer, ed., *Cambridge Companion*, 417–47.
2. Maurice A. Finocchiaro, *Retrying Galileo: 1633–1992* (Berkeley: University of California Press, 2007). Douai may have been typical in teaching heliocentrism from 1755: see Michael Sharratt, 'Galileo's "Rehabilitation" ', in *C&G* 323–39, at 328.
3. The title, we may note, was altered by the publisher (*OG* xvii 370), so the implicit reference to Tartaglia was not Galileo's. Galileo found the title plebeian, where his original, now lost, was noble. On Tartaglia and his influence on Galileo, see Mary J. Henninger-Voss, 'How the "New Science" of Cannons Shook Up the Aristotelian Cosmos', *Journal of the History of Ideas* 63 (2002), 371–97.
4. *OG* xiii 420, 427, xiv 102; see also 372.
5. Maurice A. Finocchiaro, 'Aspects of the Controversy about Galileo's Trial', in *LCF* 491–511, at 501–2. Galileo tried to measure stellar parallax and failed: see Graney, 'But Still, It Moves'. Graney is broadly right in thinking that Galileo 'would have assumed he was seeing the full disk of a star, much as he saw the full disk of Mars or Jupiter' (265): see *Comets* 325–6. Galileo had already tried and failed to measure stellar parallax with regard to the new star of 1604, however, so this new failure will not have been unexpected. Attempts to measure changes in the altitude

of the pole star (*OG* xii 253–4) also amounted to an attempt to measure stellar parallax. On eastward deviation, Domenico Bertoloni Meli, 'St Peter and the Rotation of the Earth: The Problem of Fall Around 1800', in P.M. Harman and Alan E. Shapiro, eds, *The Investigation of Difficult Things* (Cambridge: Cambridge University Press, 1992), 421–47.

6. *OG* v 195, xii 34.

7. So too the supporters of Tycho accepted that their argument against Copernicanism was bound to be from physics rather than astronomy: *OG* xvii 363.

8. Favaro, *Adversaria galilæiana*, 13–18.

9. For a sharply different view, see Thomas S. Kuhn, 'Mathematical versus Experimental Traditions in the Development of Physical Science' [1976], in Kuhn, *The Essential Tension* (Chicago: University of Chicago Press, 1977), 31–65. On Descartes as an experimental scientist, see Jed Z. Buchwald, 'Descartes's Experimental Journey past the Prism and through the Invisible World to the Rainbow', *Annals of Science* 65 (2008), 1–46.

10. *OG* xiii 419.

11. The Redondi thesis has been much debated: see, for example, Vincenzo Ferrone and Massimo Firpo, 'From Inquisitors to Microhistorians: A Critique of Pietro Redondi's *Galileo eretico*', *Journal of Modern History* 58 (1986), 485–524. The only author I know to adopt it wholesale is Michael White, *Galileo Antichrist* (London: Weidenfeld and Nicolson, 2007).

12. Descartes claimed his own atomism was entirely compatible with transubstantiation: *OG* xviii 318; Peter Dear, 'The Church and the New Philosophy', in Pumfrey, Rossi and Slawinski, eds, *Science, Culture, and Popular Belief*, 119–39, at 132–3. But this did not prevent his being condemned by the Inquisition in 1663: Jean-Robert Armogathe and Vincent Carraud, 'La première condamnation des Œuvres de Descartes, d'après des documents inédits aux Archives du Saint-Office', *Nouvelles de la république des lettres* 2001/II, 103–37.

13. See, for example, *OG* xvi 215–16.

14. There are passing references to attendance at Mass (*OG* xvi 163), to an invitation to attend a Church service to do business there (xiv 107), and to a religious book sent to him by a friend, though in lieu of a secular book that Galileo had requested (xvi 103, 175).

15. *OG* xi 150.

16. For a critique of Biagioli's book, see Michael H. Shank, 'How Shall We Practice History: The Case of Biagioli's *Galileo Courtier*', *Early Science and Medicine* 1 (1996), 106–50. Biagioli's book develops a line of argument well expressed in Westfall, 'Science and Patronage'. Both assume that Galileo was pursuing status, and that status was defined by patrons. The alternative view is that Galileo was pursuing – rather clumsily and awkwardly to be sure – truth. See, for example, Cardi's advice to him to write the truth without trying to ingratiate himself and without being afraid of the consequences: *OG* xi 502. It is also worth remarking that Galileo was not very good at performing the part of a citizen of the republic of letters – one could not write a book about Galileo's correspondence comparable to Lisa Jardine's *Erasmus, Man of Letters* (Princeton: Princeton University Press, 1994). Galileo's place in the republic of letters depended on others acting on his behalf: see Torrini, 'Galileo e la repubblica degli scienziati'; Stéphane Garcia, 'Élie Diodati – Galilée: La rencontre de deux logiques', in *LCF* 883–92; and *EDG* 259–65.

17. For a negative assessment of Brecht's play, see William R. Shea and Mariano Artigas, *Galileo Observed* (Sagamore Beach: Science History Publications, 2006), 53–84.

18. Hugh Rorrison, 'Commentary', in Bertolt Brecht, *Life of Galileo*, ed. Hugh Rorrison (London: Methuen, 2006), xx–xli, at xx.

19. Julien Offray de La Mettrie, *L'homme machine*, ed. Aram Vartanian (Princeton: Princeton University Press, 1960), 11, 12.

Bibliography

Adler, Carl G., and Byron L. Coulter, 'Galileo and the Tower of Pisa Experiment', *American Journal of Physics* 46 (1978), 199–201.

Alexander, Amir, 'Lunar Maps and Coastal Outlines: Thomas Hariot's Mapping of the Moon', *SHPS* 29 (1998), 345–68.

Ariew, Roger, 'The Initial Response to Galileo's Lunar Observations', *SHPS* 32 (2001), 571–81 and 749.

Ariotti, Piero, 'From the Top to the Foot of the Mast on a Moving Ship', *Annals of Science* 28 (1972), 191–203.

Armogathe, Jean-Robert and Vincent Carraud, 'La première condamnation des Œuvres de Descartes, d'après des documents inédits aux Archives du Saint-Office', *Nouvelles de la république des lettres* 2001/II, 103–37.

Artigas, Mariano, Rafael Martínez and William R. Shea, 'New Light on the Galileo Affair', in *C&G* 213–33.

Bacon, Francis, *The Philosophical Works of Francis Bacon*, ed. James Spedding (5 vols, London: Longman, 1861).

Bald, R. C., *John Donne: A Life* (Oxford: Clarendon Press, 1970).

Baldini, Ugo, 'The Development of Jesuit "Physics" in Italy, 1550–1700: A Structural Approach', in Constance Blackwell and Sachiko Kusukawa, eds, *Philosophy in the Sixteenth and Seventeenth Centuries: Conversations with Aristotle* (Farnham: Ashgate, 1999), 248–79.

—— *Legem impone subactis: Studi su filosofia e scienza dei Gesuiti in Italia, 1540–1632* (Rome: Bulzoni, 1992).

Baroncini, Gabriele, 'Sulla galileiana *esperienza sensata*', *Studi secenteschi* 25 (1984), 147–72.

Barra, Mario, 'Galileo Galilei e la probabilità', in *LCF* 101–18.

Bedini, Silvio A. *The Pulse of Time: Galileo Galilei, the Determination of Longitude, and the Pendulum Clock* (Florence: Olschki, 1991).

Bellinati, Claudio, 'Integrazioni e correzioni alle pubblicazioni di Antonio Favaro, studioso di Galileo Galilei', in Santinello, ed., *Galileo e la cultura padovana*, 275–90.

Beretta, Francesco, 'The Documents of Galileo's Trial: Recent Hypotheses and Historical Criticism', in *C&G* 191–212.

—— 'Galileo, Urban VIII, and the Prosecution of Natural Philosophers', in *C&G* 234–61.

—— 'Un nuovo documento sul processo di Galileo Galilei: La lettera di Vincenzo Maculano del 22 aprile 1633', *Nuncius* 16 (2001), 629–41.

—— 'Le procès de Galilée et les Archives du Saint-Office', *Revue des sciences philosophiques et théologiques* 83 (1999), 441–90.

—— 'Urbain VIII Barberini protagoniste de la condamnation de Galilée', in *LCF* 549–73.

Bertoloni Meli, Domenico, 'St Peter and the Rotation of the Earth: The Problem of Fall Around 1800', in P.M. Harman and Alan E. Shapiro, eds, *The Investigation of Difficult Things* (Cambridge: Cambridge University Press, 1992), 421–47.

—— *Thinking with Objects: The Transformation of Mechanics in the Seventeenth Century* (Baltimore: Johns Hopkins University Press, 2006).

Besomi, Ottavio, and Michele Camerota, *Galileo e il Parnaso Tychonico* (Florence: Olschki, 2000).

Biagioli, Mario, *Galileo Courtier: The Practice of Science in the Culture of Absolutism* (Chicago: University of Chicago Press, 1993).

—— *Galileo's Instruments of Credit: Telescopes, Images, Secrecy* (Chicago: University of Chicago Press, 2006).

—— 'Scientific Revolution, Social Bricolage, and Etiquette', in Roy Porter and Mikuláš Teich, eds, *The Scientific Revolution in National Context* (Cambridge: Cambridge University Press, 1992), 11–54.

Bianchi, Luca, 'Agostino Oreggi, qualificatore del *Dialogo*, e i limiti della conoscenza scientifica', in *LCF* 575–84.

Black, Christopher F., *The Italian Inquisition* (New Haven, Conn.: Yale University Press, 2009).

Blackwell, Richard J., *Behind the Scenes at Galileo's Trial: Including the First English Translation of Melchior Inchofer's 'Tractatus syllepticus'* (Notre Dame: University of Notre Dame Press, 2006).

—— *Galileo, Bellarmine, and the Bible: Including a Translation of Foscarini's 'Letter on the Motion of the Earth'* (Notre Dame: University of Notre Dame Press, 1991).

Bordiga, Giovanni, *Giovanni Battista Benedetti* [1926] (Venice: Istituto veneto di scienze, lettere ed arti, 1985).

Brecht, Bertolt, *Life of Galileo*, ed. Hugh Rorrison (London: Methuen, 2006).

Bredekamp, Horst, 'Gazing Hands and Blind Spots: Galileo as Draftsman', *Science in Context* 13 (2000), 423–62.

Bucciantini, Massimo, *Contro Galileo: Alle origini dell'affaire* (Florence: Olschki, 1995).

—— *Galileo e Keplero* (Turin: Einaudi, 2007).

—— 'Novità celesti e teologia', in *LCF* 795–808.

Buchwald, Jed Z., 'Descartes's Experimental Journey past the Prism and through the Invisible World to the Rainbow', *Annals of Science* 65 (2008), 1–46.

Butler, George F., 'Milton's Meeting with Galileo: A Reconsideration', *Milton Quarterly* 39 (2005), 132–9.

Büttner, Jochen, Peter Damerow and Jürgen Renn, 'Traces of an Invisible Giant: Shared Knowledge in Galileo's Unpublished Treatises', in *LCF* 183–201.

Camerota, Michele, 'Borro, Girolamo', in *New Dictionary of Scientific Biography* (New York: Scribner, 2007).

—— 'Buonamici, Francesco', in *New Dictionary of Scientific Biography* (New York: Scribner, 2007).

—— *Galileo Galilei e la cultura scientifica nell'età della controriforma* (Rome: Salerno Editrice, 2004).

—— 'Galileo, Lucrezio, e l'atomismo', in M. Beretta and F. Citti, eds, *Lucrezio, la natura e la scienza* (Florence: Olschki, 2008), 141–75.

Camerota, Michele, and Mario Helbing, 'Galileo and Pisan Aristotelianism: Galileo's *De motu antiquiora* and the *Quaestiones de motu elementorum* of the Pisan Professors', *Early Science and Medicine* 5 (2000), 319–65.

Camiz, Francha Trinchieri, 'The Roman "Studio" of Francesco Villamena', *Burlington Magazine* 136 (1994), 506–16.

Campanella, Tommaso, *Apologia pro Galileo*, ed. Michel-Pierre Lerner (Pisa: Scuola normale superiore, 2006).

—— *A Defense of Galileo*, trans. Richard J. Blackwell (Notre Dame: University of Notre Dame Press, 1994).

—— *Opuscoli astrologici*, ed. Germana Ernst (Milan: Rizzoli, 2003).

Carroll, William E., 'Galileo Galilei and the Myth of Heterodoxy', in John Brooke and Ian Maclean, eds, *Heterodoxy in Early Modern Science and Religion* (Oxford: Oxford University Press, 2005), 115–44.

Carugo, Adriano, and Alistair C. Crombie, 'The Jesuits and Galileo's Ideas of Science and Nature', *Annali dell'Istituto e Museo di storia della scienza di Firenze* 8 (1983), 3–68.

Castagnetti, Giuseppe, and Michele Camerota, 'Antonio Favaro and the *Edizione Nazionale* of Galileo's Works', *Science in Context* 13 (2000), 357–62.

—— 'Raffaello Caverni and His *History of the Experimental Method in Italy*', *Science in Context* 13 (2000), 327–39.

Cerbu, Thomas, 'Melchior Inchofer, "un homme fin et rusé" ', in *LCF* 587–611.

Cesi, Federico, 'Del natural desiderio di sapere', in M. L. Altieri Biagi and B. Basile, eds, *Scienziati del Seicento* (Milan: Rizzoli Editore, 1969), 53–92.

Chalmers, Alan, 'Galileo's Telescopic Observations of Venus and Mars', *British Journal for the Philosophy of Science* 36 (1985), 175–84.

Chalmers, Alan, and Richard Nicholas, 'Galileo on the Dissipative Effect of a Rotating Earth', *SHPS* 14 (1983), 315–40.

Chua, Daniel, 'Vincenzo Galilei, Modernity and the Division of Nature', in Suzanna Clark and Alexander Rehding, eds, *Music Theory and Natural Order from the Renaissance to the Early Twentieth Century* (Cambridge: Cambridge University Press, 2006), 17–29.

Cioni, Michele, *I documenti galileiani del S. Uffizio di Firenze* [1908] (Florence: Giampiero Pagnini, 1996).

Clark, Stuart, *Vanities of the Eye: Vision in Early Modern European Culture* (Oxford: Oxford University Press, 2007).

Clavelin, Maurice, 'Galilée astronome philosophe', in *LCF* 19–40.

—— *Galilée copernicien* (Paris: Albin Michel, 2004).

—— *The Natural Philosophy of Galileo: Essay on the Origins and Formation of Classical Mechanics* (Cambridge, Mass.: MIT Press, 1974).

Conti, Lino, 'Francesco Stelluti, il copernicanesimo dei Lincei e la teoria galileiana delle maree', in Carlo Vinti, ed., *Galileo e Copernico* (Assisi: Porziuncola, 1990), 141–236.

Cooper, Lane, *Aristotle, Galileo and the Tower of Pisa* (Ithaca: Cornell University Press, 1935).

Copernicus, Nicolaus, *On the Revolutions*, trans. E. Rosen (Baltimore: Johns Hopkins University Press, 1992).

Coulter, Byron L., and Carl G. Adler, 'Can a Body Pass a Body Falling through the Air?' *American Journal of Physics* 47 (1979), 841–6.

Cozzi, Gaetano, 'Galileo Galilei, Paolo Sarpi, e la società veneziana', in Cozzi, *Paolo Sarpi tra Venezia e l'Europa* (Turin: Einaudi, 1979), 135–234.

Cremonini, Cesare, *Le orazioni*, ed. Antonino Poppi (Padua: Antenore, 1998).

Crinò, Anna Maria, 'An Unpublished Letter on the Theme of Religion from Count Lorenzo Magalotti to the Honourable Robert Boyle in 1672', *Journal of the Warburg and Courtauld Institutes* 45 (1982), 271–8.

Crombie, Alistair C., *Styles of Scientific Reasoning* (3 vols, London: Duckworth, 1994).

Crombie, Alistair C., ed., *Scientific Change* (London: Heinemann, 1963).

Crombie, Alistair C., and Adriano Carugo, review of Wallace, *Times Literary Supplement*, 22 November 1985, 1319–20.

Damerow, Peter, Gideon Freudenthal, Peter McLaughlin and Jürgen Renn, *Exploring the Limits of Preclassical Mechanics: A Study of Conceptual Development in Early Modern Science: Free Fall and Compounded Motion in the Work of Descartes, Galileo, and Beeckman* (2nd rev. edn, New York: Springer, 2004).

d'Anania, Giovanni Lorenzo, *L'universale fabrica del mondo* (Venice: Aniello san Vito, 1576).

Daston, Lorraine, 'Baconian Facts, Academic Civility, and the Prehistory of Objectivity', in Allan Megill, ed., *Rethinking Objectivity* (Durham, N.C.: Duke University Press, 1994), 37–63.

Davidson, Nicholas, 'Unbelief and Atheism in Italy: 1500–1700', in Michael Hunter and David Wootton, eds, *Atheism from the Reformation to the Enlightenment* (Oxford: Oxford University Press, 1992), 55–86.

Dear, Peter, 'The Church and the New Philosophy', in S. Pumfrey, P. L. Rossi and M. Slawinski, eds, *Science, Culture, and Popular Belief in Renaissance Europe* (Manchester: Manchester University Press, 1991), 119–39.

—— 'Jesuit Mathematical Science and the Reconstitution of Experience in the Early Seventeenth Century', *SHPS* 18 (1987), 133–75.

—— 'Narratives, Anecdotes, and Experiments: Turning Experiences into Science in the Seventeenth Century', in Dear, ed., *Literary Structure of Scientific Argument* (Philadelphia: University of Pennsylvania Press, 1991), 135–63.

delle Colombe, Lodovico, *Risposte piacevoli* (Florence: Giovanni Antonio Caneo, 1608).

Descartes, René, *Les principes de la philosophie* (Paris: Michel Bobin, 1668).

de Vivo, Filippo, *Information and Communication in Venice: Rethinking Early Modern Politics* (Oxford: Oxford University Press, 2007).

de Waard, C., *L'expérience barométrique, ses antécédents et ses explications* (Thouars: Imprimerie Nouvelle, 1936).

Dinis, Alfredo, 'Giovanni Battista Riccioli and the Science of His Time', in Feingold, ed., *Jesuit Science*, 195–224.

Dollo, Corrado, *Galileo Galilei e la cultura della tradizione* (Soveria Mannelli: Rubbettino, 2003).

Donne, John, *The Epithalamions, Anniversaries and Epicedes*, ed. W. Milgate (Oxford: Clarendon Press, 1978).

—— *Ignatius His Conclave* (London: Richard More, 1611).

Dooley, Brendan, *Morandi's Last Prophecy and the End of Renaissance Politics* (Princeton: Princeton University Press, 2002).

Ducheyne, Steffen, 'Galileo's Interventionist Notion of "Cause" ', *Journal of the History of Ideas* 67 (2006), 443–64.

Duplessis-Mornay, Philippe, *Mémoires et correspondance* (12 vols, Paris: Treuttel et Würz, 1824–5).

Drake, Stillman, *Essays on Galileo and the History and Philosophy of Science* (3 vols, Toronto: University of Toronto Press, 1999).

—— *Galileo at Work: His Scientific Biography* (Chicago: University of Chicago Press, 1978).

Drake, Stillman, and Charles Donald O'Malley, eds, *The Controversy on the Comets of 1618* (Philadelphia: University of Pennsylvania Press, 1961).

Drake, Stillman, and I. E. Drabkin, eds, *Mechanics in Sixteenth-Century Italy: Selections from Tartaglia, Benedetti, Guido Ubaldo and Galileo* (Madison: University of Wisconsin Press, 1969).

Dupré, Sven, 'Ausonio's Mirrors and Galileo's Lenses: The Telescope and Sixteenth-Century Practical Optical Knowledge', *Galilæana* 2 (2005), 145–180.

Erlichson, Herman, 'Galileo and High Tower Experiments', *Centaurus* 36 (1993), 33–45.

Ernst, Germana, 'Astrology, Religion, and Politics in Counter-Reformation Rome', in S. Pumfrey, P. L. Rossi and M. Slawinski, eds, *Science, Culture, and Popular Belief in Renaissance Europe* (Manchester: Manchester University Press, 1991), 249–73.

Ernst, Germana, and Eugenio Canone, 'Una lettera ritrovata: Campanella a Peiresc, 19 giugno 1636', *Rivista di storia della filosofia* 49 (1994), 353–66.

Fahie, John J., *Memorials of Galileo Galilei, 1564–1642: Portraits and Paintings, Medals and Medallions, Busts and Statues, Monuments, and Mural Inscriptions* (Leamington: privately printed, 1929).

Fantoli, Annibale, 'The Disputed Injunction and Its Role in Galileo's Trial', in *C&G* 117–49.

—— *Galileo: For Copernicanism and for the Church* (3rd rev. edn, Vatican City: Vatican Observatory Publications, 2003).

Favaro, Antonio, *Adversaria galilæiana*, ed. Lucia Rossetti and Maria Laura Soppelsa (Trieste: Lint, 1992).

—— *Amici e corrispondenti di Galileo* (3 vols, Florence: Salimbeni, 1983).

—— 'Documenti inediti per la storia dei manoscritti galileiani', *Bullettino di bibliografia e di storia delle scienze matematiche e fisiche* 18 (1885), 1–112.

—— *Galileo Galilei a Padova* (Padua: Antenore, 1968).

—— *Galileo Galilei e lo studio di Padova* [1883] (2 vols, Padua: Antenore, 1966).

—— 'La libreria di Galileo Galilei descritta ed illustrata', *Bullettino di bibliografia e di storia delle scienze matematiche e fisiche* 19 (1886), 219–93.

—— *Miscellanea galileiana inedita* (Venice: Giuseppe Antonelli, 1887).

—— *Scampoli galileiani*, ed. Lucia Rossetti and Maria Laura Soppelsa (2 vols, Trieste: Lint, 1992).

Favino, Federica, 'A proposito dell'atomismo di Galileo: Da una lettera di Tommaso Campanella ad uno scritto di Giovanni Ciampoli', *Bruniana e campanelliana* 3 (1997), 265–82.

—— 'Un caso di censura postuma: La "filosofia naturale" di Giovanni Ciampoli', in *I primi Lincei e il Sant'Uffizio* (Rome: Bardi, 2005), 141–56.

—— ' "Quel petardo di mia fortuna": Reconsiderando la "caduta" di Giovan Battista Ciampoli', in *LCF* 863–82.

—— 'Scetticismo ed empirismo: Ciampoli Linceo', in Andrea Battistini, Gilberto de Angelis and Giuseppe Olmi, eds, *All'origine della scienza moderna: Federico Cesi e l'Accademia dei Lincei* (Bologna: Il Mulino, 2007), 175–202.

Febvre, Lucien, *The Problem of Unbelief in the Sixteenth Century* [1942] (Cambridge, Mass.: Harvard University Press, 1982).

Feingold, Mordechai, 'The Grounds for Conflict: Grienberger, Grassi, Galileo, and Posterity', in Feingold, ed., *The New Science and Jesuit Science: Seventeenth-Century Perspectives* (Dordrecht: Kluwer, 2003), 121–57.

—— 'Jesuits: Savants', in Feingold, ed., *The New Science and Jesuit Science*, 1–45.

Feingold, Mordechai, ed., *Jesuit Science and the Republic of Letters* (Cambridge, Mass.: MIT Press, 2003).

Feldhay, Rivka, *Galileo and the Church: Political Inquisition or Critical Dialogue?* (Cambridge: Cambridge University Press, 1995).

Ferrone, Vincenzo, and Massimo Firpo, 'From Inquisitors to Microhistorians: A Critique of Pietro Redondi's *Galileo eretico*', *Journal of Modern History* 58 (1986), 485–524.

Feyerabend, Paul K., *Against Method* (London: Verso, 1975, 2nd edn 1988, 3rd edn 1993, 4th edn 2010).

—— *Farewell to Reason* (London: Verso, 1987).

—— first reply to Whitaker, *Science* 209 (1980), 544.

—— second reply to Whitaker, *Science* 211 (1981), 876–7.

Findlen, Paula, *Possessing Nature: Museums, Collecting, and Scientific Culture in Early Modern Italy* (Berkeley: University of California Press, 1994).

Finocchiaro, Maurice A., 'Aspects of the Controversy about Galileo's Trial', in *LCF* 491–511.

—— *The Galileo Affair: A Documentary History* (Berkeley: University of California Press, 1989).

—— 'Galileo's Copernicanism and the Acceptability of Guiding Assumptions', in Arthur Donovan, Larry Laudan and Rachel Laudan, eds, *Scrutinizing Science* (Baltimore: Johns Hopkins University Press, 1988), 49–67.

—— 'Physical-Mathematical Reasoning: Galileo on the Extruding Power of Terrestrial Rotation', *Synthèse* 134 (2003), 217–44.

—— *Retrying Galileo, 1633–1992* (Berkeley: University of California Press, 2007).

Fioravanti, Leonardo, *Il reggimento della peste* (Venice: heirs of Melchior Sessa, 1571).

Frajese, Vittorio, *Sarpi scettico: Stato e chiesa a Venezia tra Cinque e Seicento* (Bologna: Il Mulino, 1994).

Fredette, Raymond, 'Galileo's *De motu antiquiora*: Notes for a Reappraisal', in *LCF* 165–82.

Galilei, Galileo, *Cause, Experiment and Science: A Galilean Dialogue Incorporating a New English Translation of Galileo's 'Bodies That Stay Atop Water, or Move in It'*, trans. Stillman Drake (Chicago: University of Chicago Press, 1981).

—— *Contro il portar la toga* (Pisa: ETS, 2005).

—— *Dialogo sopra i due massimi sistemi del mondo, tolemaico e copernicano*, ed. O. Besomi and M. Helbing (2 vols, Padua: Antenore, 1998).

—— *Dialogue Concerning the Two Chief World Systems – Ptolemaic and Copernican*, trans. Stillman Drake (Berkeley: University of California Press, 1953, 2nd rev. edn 1966).

—— *Discorso delle comete*, ed. Ottavio Besomi and Mario Helbing (Rome: Antenore, 2002).

—— *Discoveries and Opinions of Galileo: Including 'The Starry Messenger' (1610), 'Letter to the Grand Duchess Christina' (1615), and Excerpts from 'Letters on Sunspots' (1613), 'The Assayer' (1623)*, trans. Stillman Drake (New York: Anchor Books, 1990).

—— *Galileo against the Philosophers in His 'Dialogue of Cecco di Ronchitti' (1605) and 'Considerations of Alimberto Mauri' (1606)*, trans. Stillman Drake (Los Angeles: Zeitlin and Ver Brugge, 1976).

—— *Galileo's Early Notebooks: The Physical Questions*, trans. William A. Wallace (Notre Dame: University of Notre Dame Press, 1977).

—— *Galileo's Logical Treatises*, trans. William A. Wallace (Dordrecht: Kluwer Academic Publishers, 1992).

—— *Lettera a Cristina di Lorena*, ed. Mauro Pesce (Genoa: Marietti, 2000).

—— A Long-Lost Letter from Galileo to Peiresc on a Magnetic Clock, trans. Stillman Drake (Norwalk, Conn.: Burndy Library, 1967).

—— Le meccaniche, ed. Romano Gatto (Florence: Olschki, 2002).

—— 'On Motion', and 'On Mechanics': Comprising 'De motu' (ca. 1590) and 'Le meccaniche' (ca. 1600), ed. and trans. I. E. Drabkin and Stillman Drake (Madison: University of Wisconsin Press, 1960).

—— Operations of the Geometric and Military Compass, trans. Stillman Drake (Washington, D.C.: Burndy Library, 1978).

—— Le opere di Galileo Galilei: Edizione nazionale, ed. A. Favaro (20 vols in 21, 4th edn, Florence: G. Barbèra, 1968).

—— Il saggiatore, ed. Ottavio Besomi and Mario Helbing (Rome: Antenore, 2005).

—— Scienza e religione: Scritti copernicani, ed. Massimo Bucciantini and Michele Camerota (Rome: Donzelli, 2009).

—— Sidereus nuncius (Venice: Thomas Baglioni, 1610).

—— Sidereus nuncius: Le messager céleste, ed. Isabelle Pantin (Paris: Les Belles Lettres, 1992).

—— Sidereus nuncius, or The Sidereal Messenger, ed. and trans. Albert van Helden (Chicago: University of Chicago Press, 1989).

—— 'The Sensitive Balance', in Laura Fermi and Gilberto Bernadini, Galileo and the Scientific Revolution (New York: Dover, 1961).

—— Two New Sciences, trans. Stillman Drake (Madison: University of Wisconsin Press, 1974).

Galilei, Maria Celeste, To Father: The Letters of Sister Maria Celeste to Galileo, 1623–1633, trans. Dava Sobel (London: Fourth Estate, 2001).

Galilei, Michelagnolo, Il primo libro d'intavolatura di liuto (Munich: privately printed, 1620).

Galilei, Vincenzo, Dialogue on Ancient and Modern Music, ed. Claude V. Palisca (New Haven, Conn.: Yale University Press, 2003).

Galluzzi, Paolo, 'I sepolcri di Galileo', in Lunardi and Sabbatini, eds, Una casa per memoria, 203–55.

—— 'The Sepulchers of Galileo: The "Living" Remains of a Hero of Science', in Machamer, ed., Cambridge Companion, 417–47.

Galluzzi, Paolo, and Maurizio Torrini, eds, Le opere dei discepoli di Galileo Galilei: Carteggio (2 vols, Florence: Giunti-Barbèra, 1975–84).

Garcia, Stéphane, Élie Diodati et Galilée: Naissance d'un réseau scientifique dans l'Europe du XVIIe siècle (Florence: Olschki, 2004).

—— 'Élie Diodati – Galilée: La rencontre de deux logiques', in LCF 883–92.

—— 'Galileo's Relapse: On the Publication of the Letter to the Grand Duchess Christina (1636)', in C&G 265–78.

—— 'Peiresc, Bernegger et Diodati: Cinq lettres inédites en rapport avec Galilée', Galilæana 6 (2009), 219–33.

Garrard, Mary D., Artemisia Gentileschi (Princeton: Princeton University Press, 1989).

Gaukroger, Stephen, The Emergence of a Scientific Culture (Oxford: Oxford University Press, 2006).

Geymonat, Ludovico, Galileo Galilei: A Biography and Inquiry into His Philosophy of Science (New York: McGraw-Hill, 1965).

Gilbert, Neal W., Renaissance Concepts of Method (New York: Columbia University Press, 1963).

Gilbert, William, De magnete (New York: Dover, 1958). [English translation]

Gingerich, Owen, The Book Nobody Read: Chasing the Revolutions of Nicolaus Copernicus (New York: Penguin Books, 2005).

—— 'The Curious Case of the M-L Sidereus nuncius', Galilæana 6 (2009), 141–65.

—— 'The Galileo Sunspot Controversy', Journal for the History of Astronomy 34 (2003), 77–8.

Gingerich, Owen, and Albert van Helden, 'From Occhiale to Printed Page: The Making of Galileo's Sidereus nuncius', Journal for the History of Astronomy 34 (2003), 251–67.

Goldstein, Bernard R., 'Galileo's Account of Astronomical Miracles in the Bible: A Confusion of Sources', Nuncius 5 (1990), 3–16.

Graney, Christopher M., 'But Still, It Moves: Tides, Stellar Parallax, and Galileo's Commitment to the Copernican Theory', Physics in Perspective 10 (2008), 258–68.

Greco, Vincenzo, Giuseppe Molesini and Franco Quercioli, 'Optical Tests of Galileo's Lenses', Nature 358 (1992), 101.

Greenblatt, Stephen, *Shakespearean Negotiations* (Berkeley: University of California Press, 1988).

Greene, William C., ' "God" in Plato's Theology', *Classical Weekly* 35 (1942), 220.

Grego, Peter, *The Moon and How to Observe It* (London: Springer, 2005).

Grendler, Paul F., *The Universities of the Italian Renaissance* (Baltimore: Johns Hopkins University Press, 2002).

Grillo, Francesco, *Tommaso Campanella nell'arte di Andrea Sacchi e Nicola Poussin* (Cosenza: Edizioni Pellegrini, 1979).

Guerrini, Luigi, *Galileo e la polemica anticopernicana a Firenze* (Florence: Edizioni Polistampa, 2009).

Hamou, Philippe, ' "La nature est inexorable": Pour une reconsidération de la contribution de Galilée au problème de la connaissance', *Galilæana* 5 (2008), 149–77.

Harris, Ann Sutherland, 'Letter to the Editor', *Art Bulletin* 59 (1977), 304–7.

Headley, John M., *Tommaso Campanella and the Transformation of the World* (Princeton: Princeton University Press, 1997).

Helbing, Mario Otto, 'Galileo e le *Questioni meccaniche* attribuite ad Aristotele', in *LCF* 217–36.

Henninger-Voss, Mary J., 'How the "New Science" of Cannons Shook Up the Aristotelian Cosmos', *Journal of the History of Ideas* 63 (2002), 371–97.

—— 'Working Machines and Noble Mechanics: Guidobaldo Del Monte and the Translation of Knowledge', *Isis* 91 (2000), 233–59.

Hirschman, Albert O., *Exit, Voice, and Loyalty* (Boston, Mass.: Harvard University Press, 1970).

Hobbes, Thomas, *The Correspondence of Thomas Hobbes,* ed. Noel Malcolm (2 vols, Oxford: Oxford University Press, 1994).

Hobson, Anthony, 'A Sale by Candle in 1608', *The Library* 26 (1971), 215–33.

Hofstadter, Dan, *The Earth Moves* (New York: Norton, 2009).

Huemer, Frances, 'Il dipinto di Colonia', in *Il Cannochiale e il Pennello*, ed. Lucia Tongiorgi and Alessandro Tossi (Florence: Giunti, 2009), 60–70.

Huygens, Christiaan, *The Pendulum Clock* (Ames: Iowa State University Press, 1986).

Inchofer, Melchior, *La monarchie des solipses* (Amsterdam: Herman Uytwerf, 1722).

Jardine, Lisa, *Erasmus, Man of Letters* (Princeton: Princeton University Press, 1994).

Jardine, Nicholas, 'Galileo's Road to Truth and the Demonstrative Regress', *SHPS* 7 (1976), 277–318.

Johansen, Thomas Kjeller, *Plato's Natural Philosophy* (Cambridge: Cambridge University Press, 2004).

Johns, Adrian, *The Nature of the Book* (Chicago: University of Chicago Press, 1998).

Johnson, Francis R., and Sanford V. Larkey, 'Thomas Digges, the Copernican System, and the Idea of the Infinity of the Universe in 1576', *Huntington Library Bulletin* 5 (1934), 69–117.

Jütte, Robert, *A History of the Senses* (Cambridge: Polity, 2005).

Kepler, Johannes, *Discussion avec le messager céleste*, ed. Isabelle Pantin (Paris: Les Belles Lettres, 1993).

—— *Kepler's Conversation with Galileo's Sidereal Messenger*, trans. Edward Rosen (New York: Johnson Reprint Corporation, 1965).

Kelter, Irving A., 'The Refusal to Accommodate: Jesuit Exegetes and the Copernican System', in *C&G* 38–53.

Koestler, Arthur, *The Sleepwalkers* (London: Hutchinson, 1959).

Koyré, Alexandre, 'Galilée et l'expérience de Pise. À propos d'une légende', *Annales de l'Université de Paris* 12 (1937), 441–53.

—— *Galileo Studies* (Atlantic Highlands: Humanities Press, 1978).

—— *Metaphysics and Measurement* (London: Chapman Hall, 1968).

Kuhn, Thomas S., *The Copernican Revolution* (Cambridge, Mass.: Harvard University Press, 1957).

—— *The Essential Tension: Selected Studies in Scientific Tradition and Change* (Chicago: University of Chicago Press, 1977).

Laird, W. Roy, 'Archimedes among the Humanists', *Isis* 82 (1991), 628–38.

—— 'Giuseppe Moletti's *Dialogue on Mechanics* (1576)', *Renaissance Quarterly* 40 (1987), 209–23.

—— 'Renaissance Mechanics and the New Science of Motion', in *LCF* 255–67.

La Mettrie, Julien Offray de, *L'homme machine*, ed. Aram Vartanian (Princeton: Princeton University Press, 1960).

Lattis, James M., *Between Copernicus and Galileo: Christoph Clavius and the Collapse of Ptolemaic Cosmology* (Chicago: University of Chicago Press, 1994).

Laven, Mary, *Virgins of Venice* (London: Penguin Books, 2003).

Lechner, George S., 'Letter to the Editor', *Art Bulletin* 59 (1977), 307–9.

—— 'Tommaso Campanella and Andrea Sacchi's Fresco of Divina Sapienza in the Palazzo Barberini', *Art Bulletin* 58 (1976), 97–108.

Lerner, Michel-Pierre, *Le monde des sphères* (2nd edn, 2 vols, Paris: Les Belles Lettres, 2008).

Liebreich, Karen, *Fallen Order: A History* (London: Atlantic Books, 2004).

Lindberg, David C., 'Alhazen's Theory of Vision and Its Reception in the West', *Isis* 58 (1967), 321–41.

Lindberg, David C., and Nicholas H. Steneck, 'The Sense of Vision and the Origins of Modern Science', in Alan G. Debus, ed., *Science, Medicine and Society in the Renaissance* (2 vols, London: Heinemann, 1972), i, 29–45.

Lunardi, Roberto, and Oretta Sabbatini, eds, *Una casa per memoria*, vol. iii of *Il rimembrar delle passate cose* (Florence: Edizioni Polistampa, 2009).

Machamer, Peter K., 'Feyerabend and Galileo: The Interaction of Theories and the Reinterpretation of Experience', *SHPS* 4 (1973), 1–46.

—— 'Galileo's Machines, His Mathematics, and His Experiments', in Machamer, ed., *Cambridge Companion*, 53–79.

Machamer, Peter K., ed., *The Cambridge Companion to Galileo* (Cambridge: Cambridge University Press, 1998).

MacLachlan, James, 'A Test of an "Imaginary" Experiment of Galileo's', *Isis* 64 (1973), 374–9.

McMullin, Ernan, 'Galilean Idealization', *SHPS* 16 (1985), 247–73.

—— 'Galileo's Theological Venture', in *C&G* 88–116.

McMullin, Ernan, ed., *Galileo, Man of Science* (New York: Basic Books, 1967).

—— *The Church and Galileo* (Notre Dame: University of Notre Dame Press, 2005).

McTighe, Thomas P., 'Galileo's "Platonism": A Reconsideration', in McMullin, ed., *Galileo, Man of Science*, 365–87.

Maffioli, Cesare S., *Out of Galileo: The Science of Waters, 1628–1718* (Rotterdam: Erasmus Publishing, 1994).

Magalotti, Lorenzo, *Delle lettere familiari* (2 vols, Florence: SAR, 1769).

Maschietto, Ludovico, 'Girolamo Spinelli e Benedetto Castelli', in Santinello, ed., *Galileo e la cultura padovana*, 453–67.

Massa, Daniel, 'Giordano Bruno and the Top-Sail Experiment', *Annals of Science* 30 (1973), 201–11.

Mayer, Thomas F., 'The Roman Inquisition's Precept to Galileo (1616)', *British Journal for the History of Science* (online pre-print, 2009).

Mersenne, Marin, *Les nouvelles pensées de Galilée*, ed. Pierre Costabel and Michel-Pierre Lerner (2 vols, Paris: Vrin, 1973).

Middleton, W. E. Knowles, *The Experimenters: A Study of the Accademia del Cimento* (Baltimore: Johns Hopkins University Press, 1971).

Milani, Marisa, 'Il "Dialogo in perpuosito de la stella nuova" di Cecco di Ronchitti da Brugine', *Giornale storico della letteratura italiana* 170 (1993), 66–86.

—— 'Galileo Galilei e la letteratura pavana', in Santinello, ed., *Galileo e la cultura padovana*, 193–217.

Miller, David Marshall, 'The Thirty Years War and the Galileo Affair', *History of Science* 46 (2008), 49–74.

Miller, P. N., 'Description Terminable and Interminable', in Gianna Pomata and Nancy G. Siraisi, eds, *Historia* (Cambridge, Mass.: MIT Press, 2005), 355–97.

Milton, John, *Paradise Lost* (rev. edn, London: S. Simmons, 1674).

Monconys, Balthasar de, *Les voyages de Balthasar de Monconys: Documents pour l'histoire de la science*, ed. Charles Henry (Paris: A. Hermann, 1887).

315

Montesinos, José, and Carlos Solís, eds, *Largo campo di filosofare: Eurosymposium Galileo 2001* (La Orotava: Fundación Canaria Orotava de Historia de la Ciencia, 2001).

Moody, Ernest A., 'Galileo and Avempace: The Dynamics of the Leaning Tower Experiment', *Journal of the History of Ideas* 12 (1951), 163–93, 375–422.

Moscovici, Serge, *L'expérience du mouvement: Jean-Baptiste Baliani, disciple et critique de Galilée* (Paris: Hermann, 1967).

Moss, Jean Dietz, *Novelties in the Heavens: Rhetoric and Science in the Copernican Controversy* (Chicago: University of Chicago Press, 1993).

Muir, Edward, *The Culture Wars of the Late Renaissance* (Boston: Harvard University Press, 2007).

Nardi, Giovanni, *De igne subterraneo* (Florence: A. Massa et L. Landis, 1641).

Naudæana et Patiniana (Paris: Florentin et Pierre Delaulne, 1701).

Naylor, R. H., 'Galileo, Copernicanism and the Origins of the New Science of Motion', *British Journal for the History of Science* 36 (2003), 151–81.

—— 'Galileo's Experimental Discourse', in David Gooding, Trevor Pinch and Simon Schaffer, eds, *The Uses of Experiment* (Cambridge: Cambridge University Press, 1989), 117–34.

—— 'Galileo's Physics for a Rotating Earth', in *LCF* 337–55.

—— 'Galileo's Tidal Theory', *Isis* 98 (2007), 1–22.

Owen, G. E. L., ' "Tithenai ta Phainomena" ' [1961], in Jonathan Barnes, Malcolm Schofield and Richard Sorabji, eds, *Articles on Aristotle 1: Science* (London: Duckworth, 1975), 14–34.

Pagano, Sergio M., ed., *I documenti vaticani del processo di Galileo Galilei (1611–1741)* (Vatican City: Archivio segreto vaticano, 2009).

Palisca, Claude V., 'Vincenzo Galilei, scienziato sperimentale, mentore del figlio Galileo', *Nuncius* 15 (2000), 497–514.

Palmieri, Paolo, 'Galileo Did Not Steal the Discovery of Venus' Phases: A Counterargument to Westfall', in *LCF* 433–44.

—— 'Galileo and the Discovery of the Phases of Venus', *Journal for the History of Astronomy* 32 (2001), 109–29.

—— 'Mental Models in Galileo's Early Mathematization of Nature', *SHPS* 34 (2003), 229–64.

—— 'A New Look at Galileo's Search for Mathematical Proofs', *Archive for History of Exact Sciences* 60 (2006), 285–317.

—— *Reenacting Galileo's Experiments: Rediscovering the Techniques of Seventeenth-Century Science* (Lewiston, N.Y.: Edwin Mellen Press, 2008).

—— 'Re-Examining Galileo's Theory of Tides', *Archive for History of Exact Sciences* 53 (1998), 223–375.

—— ' "Spuntar lo scoglio più duro": Did Galileo Ever Think the Most Beautiful Thought Experiment in the History of Science?' *SHPS* 36 (2005), 223–40.

Panofsky, Erwin, 'Galileo as a Critic of the Arts', *Isis* 47 (1956), 3–15.

Pantin, Isabelle, ' "Dissiper les ténèbres qui restent encore à percer": Galilée, l'église conquérante et la république des philosophes', in Alain Mothu, ed., *Révolution scientifique et libertinage* (Turnhout: Brepols, 2000), 11–34.

Parrott, David, 'The Mantuan Succession, 1627–31', *English Historical Review* 112 (1997), 20–65.

Pedersen, Olaf, 'Galileo's Religion', in G. V. Coyne, M. Heller and J. Życiński, eds, *The Galileo Affair: A Meeting of Faith and Science* (Vatican City: Specola Vaticana, 1985), 75–102.

Pesce, Mauro, 'Le redazioni originali della lettera "copernicana" di G. Galilei a B. Castelli', *Filologia e critica* 17 (1992), 394–417.

Picanyol, Leodegario, *Le scuole pie e Galileo Galilei* (Rome: PP. Scolopi di S. Pantaleo, 1942).

Pin, Corrado, 'Progetti e abbozzi sarpiani sul governo dello stato "in questi nostri tempi turbolenti" ', in Paolo Sarpi, *Della potestà de' prencipi* (Venice: Marsilio, 2006), 89–120.

Plutarch, 'Life of Marcellus', in *Lives* (London: Loeb, 1917).

Poppi, Antonino, *Cremonini e Galilei inquisiti a Padova nel 1604* (Padua: Antenore, 1992).

—— 'Una implicita ritrattazione di Antonio Favaro sulla licenza di stampa del *Sidereus nuncius*', *Atti e memorie dell'Accademia patavina di scienze, lettere ed arti già Accademia dei Ricovrati: Memorie della classe di scienze morali, lettere ed arti*, 110 (1998), 99–105.

Purnell, Frederick, Jr, 'Jacopo Mazzoni and Galileo', *Physis* 14 (1972), 273–94.

Redondi, Pietro, 'From Galileo to Augustine', in Machamer, ed., *Cambridge Companion*, 175–210.
—— *Galileo Heretic*, trans. Raymond Rosenthal (London: Allen Lane, 1988).
—— 'La nave di Bruno e la pallottola di Galileo: Uno studio di iconografia della fisica', in Adriano Prosperi, ed., *Piacere del testo* (Rome: Bulzoni, 2001), 285–363.
—— 'Vent'anni dopo', in Redondi, *Galileo eretico* (3rd edn, Turin: Einaudi, 2004), 467–85.
Reeves, Eileen, *Galileo's Glassworks: The Telescope and the Mirror* (Cambridge, Mass.: Harvard University Press, 2008).
—— *Painting the Heavens: Art and Science in the Age of Galileo* (Princeton: Princeton University Press, 1997).
Renn, Jürgen, and Matteo Valleriani, 'Galileo and the Challenge of the Arsenal', *Nuncius* 16 (2001), 481–503.
Renn, Jürgen, Peter Damerow and Simone Rieger, 'Hunting the White Elephant: When and How Did Galileo Discover the Law of Fall?' *Science in Context* 13 (2000), 299–419.
Reston, James, *Galileo: A Life* (London: Cassell, 1994).
Reynolds, Anne, 'Galileo Galilei's Poem "Against Wearing the Toga" ', *Italica* 59 (1982), 330–41.
Ricci, Saverio, 'I Lincei e le novità celesti prima del *Nuncius sidereus*', in Massimo Bucciantini and Maurizio Torrini, eds, *La diffusione del copernicanesimo in Italia (1543–1610)* (Florence: Olschki, 1997), 221–36.
Righini Bonelli, Maria Luisa, and William Shea, *Galileo's Florentine Residences* (Florence: Istituto e Museo di storia della scienza, 1979).
Rist, J. M., *Epicurus: An Introduction* (Cambridge: Cambridge University Press, 1972).
Rocco, Antonio, *L'Alcibiade fanciullo a scola* (Rome: Salerno Editrice, [1988]).
Ronchi, Vasco, 'The Influence of the Early Development of Optics on Science and Philosophy', in McMullin, ed., *Galileo, Man of Science*, 195–206.
Ronconi, G., 'Paolo Gualdo e Galileo', in Santinello, ed., *Galileo e la cultura padovana*, 375–88.
Rosen, Edward, 'Galileo the Copernican', in Carlo Maccagni, ed., *Saggi su Galileo Galilei* (Florence: Barbèra, 1967), 181–92.
Rowland, Ingrid D., *The Scarith of Scornello: A Tale of Renaissance Forgery* (Chicago: University of Chicago Press, 2004).
Ruby, Jane E., 'The Origins of Scientific "Law" ', *Journal of the History of Ideas* 47 (1986), 341–59.
Russo, Lucio, *The Forgotten Revolution: How Science Was Born in 300 BC and Why It had To Be Reborn* (Berlin: Springer, 2004).
Santinello, Giovanni, ed., *Galileo e la cultura padovana* (Padua: CEDAM, 1992).
Sarpi, Paolo, *Lettere ai gallicani*, ed. Boris Ulianich (Wiesbaden: Franz Steiner Verlag, 1961).
—— *Lettere ai protestanti*, ed. Mario D. Busnelli (2 vols, Bari: Laterza, 1931).
—— *Pensieri naturali, metafisici e matematici*, ed. Luisa Cozzi and Liberio Sosio (Milan: Riccardo Ricciardi, 1996).
Schaffer, Simon, 'Glass Works: Newton's Prisms and the Uses of Experiment', in David Gooding, Trevor Pinch and Simon Schaffer, eds, *The Uses of Experiment* (Cambridge: Cambridge University Press, 1989), 67–104.
Schemmel, Matthias, 'England's Forgotten Galileo', in *LCF* 269–80.
Schiavo, A., 'Notizie riguardanti la Badia di Passignano estratte dai fondi dell'Archivio di Stato di Firenze', *Benedictina* 4 (1955), 31–92.
Schmitt, Charles B., 'Experience and Experiment: A Comparison of Zabarella's View with Galileo's in *De motu*', *Studies in the Renaissance* 16 (1969), 80–138.
Schmitt, Charles B., and Charles Webster, 'Harvey and M. A. Severino', *Bulletin of the History of Medicine* 45 (1971), 49–75.
Scott, John Beldon, 'Galileo and Urban VIII: Science and Allegory at Palazzo Barberini', in Lorenza Mochi Onori, Sebastian Schütze and Francesco Solinas, eds, *I Barberini e la cultura Europea del Seicento* (Rome: De Luca Editori, 2007), 127–36.
—— *Images of Nepotism: The Painted Ceilings of Palazzo Barberini* (Princeton: Princeton University Press, 1991).
Settle, Thomas B., 'Experimental Sense in Galileo's Early Works and Its Likely Sources', in *LCF* 831–50.

—— 'An Experiment in the History of Science', *Science* 133 (1961), 19–23.

—— 'Galileo and Early Experimentation', in Rutherford Aris, H. Ted Davis and Roger H. Stuewer, eds, *The Springs of Scientific Creativity* (Minneapolis: University of Minnesota Press, 1983), 3–20.

Shank, Michael H., 'How Shall We Practice History: The Case of Biagioli's *Galileo Courtier*', *Early Science and Medicine* 1 (1996), 106–50.

—— 'Setting the Stage: Galileo in Tuscany, the Veneto, and Rome', in *C&G* 57–87.

Shapiro, Barbara J., *A Culture of Fact: England, 1550–1720* (Ithaca: Cornell University Press, 2000).

Sharratt, Michael, *Galileo: Decisive Innovator* (Cambridge: Cambridge University Press, 1994).

—— 'Galileo's "Rehabilitation" ', in *C&G* 323–39.

Shea, William R., 'Galileo the Copernican', in *LCF* 41–60.

Shea, William R., and Mariano Artigas, *Galileo in Rome: The Rise and Fall of a Troublesome Genius* (Oxford: Oxford University Press, 2003).

—— *Galileo Observed: Science and the Politics of Belief* (Sagamore Beach: Science History Publications, 2006).

Sizi [or Sizzi], Francesco, *Dianoia*, trans. Clelia Pighetti (Florence: Barbèra, 1964).

Smith, Logan Pearsall, *The Life and Letters of Sir Henry Wotton* (2 vols, Oxford: Clarendon Press, 1907).

Sobel, Dava, *Galileo's Daughter: A Drama of Science, Faith and Love* (London: Fourth Estate, 1999).

Speller, Jules, *Galileo's Inquisition Trial Revisited* (Frankfurt: Peter Lang, 2008).

Spini, Giorgio, *Galileo, Campanella, e il 'Divinus poeta'* (Bologna: Il Mulino, 1996).

—— 'The Rationale of Galileo's Religiousness', in Carlo L. Golino, ed., *Galileo Reappraised* (Berkeley: University of California Press, 1966), 44–66.

—— *Ricerca dei libertini* (2nd rev. edn, Florence: La Nuova Italia, 1983).

Stabile, Giorgio, 'Il concetto di esperienza in Galileo e nella scuola galileiana', in *Experientia: X colloquio internazionale* (Florence: Olschki, 2002), 217–41.

Stella, Aldo, 'Galileo, il circolo culturale di Gian Vincenzo Pinelli e la "patavina libertas" ', in Santinello, ed., *Galileo e la cultura padovana*, 325–44.

Stevens, Henry, *Thomas Hariot the Mathematician, the Philosopher, and the Scholar* (London: privately printed, 1900).

Strocchia, Sharon T., *Nuns and Nunneries in Renaissance Florence* (Baltimore, Md: Johns Hopkins University Press, 2009).

Sutton, Robert B., 'The Phrase "Libertas philosophandi" ', *Journal of the History of Ideas* 14 (1953), 310–16.

Sylla, Edith, 'Galileo and the Oxford *Calculatores*', in William A. Wallace, ed., *Reinterpreting Galileo* (Washington, D.C.: Catholic University of America Press, 1986), 53–108.

Tarde, Jean, *À la rencontre de Galilée: Deux voyages en Italie* (Geneva: Slatkine, 1984).

Thomason, Neil, 'Sherlock Holmes, Galileo, and the Missing History of Science', *PSA 1994: Proceedings of the Biennial Meeting of the Philosophy of Science Association*, i, 323–33.

Tobert, Jane T., 'Peiresc and Censorship: The Inquisition and the New Science, 1610–1637', *Catholic Historical Review* 89 (2003), 24–38.

Topper, David, 'Colluding with Galileo', *Journal for the History of Astronomy* 34 (2003), 75–7.

Torrini, Maurizio, 'Galileo e la repubblica degli scienziati', in *LCF* 783–94.

Torrini, Maurizio, et al., *Il diavolo e il diavoletto* (Montevarchi: Accademia valdarnese del Poggio, 1997).

Valleriani, Matteo, 'A View on Galileo's *Ricordi autografi*: Galileo Practitioner in Padua', in *LCF* 281–91.

van Helden, Albert, 'Galileo and Scheiner on Sunspots: A Case Study in the Visual Language of Astronomy', *Proceedings of the American Philosophical Society* 140 (1996), 358–96.

—— 'Galileo, Telescopic Astronomy, and the Copernican System', in René Taton and Curtis Wilson, eds, *Planetary Astronomy from the Renaissance to the Rise of Astrophysics: Part A: Tycho Brahe to Newton* (Cambridge: Cambridge University Press, 1989), 81–105.

—— 'The Invention of the Telescope', *Transactions of the American Philosophical Society* 67:4 (1977), 1–67.

Vasari, Giorgio, *The Lives of the Artists*, trans. Julia Conway Bondanella and Peter Bondanella (Oxford: Oxford University Press, 1991).

Vérin, Hélène, 'Galilée et Antoine de Ville: Un courier sur l'idée de matière', in *LCF* 307–21.

Vickers, Brian, 'Epideictic Rhetoric in Galileo's *Dialogo*', *Annali dell'Istituto e Museo di storia della scienza* 8 (1983), 69–102.

Walker, D. P., 'Some Aspects of the Musical Theory of Vincenzo Galilei and Galileo Galilei', *Proceedings of the Royal Musical Association* 100 (1974), 33–47.

—— *Spiritual and Demonic Magic from Ficino to Campanella* (London: Warburg Institute, 1958).

Wallace, William A., *Domingo de Soto and the Early Galileo* (Aldershot: Ashgate, 2004).

—— *Galileo and His Sources: The Heritage of the Collegio Romano in Galileo's Science* (Princeton: Princeton University Press, 1984).

—— 'Galileo's Pisan Studies in Science and Philosophy', in Machamer, ed., *Cambridge Companion*, 27–52.

Weinberg, Bernard, 'The Accademia degli Alterati and Literary Taste from 1570 to 1600', *Italica* 31 (1954), 207–14.

Westfall, Richard S., 'Floods along the Bisenzio: Science and Technology in the Age of Galileo', *Technology and Culture* 30 (1989), 879–907.

—— 'Science and Patronage: Galileo and the Telescope', *Isis* 76 (1985), 11–30.

Westfall, Richard S., ed., *Essays on the Trial of Galileo* (Vatican City: Vatican Observatory Publications, 1989).

Westman, Robert S., 'The Copernicans and the Churches', in David C. Lindberg and Ronald L. Numbers, eds, *God and Nature* (Berkeley: University of California Press, 1986), 76–113.

Whitaker, Ewan A., 'Galileo's Lunar Observations', *Science* 208 (1980), 446.

—— 'Galileo's Lunar Observations and the Dating of the Composition of the *Sidereus nuncius*', *Journal for the History of Astronomy* 9 (1978), 155–69.

—— 'Lunar Topography: Galileo's Drawings', *Science* 210 (1980), 136.

White, Michael, *Galileo Antichrist* (London: Weidenfeld and Nicolson, 2007).

Wilding, Nick, 'Galileo's Idol: Gianfrancesco Sagredo Unveiled', *Galilæana* 3 (2006), 229–45.

—— 'The Return of Thomas Salusbury's *Life of Galileo* (1664)', *British Journal for the History of Science* 41 (2008), 1–25.

Willach, Rolf, *The Long Route to the Invention of the Telescope* (Philadelphia: American Philosophical Society, 2008).

Wilson, Catherine, *The Invisible World: Early Modern Philosophy and the Invention of the Microscope* (Princeton N.J.: Princeton University Press, 1995).

Winkler, Mary G., and Albert van Helden, 'Representing the Heavens: Galileo and Visual Astronomy', *Isis* 83 (1992), 195–217.

Wisan, Winifred L., 'Galileo and the Process of Scientific Creation', *Isis* 75 (1984), 269–86.

—— 'Galileo's *De systemate mundi* and the New Mechanics', in Paolo Galluzzi, ed., *Novità celesti e crisi del sapere* (Florence: Giunti-Barbèra, 1984), 41–9.

Wootton, David, 'Accuracy and Galileo: A Case Study in Quantification and the Scientific Revolution', *Journal of the Historical Society* 10 (2010), 43–55.

—— *Bad Medicine: Doctors Doing Harm Since Hippocrates* (Oxford: Oxford University Press, 2006).

—— 'New Light on the Composition and Publication of the *Sidereus nuncius*', *Galilæana* 6 (2009), 61–78.

—— *Paolo Sarpi: Between Renaissance and Enlightenment* (Cambridge: Cambridge University Press, 1983).

Index